Momentum Wave Functions–1982
(Adelaide, Australia)

AIP Conference Proceedings
Series Editor: Hugh C. Wolfe
Number 86

Momentum Wave Functions–1982
(Adelaide, Australia)

Editor
Erich Weigold
Flinders University of South Australia

American Institute of Physics
New York   1982

Copying fees: The code at the bottom of the first page of each article in this volume gives the fee for each copy of the article made beyond the free copying permitted under the 1978 US Copyright Law. (See also the statement following "Copyright" below). This fee can be paid to the American Institute of Physics through the Copyright Clearance Center, Inc., Box 765, Schenectady, N.Y. 12301.

L.C. Catalog Card No. 82-072375
ISBN 0-88318-185-1
DOE CONF- 820245

PREFACE

The elucidation of atomic, molecular, and nuclear structure from the study of single particle momentum distributions is a fundamental and potentially powerful tool in both chemistry and physics. These studies are carried out experimentally with knockout reactions: (e,2e), Compton scattering and positron annihilation in atomic and molecular physics; (p,2p), (p,pn), (γ,p), (e,e'p) and (e,e'n) in nuclear physics. In addition, in nuclear physics reactions such as (p,pα) or (α,2α) permit studies of multi-particle cluster momentum distributions. Some of these processes were discussed at the first Workshop on Momentum Wave Functions in Atomic, Molecular and Nuclear Systems held in 1976 at Indiana University, and a highly successful exchange of information took place between nuclear physicists, atomic and molecular physicists and chemists. Since then both experimental and theoretical research has delineated the conditions under which structure information can be obtained from knockout processes and a large amount of new information has been accumulated. This second meeting held at Flinders University, Adelaide, South Australia from 18-24 February 1982, was organised both to consolidate the advances already made and to set new directions for future research.

Since the workshop was interdisciplinary in nature it was structured to allow as much exchange of information as possible. Each scientific session consisted of several indepth contributions followed by open discussion and panel discussions. In addition several social and recreational events were organised in order to facilitate the further exchange of information.

The workshop/seminar was held under the auspices of the US/Australia Agreement for Scientific and Technical Cooperation, administered by the Australian Department of Science and Technology and the National Science Foundation of the U.S.A.

Erich Weigold

TABLE OF CONTENTS

MOMENTUM DISTRIBUTIONS : OPENING REMARKS

Erich Weigold
Institute for Atomic Studies, The Flinders University of South Australia,
Adelaide, S.A.

The common theme of this interdisciplinary workshop is single particle momentum distributions in manybody systems. For the purposes of the workshop these systems have been restricted to atoms, molecules and nuclei, with only an occasional excursion into solids. Therefore a variety of research fields are represented at the workshop:- quantum chemistry, atomic and molecular physics, and nuclear physics being the principal ones.

Despite the varied background of the participants there is one system we all understand, namely atomic hydrogen. It therefore provides an appropriate example which I can use to demonstrate how a knockout reaction can give a direct measurement of the single particle momentum distribution. I will be extremely simply in what I say, since the various complexities will be fully elaborated in later discussions.

The problem of the hydrogen atom has played a central role in the development of quantum mechanics, beginning with Bohr's daring speculations. It was also the first problem tackled by Schrödinger with his new wave mechanics and similarly it was used by Heisenberg in his first papers as a prime example of the success of quantum mechanics. It has always played a central role in the teaching of quantum physics and has served as a most important heuristic tool, shaping our intuition and inspiring many expositions.

The Schrödinger equation for the hydrogen atom is usually solved in the position representation, the solution to the equation being the wave functions $\psi_{n\ell m}(\underline{r})$. The simplest physical observable that can be derived from ψ is $|\psi_{n\ell m}(\underline{r})|^2$, which gives the probability density of finding the electron at a position $\underline{r}$ with respect to the centre of mass of the atom. However, although $|\psi_{n\ell m}(\underline{r})|^2$ is stated to be an observable it has never been directly observed. The standard texts show only calculated values of $|\psi|^2$ and discuss it at most by means of thought experiments.

If Schrödinger's equation is solved in the momentum representation instead of the coordinate representation, the absolute square of the corresponding momentum state wave function $\phi_{n\ell m}(\underline{p})$ would give the momentum probability distribution of the electron in the state defined by the quantum numbers n, ℓ and m. Lohmann and Weigold[1] have recently reported the first direct measurement of this observable for the ground state of the hydrogen atom using the (e,2e) reaction.

Figure 1 illustrates three different types of collisions which can take place in the (e,2e) reaction on atomic hydrogen, which is a three body problem. In the first case the incident electron can be considered as interacting mainly with the ion core (proton in this case). In general this will lead to emitted electrons (B) of low energy and momentum. They will be emitted in the backward direction as well as the forward direction. Such collisions, which give rise to recoil terms in the full collision amplitude, will be discussed by H. Ehrhardt in his talk. It is obvious that in such collisions information on the

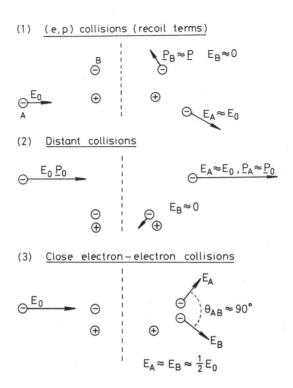

(1) (e,p) collisions (recoil terms)

$P_B \approx P$ $E_B \approx 0$

E_0
A

$P_B \approx P$ $E_B \approx 0$

$E_A \approx E_0$

(2) Distant collisions

$E_0 P_0$

$E_A \approx E_0, P_A \approx P_0$

$E_B \approx 0$

(3) Close electron-electron collisions

E_0

E_A

$\theta_{AB} \approx 90°$

E_B

$E_A \approx E_B \approx \frac{1}{2} E_0$

Fig. 1 Schematic representation of several types of collisions in the electron impact ionization of atomic hydrogen. The subscripts o, A and B denote respectively the incident, "scattered" and "ejected" electrons.

electron momentum distribution in the target is severely masked.

The second type of collision is one where the incident electron makes a distant collision with the target. In this case the electron is scattered into a very forward angle with little energy loss, inter-acting with the target as a whole. At high energies this type of collision can be shown[2] to be equivalent to the absorption of a photon of energy $h\nu = E_o - E_A$, and hence these (e,2e) collisions are sometimes called dipole (e,2e). Again we will not be able to extract any information on the momentum distribution of the ejected electron in its bound state.

Close electron-electron collisions are illustrated by the third type of collision in figure 1, in which the residual ion (proton) plays essentially the role of a spectator. If we define the momentum transferred by the scattered electron by

$$\underline{K} = \underline{P}_o - \underline{P}_A , \qquad (1)$$

where $p_A \geqslant p_B$ by convention, it is obvious that we need to maximise K to ensure close electron-electron collisions. In addition to high energy this implies that we must choose $K = |p_o - p_A| = |p_o - p_B|$ since the "scattered" and ejected electron are indistinguishable. Thus we choose symmetric kinematics ($E_A = E_B$, $p_A = p_B$ and $\theta_A = \theta_B$) and high energies.

The reason why this third category of collisions leads to information on the momentum distribution of the struck electron can be readily seen from semiclassical arguments. If the proton plays a spectator role (except for momentum conservation) the (e,2e) cross section can be simply described by the electron-electron collision cross section multiplied by the probability of the electron having the momentum $p = p_o - p_A - p_B$ required by momentum conservation.

The first factor is simply the Rutherford cross section $4K^{-4}$ and the second is of course $|\phi_{1s}(p)|^2$, and thus in atomic units

$$\sigma_{c\ell}(p,E) \sim \frac{4}{K^4}|\phi_{1s}(p)|^2 \tag{2}$$

In noncoplanar symmetric kinematics, where $\theta_A = \theta_B = \theta$ is fixed and the kinematic variable is the out of plane azimuthal angle $\phi = \pi - (\phi_A - \phi_B)$, K is independent of p (i.e. ϕ) for fixed E_o, E_A, E_B, and the cross section should be directly proportional to $|\phi_{1s}(p)|^2$.

We must of course allow for the indistinguishability of electrons and the fact that the electron is bound by an energy $\varepsilon = E_o - E$, where $E = E_A + E_B$ is the total energy. If the electron energies are high enough we can describe the continuum electron waves by plane waves and we can then use the plane wave impulse approximation (PWIA) to describe symmetric (e,2e) collisions[3]. In this approximation the cross section is given by[1]

$$\sigma = (2\pi)^4 \frac{p_A p_B}{p_o} f_{ee} |\phi_{1s}(p)|^2 , \tag{3}$$

where f_{ee} is the antisymmetrized two electron Mott scattering factor, which for symmetric kinematics is given by

$$f_{ee} = \frac{1}{4\pi^4} \frac{2\pi\eta}{(e^{2\pi\eta}-1)} \frac{1}{K^4} \sim \frac{1}{4\pi^4} \frac{1}{K^4} \tag{4}$$

The factor involving $\eta = |p_A - p_B|^{-1}$ is the Gamow factor which determines the particle density in Coulomb scattering. Since η is very small and nearly constant at high energies, the Gamow factor is very close to unity at all angles for which the cross section is measurable. For noncoplanar symmetric kinematics K is independent of ϕ and the cross section is again simply proportional to $|\phi_{1s}(p)|^2$.

Lohmann and Weigold[1] used the noncoplanar symmetric (e,2e) technique to measure the shape of the (e,2e) cross section on hydrogen at a number of energies. The momentum distributions obtained by them at 400, 800 and 1200eV are shown in figure 2, normalized to the momentum distribution $|\phi_{1s}(p)|^2 = 8\pi^{-2}(1+p^2)^{-4}$ given by the Schrödinger equation (solid line).

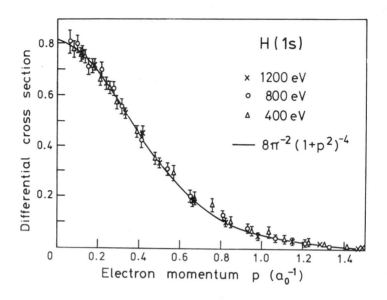

Fig. 2 The 400eV, 800eV and 1200eV noncoplanar symmetric momentum profiles for atomic hydrogen compared with the calculated momentum distribution $|\phi_{1s}(p)|^2$ (solid line).

These data provide the first direct experimental demonstration of the interpretation of wave functions as probability amplitudes in the simplest case, namely atomic hydrogen. It is a case which of course is familiar to everyone in the workshop.

Later in the workshop we will look at the (e,2e) reaction in much more detail and see it applied to the study of momentum distributions in atoms and molecules, and we will even hear some preliminary results on solids. We will learn of the successes of the analogous and historically older (p,2p) reaction in nuclear physics and the use of other techniques such as (γ,p) in obtaining nuclear momentum distributions. We will also hear of the problems and limitations of the various techniques and hopefully there will be many suggestions about where future work should take us. Finally, let me not ignore the fact that since the overall aim is to understand the detailed structure of the various manybody systems, we need to discuss the current status of structure theories as well as our understanding of the reactions we use to obtain the momentum space structure information. Based on the success of the first workshop at Indiana University[4] in 1976, I am certain we will learn much from each other once we understand each other's language. I look forward to a very fruitful workshop.

REFERENCES

1. B. Lohmann and E. Weigold, Phys. Lett. 86A, 139 (1981).
2. C.E. Brion, Radiat. Res. 64, 37 (1975).
3. I.E. McCarthy and E. Weigold, Phys. Rep. 27C, 275 (1976).
4. D.W. Devins, Editor "Momentum Wave Functions - 1976", AIP Conference Proceedings No. 36 (AIP Press, 1977).

THEORY OF THE (e,2e) REACTION

I.E. McCarthy

Institute for Atomic Studies,
The Flinders University of South Australia,
Bedford Park, S.A. 5042, Australia.

At the 1976 Indiana meeting it was shown[1], with the aid of some early examples, that the (e,2e) reaction is capable of yielding reliable information about two aspects of atomic and molecular structure, namely momentum profiles of single-particle orbitals, and spectroscopic factors for components of such orbitals split among final-ion eigenstates by electron-electron correlations. To obtain this information one must choose favorable kinematics for the reaction. It is possible to find different kinematic regions where quite-sophisticated reaction theories are severely tested, even in cases where the structure aspect of the reaction is clear and easy to calculate.

In this talk I want first to discuss just what has been found out about structure and structure calculations and how reliable this information is. At the end I will say something about the state of the art with respect to reaction theory.

The binary encounter approximation

The (e,2e) experiment involves complete knowledge of the momenta and energies of four bodies: an incident electron ($\vec{k}_0$), a target atom or molecule in its ground state $|g\rangle$, two outgoing electrons ($\vec{k}_A, \vec{k}_B$) and the residual ion in an eigenstate $|f\rangle$ (which may be in the continuum).

The amplitude M_f for the reaction is written

$$M_f = \langle \vec{k}_A \vec{k}_B (f|T|g\rangle|\vec{k}_0\rangle \quad , \tag{1}$$

where T is whatever operator is necessary to make M_f exact.

The binary encounter approximation amounts to the assumption that T is a three-body operator involving the coordinates only of the center of mass of the residual ion and the electrons 1 and 2, one of which is the incident electron, the other being initially bound. (Antisymmetry will be implicit throughout the discussion). With this approximation the ion state vector $|f\rangle$ commutes with T.

$$M_f = \langle \vec{k}_A \vec{k}_B |T| (f|g\rangle \vec{k}_0\rangle \quad . \tag{2}$$

The (e,2e) amplitude is a momentum transform of the generalized overlap amplitude $(f|g\rangle$, which depends only on the structure of the target and ion. The differential cross section may be considered as a measure of this function via a profile of recoil momenta q.

$$\vec{q} = \vec{k}_0 - \vec{k}_A - \vec{k}_B \quad . \tag{3}$$

If we can find a kinematic situation where the momentum transform is simple, then the experiment is a direct measurement of the structure properties. In fact there is a region easily accessible in the laboratory where it amounts to a Fourier transform with respect to q and multiplication by the Mott scattering amplitude for two electrons with appropriate momenta.

If we choose a case where $(f|g>$ is simple, then the experiment can be used to investigate the reaction transform or mechanism. An example is the hydrogen atom[2] where $(f|g>$ is the 1s orbital. Helium is nearly as simple, but (e,2e) already provides in this case some interesting structure information[3].

Structure effects

The generalized overlap amplitude is the main product of a perturbation treatment of the one-body Green's function[4]. Such a treatment includes some relaxation effects and is valid for ion states $|f)$ in the continuum. I will not discuss details of quantum chemistry calculations.

In order to illustrate the principles of structure effects I will consider $|g>$ and $|f)$ as wave functions determined by a configuration-interaction calculation using target Hartree-Fock wave functions $|\alpha>$ as a basis.

$$|g> = \Sigma_\alpha a_\alpha^{(g)} |\alpha> , \qquad (4)$$

$$|f) = \Sigma_{j\beta} t_{j\beta}^{(f)} C_{jr\beta} \psi_j^+ |\beta> . \qquad (5)$$

In (5) we are considering the ion as a linear combination with coefficients $t_{j\beta}^{(f)}$ of configurations with a hole in the orbital j of the target determinant $|\beta>$. $C_{jr\beta}$ is a Clebsch-Gordan coefficient ensuring that all configurations in the sum belong to the representation r of the point group of the target. For atoms, where this is the rotation group, it ensures that they all have the same angular momentum and parity.

The use of the same orthonormal basis for target and ion simplifies the overlap amplitude. Including the effects of antisymmetrization we have

$$(f|g> = n_r^{\frac{1}{2}} \Sigma_{j\alpha} a_\alpha^{(g)} t_{j\alpha}^{(f)} C_{jr\alpha} \psi_j , \qquad (6)$$

where n_r is the degeneracy of the representation r.

In nearly every case the target state is dominated by the Hartree-Fock ground state, so that $a_0^{(g)}$ is close to 1 and all other $a_\alpha^{(g)}$ are small. Whatever the reaction transform T, eq. (2) shows that if the binary encounter approximation is valid the (e,2e) amplitude is proportional to the magnitude of $(f|g>$. The effects of $a_\alpha^{(g)}$ for $\alpha \neq 0$ are therefore small, and negligible except in the case where $C_{jr\alpha}$ implies a selection rule forbidding the term in $a_0^{(g)}$.

Ion structure

We first consider the case where $a_0^{(g)}$ is allowed.

$$(f|g> = n_r^{\frac{1}{2}} a_0^{(g)} \Sigma_j t_{jo}^{(f)} \psi_j . \qquad (7)$$

In most experiments only relative cross sections are observed. Since $[a_0^{(g)}]^2$ multiplies all cross sections, it is irrelevant. The sum in (7) over orbitals belonging to the same representation reduces to a single orbital ψ_i, the characteristic orbital, if we choose the appropriate potential. The momentum profile $P_i^{(f)}(q)$ for a function ψ_i is proportional to the spectroscopic factor

$$S_i^{(f)} = [t_{io}^{(f)}]^2 . \tag{8}$$

Orthonormality and closure properties of state vectors result in the sum rule

$$\Sigma_f \, S_i^{(f)} = 1 \, , \qquad f \in r \, , \tag{9}$$

and in the expression for the orbital energy

$$E_i = \langle \psi_i | H | \psi_i \rangle = \Sigma_f \, S_i^{(f)} \, E_f \, , \qquad f \in r \, , \tag{10}$$

where E_f is the energy of each ion state f belonging to the representation r. Note that these conclusions depend only on the binary encounter approximation and the slow variation of the reaction transform with energy over the range of E_f. Their validity may be tested by seeing if the ratio of cross sections $P_i^{(f)}(q)$ depends on total energy E. It has been experimentally verified that there is essentially no energy dependence for E > 400eV in the cases of outer and inner valence orbitals for atoms and molecules, either of the shape of the $P_i^{(f)}$ or of their ratios, up to about q = 1-2a.u.

The reaction transform

The approximations that can be made for T simplify with increasing energy E. The distorted-wave impulse approximation assumes that T is the two-electron collision operator t and that the continuum state vectors are products of elastic-scattering functions $\chi^{(\pm)}(\vec{k})$ calculated in appropriate potentials. Apart from unobserved constants we have

$$P_i^{(f)} = n_r \, S_i^{(f)} \, |\langle \chi^{(-)}(\vec{k}_A) \chi^{(-)}(\vec{k}_B) | t | \psi_i \chi^{(+)}(\vec{k}_0) \rangle|^2 . \tag{11}$$

At higher energies the elastic-scattering functions are quite-well approximated by plane waves in the outer region of the target.

$$P_i^{(f)} = n_r \, S_i^{(f)} \, f_{ee} \, \int d\Omega |\langle \vec{q} | \psi_i \rangle|^2 \, , \tag{12}$$

where f_{ee} is the Mott-scattering factor (half-off-shell)

$$f_{ee} = \frac{1}{|\vec{k}-\vec{k}'|^4} + \frac{1}{|\vec{k}+\vec{k}'|^4} - \frac{1}{|\vec{k}-\vec{k}'|^2} \frac{1}{|\vec{k}+\vec{k}'|^2} \cos\left[\eta \, \ell n \frac{|\vec{k}+\vec{k}'|^2}{|\vec{k}-\vec{k}'|^2} \right] \, ,$$

$$\tag{13}$$

$$\vec{k} = \tfrac{1}{2}(\vec{k}_0+\vec{q}), \quad \vec{k}' = \tfrac{1}{2}(\vec{k}_A-\vec{k}_B), \quad \eta = 1/2k' . \tag{14}$$

f_{ee} is almost independent of q in noncoplanar symmetric geometry. We have introduced the spherical average to take care of rotational degeneracies. For molecules we also have a vibrational average, which has been confirmed numerically in the case of H_2 to be adequately approximated by considering the nuclei in their target equilibrium positions.

Experimental determination of ion structure

Eq. (12) determines an "experimental orbital" ψ_i, which again is a property of $\langle f|g \rangle$ if it does not change with increasing energy. In fact for noncoplanar symmetric kinematics and E > 400eV neither $S_i^{(f)}$ nor ψ_i are energy dependent and $P_i^{(f)}$ is given quite well by the target Hartree-Fock orbital ψ_i. We have thus generalized the concept of an orbital to the situation where the ion orbital splits into several "satellites".

The case of argon[5] is an excellent illustration of the principles.

Fig. 1 shows the observed energy dependence of $P_i^{(f)}$.

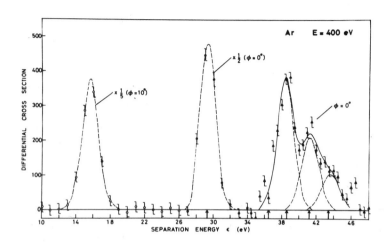

Fig. 1 Energy-dependence of the cross section for (e,2e) on
argon at E = 400eV.

Fig. 2 shows that the 15.76eV state is the 3p state. The profile is
given quite well by the Hartree-Fock 3p wave function.

Fig. 3 shows that all other states have essentially the same profile shape
and that it is the Hartree-Fock 3s shape.

The generalized-orbital concept is therefore valid both for 3p and
3s orbitals. The centroid (10) of the 3s states is in fact at the
Hartree-Fock value 35eV.

Calculation of one-body properties

Having verified that the (e,2e) reaction in noncoplanar symmetric
geometry at high enough energy gives consistent structure information we
will see how this information compares with the corresponding quantities
calculated by the methods of quantum chemistry.

The shape of the momentum profile $P_i^{(f)}(q)$ is proportional to the
spherically-averaged square of the momentum-space orbital $\langle q | \psi_i \rangle$. It is
a target one-body quantity. The best variational calculation of the
orbital ψ_i is the Hartree-Fock calculation[6]. We have seen that it coin-

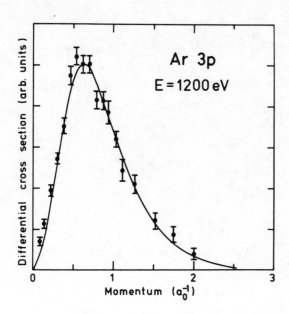

Fig. 2 Momentum profile of the 15.76eV state of argon.

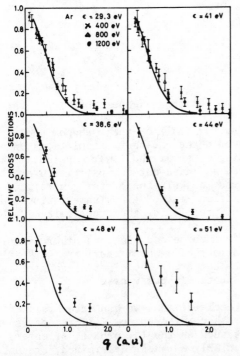

Fig. 3 Momentum profiles for the inner valence (3s) shell of argon.

cides with the experimental orbital in the case of the valence orbitals of argon.

Fig. 4 illustrates the momentum profiles for the valence orbitals of

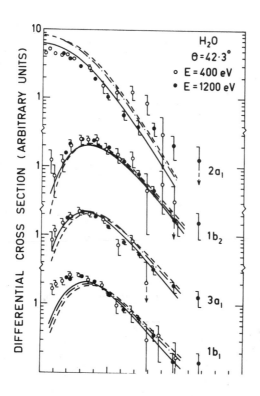

Fig. 4 Momentum profiles for the valence orbitals of water.

the water molecule[7].   Again, for small q, their shapes and ratios are independent of E.   For higher q the one-body calculations underestimate $P_1^{(f)}(q)$, the discrepancy being worse at the lower E value.   The molecular orbital calculations[8] (dashed and dash-dot curves) give overall ratios and shapes quite well.   The direct calculation of the generalized overlap amplitude by the Green's function[9] method gives some improvement, particularly in the magnitude of the profile for the inner valence orbital $2a_1$.

We can say in general from these and numerous other examples that the one-body variational calculations of quantum chemistry work well.   The Green's function method works better.   The binary-encounter interpretation of (e,2e) is confirmed in essentially every case.

The confirmation is so decisive that we can attach significance to the cases where there is a clear shape discrepancy.   This occurs only for lone-pair molecular orbitals such as the $1\sigma_1$ orbital of water.   Low-q components, which are particularly uniquely determined, are underestimated by theory.   This is attributed to the insensitivity of the variational integral to the shape of the wave function at very large distances where the potential is very small.   There is more density at large distances than given by the variational method.

Calculation of many-body properties

The spectroscopic factors (5), (8) are many-body properties which are calculated in quantum chemistry by introducing configuration interaction. There are several ways to do this.   One is analogous to the phenomenological shell model of nuclear physics.   Hamiltonian matrix elements for a restricted basis are treated as parameters, whose values are determined by fitting spectral eigenvalues.

This was done for the 3s hole representation of the argon ion, where the spectroscopic factors are very-well determined, by McCarthy, Uylings and Poppe[10], using a model space consisting of all the configurations obtainable from the orbitals 3s, 4s, 3p, 3d, 4d.   A total of 60 energy eigenvalues are fitted by varying the 30 parameters of this model space. The comparison with (e,2e) spectroscopic factors is given in Table I.   The spectroscopic factor for the lowest state is given quite well by the shell model.   For higher states the spectrum is distorted by the absence of the continuum (E_f > 43eV) from the model space.   In fact the continuum represents about 13% of the total strength.   We however do have an indication that the configuration-interaction picture is valid in principle if enough configurations can be included in the calculation by some means.

What about *ab initio* calculations for the argon ion? The most successful to date have been performed by Amos, Mitroy and Morrison[11] using a similar model space, with the interaction integrals calculated numerically using the ee Coulomb potential.   Results are also given in Table I.   Again

TABLE I

Comparison of phenomenological shell model and *ab initio* CI calculations with the results of (e,2e) for the 3s-hole representation of the argon ion.

Shell Model		CI		(e,2e)	
E_f(eV)	$S_{3s}^{(f)}$	E_f(eV)	$S_{3s}^{(f)}$	E_f(eV)	$S_{3s}^{(f)}$
29.24	1.61	28.7	.600	29.3±0.1	.53±.05
38.58	.15 ⎱ .17	36.9	.019 ⎱ .161	38.6±0.1	.23±.02
38.61	.02 ⎰	39.1	.142 ⎰		
41.20	.20			41.2±0.2	.13±.02
				43.4±0.1	.05±.01
		41.8	.075	>44	.08±.01
		>42	.095		

one can say that the agreement with (e,2e) data is promising.   Less-detailed calculations by Smid and Hansen[12] show a peculiar result.   The spectroscopic factor corresponding to the highest (unphysical) eigenvalue is 2 to 3 times that for the one-hole (lowest) state.   This makes the whole CI procedure suspect, probably because of inadequate representation of the continuum.

A variational many-body procedure is used here.   Pseudo-states are introduced into the model space with orbital shape parameters chosen to minimize the energy of a particular eigenstate. One might hope that this would mimic the continuum. However results of Fuss et al. for Krypton[13] show that as soon as the variational procedure is applied the lowest-state strength increases dramatically, i.e. it changes in the wrong direction for improving agreement with experiment. Similar results have been obtained by Glass[14] for argon. Clearly great care must be taken in such calculatio

For molecules we have Green's function calculations of spectroscopic factors (pole strengths). Since a molecule has less symmetry than an atom, fewer configurations contribute to a particular point-group representation and the configuration interaction calculation is less complex. Comparison in Table II of (e,2e) pole strengths for hydrogen bromide[15] with a

TABLE II

Electron separation energies (eV) and pole strengths (in brackets)
for the valence shells of HBr.

Orbital	(e,2e)	GF (ref. 15)	GF (ref. 16)
4	11.9 [~1]	10.91 [.944]	10.85 [.96]
		28.80 [.010]	39.01 [.013]
		29.80 [.024]	
8	15.6 [~1]	15.06 [.928]	15.00 [.950]
		31.35 [.024]	43.37 [.012]
		46.79 [.011]	43.74 [.011]
7	24.3† [.42±.03]	25.53 [.70]	22.40 [.076]
	30.3† [.58±.03]	26.50 [.053]	25.66 [.59]
		27.88 [.052]	29.90 [.25]
		28.28 [.026]	39.29 [.012]
		31.60 [.012]	56.12 [.011]
		31.82 [.025]	58.52 [.024]
		32.04 [.012]	
		34.97 [.033]	
		45.56 [.019]	
		46.90 [.026]	

† center of main peaks

Green's function calculation[16] show once again that qualitative agreement for level densities is the best that can be obtained with present computing technology.

Target correlations

The case where the term in $a_0^{(g)}$ is forbidden in eq. (6) by a selection rule is interesting, because the (e,2e) profile then depends on the $a_\alpha^{(g)}$ for $\alpha \neq 0$, i.e. on the electron-electron correlations of the target. Helium is a simple example. The excitation by (e,2e) of the unresolved n=2 states of He$^+$ can be observed. Its cross section is of the order of 1% of the allowed cross section for exciting the 1s state. This gives an idea of the order of magnitude of target correlation effects. Its momentum profile can be calculated quite well[3] if a sufficiently-detailed ground-state wave function[17] is used. This is illustrated in Fig. 5.

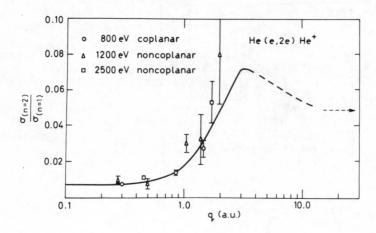

Fig. 5 Momentum profile for the excitation of the forbidden n=2
        states of He^+ by (e,2e).

Reaction mechanism - large momentum transfer

In a collision in which the final state is close to symmetric one
would expect that the two-electron collision is an important feature of
the reaction.   There is no general three-body theory for a problem involving Coulomb potentials, and one may consider the (e,2e) reaction as a means
of investigating what to expect from such a theory.   Here experiment and
theory go hand in hand.

We take the distorted-wave impulse approximation (11) as our basic
approximation.   At sufficiently-high energies the distorted waves are
adequately represented by the averaged eikonal approximation in which a
plane wave is used with a wave number shifted to allow for a constant
complex distorting potential $\overline{V}+i\overline{W}$.   In this approximation eq. (11) fac-
torizes into a ee collision factor f_{ee}(13) and a distorted-wave transform.
The validity of the approximation is shown for the 5p orbital of xenon in
Fig. 6 in comparison with absolute experimental cross sections[18].

$$P_i^{(f)} = n_r S_i^{(f)} f_{ee} |<\chi^{(-)}(\vec{k}_A)\chi^{(-)}(\vec{k}_B)|\psi_i \chi^{(+)}(\vec{k}_0)>|^2 . \quad (15)$$

In an investigation by Camilloni et al.[19] for helium, the value of the
constant potential was chosen to fit the central coplanar-symmetric (e,2e)
profile.   With this value one can choose the experimental kinematics to
keep the distorted-wave transform constant at the value $<\overline{q}|\psi_i>$ while vary-

14

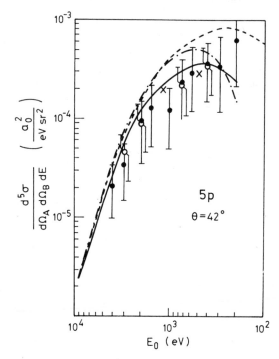

Fig. 6 Energy-dependence of the absolute cross section for the
(e,2e) reaction on the 5p state of xenon, in coplanar
symmetric geometry, $\theta = 42°$. The dash-dot curve is the
plane-wave impulse approximation, the dashed curve is the
averaged-eikonal DWIA with $\bar{V} = 10eV$, $\bar{W} = 0$. The full
curve is the same approximation with $\bar{V} = 10eV$, $\bar{W} = 7eV$ Å.

ing f_{ee}. Fig. 7 shows that f_{ee} is represented much better by the spin-
averaged square of the ee t-matrix (DWIA) rather than that of the potential
(DWBA).

Calculations in the factorized approximation have been done with distorted waves calculated in reasonable optical potentials. For neon[20]
we see in Fig. 8 that distortion accounts for the enhanced cross section
for higher q. A most-detailed calculation was done for hydrogen[21] by
Noble. Here the effect of different assumptions for the screening effect[22]
was shown to be unimportant. In fact the assumption that each distorted
wave has an effective charge of 1 worked best. The quality of the comparison with experiment is shown in Fig. 9.

Reaction mechanism - small momentum transfer

(e,2e) coincidence experiments with small momentum transfer to the
scattered electron (defined as the faster of the emerging electrons) have
been the subject of a large program of investigation by the group of
Ehrhardt[23]. These experiments all show a forward binary encounter peak
and a backward peak attributed to electron recoil after heavy-particle
knockout. The forward peak is much larger. In fact we can see from (13)

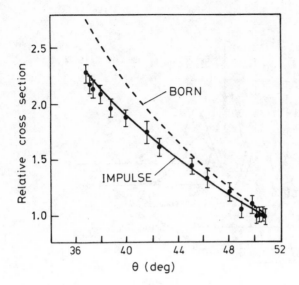

Fig. 7 The variation of f_{ee} for helium at 400eV with $\overline{q}$ = 0.7 a.u. The dashed curve is the averaged-eikonal ($\overline{V}$ = 20eV) DWBA, the full curve is the averaged-eikonal DWIA.

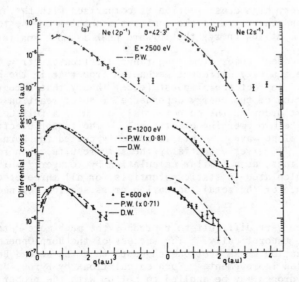

Fig. 8 Noncoplanar-symmetric (e,2e) cross sections for neon.

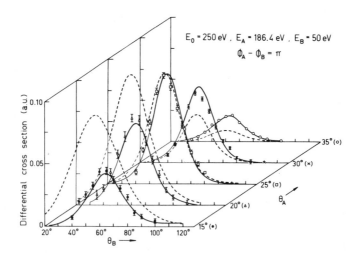

Fig. 9   The (e,2e) cross section for hydrogen.   The full curve
is the factorized DWIA, the dashed curve is the PWIA.

that it has a very high cross section in comparison with the high momentum-
transfer mode and therefore is the dominant mechanism in the total reaction
cross section.   At low energy the forward peak is much smaller.

The forward peak comes from long-range electron-electron encounters
and we may expect a very different mechanism from that in the high momentum-
transfer case.   In fact the simplest (e,2e) theory that integrates to give
a good description of the energy dependence of total reaction cross sections
is only of first order in the ee potential, but uses a full Coulomb wave for
the slower of the two emerging electrons[24].   The faster electrons are
represented by plane waves.   This is commonly called the Born-Oppenheimer
approximation.   It gives essentially the same results, even for highly-
charged target ions, as a similar calculation by Younger[25] which uses distorted waves calculated in static potentials for all three electrons.   The
energy dependence of the total ionization cross section for neon[26] is shown
in Fig. 10.

It has been very difficult to reproduce the much-more-detailed data of
Ehrhardt et al. theoretically.   The success of the Born-Oppenheimer theory
for total ionization cross sections suggests that the second Born approximation may be an improvement.   Such calculations by Byron, Joachain and
Piraux[27] for hydrogen may be applied to helium with the proper kinematic
adjustments.   They are very promising.

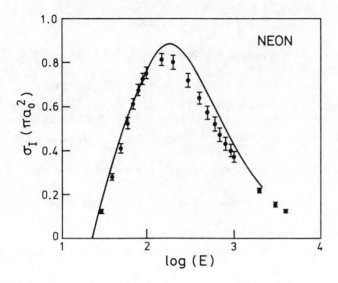

Fig. 10 Energy dependence of the total ionization cross section
for neon compared with the Born-Oppenheimer approximation.

References

1.   McCarthy, I.E. in Momentum Wave-Functions-1976, AIP (New York) 1977,p.1.

2.   Lohmann, B. and Weigold, E., Phys. Lett. 86A, 139 (1977).

3.   Dixon, A.J., McCarthy, I.E. and Weigold, E., J. Phys. B9, L195 (1976).

4.   Williams, G.R.J., McCarthy, I.E. and Weigold, E., Chem. Phys. 22, 281
     (1977).

5.   Hood, S.T., McCarthy, I.E. and Weigold, E., Phys. Rev. A11, 566 (1975).

6.   Clementi, E. and Roetti, C., Atomic Data 14, 177 (1974).

7.   Dixon, A.J., Dey, S., McCarthy, I.E., Weigold, E. and Williams, G.R.J.,
     Chem. Phys. 21, 81 (1977).

8.   Snyder, L.C. and Basch, H., Molecular Wave Functions and Properties,
     Wiley (New York), 1972.

9.   Williams, G.R.J., J. Elec. Spectrosc. 15, 985 (1979).

10. McCarthy, I.E., Uylings, P. and Poppe, R., J. Phys. B11, 3299 (1978).

11. Amos, K., Mitroy, J. and Morrison, I., private communication.

12. Smid, H. and Hansen, J.E., to be published.

13. Fuss, I., Glass, R., McCarthy, I.E., Minchinton, A. and Weigold, E.,

18

14. Glass, R., private communication.

15. Brion, C.E., McCarthy, I.E., Suzuki, I.H., Weigold, E., Williams, G.R.J., Bedford, K.L., Kunz, A.B. and Weidman, R., J. Elec. Spectrosc. (in press).

16. von Niessen, W., Cederbaum, L.S., Domcke, W. and Diercksen, G.H.F., Chem. Phys. 56, 43 (1981).

17. Joachain, C.J. and Vanderpoorten, R., Physica 46, 333 (1970).

18. Giardini-Guidoni, A., Fantoni, R., Tiribelli, R., Marconero, R., Camilloni, R. and Stefani, G., CNEN preprint 79.34 (July 1979).

19. Camilloni, R., Giardini-Guidoni, A., McCarthy, I.E. and Stefani, G., Phys. Rev. A 17, 1634 (1978).

20. Dixon, A.J., McCarthy, I.E., Noble, C.J. and Weigold, E., Phys. Rev. A17, 597 (1978).

21. Weigold, E., Noble, C.J., Hood, S.T. and Fuss, I., J. Phys. B 12, 291 (1979).

22. Rudge, M.R.H., Rev. Mod. Phys. 40, 564 (1968).

23. Ehrhardt, H., Fischer, M. and Jung, K., Universität Kaiserslautern preprint.

24. Stelbovics, A.T. and McCarthy, I.E., unpublished.

25. Younger, S.M., Phys. Rev. A 23, 1138 (1981).

26. de Heer, F.J., Jansen, R.H.J. and van der Kaay, W., J. Phys. B 12, 979 (1979).

27. Byron, F.W., Jr., Joachain, C.J. and Piraux, B., unpublished.

QUASIFREE KNOCKOUT OF NUCLEONS AND NUCLEON CLUSTERS

N. S. Chant
Department of Physics & Astronomy
University of Maryland, College Park, MD 20742

ABSTRACT

Experimental results for nucleon and nucleon cluster knockout are reviewed. Results for spectroscopic factors and momentum wave functions are discussed and limitations of the theoretical description are cited.

INTRODUCTION

In this paper I wish to discuss some current attempts to determine nucleon and nucleon cluster wave functions in nuclei by means of medium energy quasifree knockout reactions. Since this is now a rather broad subject, I would like to refer you to the reviews by Miller[1] and by McDonald,[2] particularly with reference to nucleon knockout by polarized protons, to the review of cluster knockout experiments which I gave[3] at the 1978 Conference on Clustering and to an earlier talk by Roos.[4] June Matthews will discuss (e,ep) experiments shortly and Del Devins will describe recent (p,2p) and (p,pn) studies at IUCF. Having included these references, rather than attempt a thorough review, I propose to describe briefly a variety of fairly recent experiments which, I hope, reflect the diversity of recent work.

EXPERIMENT AND ANALYSIS

In case the quasifree knockout type of experiment is still unfamiliar to some, I shall first outline the main ingredients of the experiment and the theoretical analysis using the (p,pα) reaction as an example.

As shown in Fig. 1 the emitted particles are detected in coincidence using telescopes placed on opposite sides of the incident beam, usually, though not always, in a coplanar geometry. Different residual states are identified by computing a missing mass spectrum and the corresponding kinematic loci are projected onto the proton energy axis to yield so-called "energy-sharing distributions" of

$$A(p, p\alpha)B$$

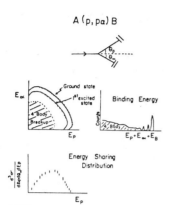

NOTES:

1) Recoil momentum $\vec{p}_B$ varies smoothly across kinematic locus.

2) For given θ_P choose θ_α such that $\vec{p}_B = 0$ somewhere on kinematic locus.

3) If B is a spectator, $-\vec{p}_B$ is momentum of alpha in target before collision.

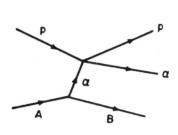

Fig. 1

$$A(a, cd)B$$

PWIA

$$\frac{d^3\sigma}{d\Omega_c \, d\Omega_d \, dE_c} = K \, S_b \, |\phi(-\vec{p}_B)|^2 \, \frac{d\sigma}{d\Omega}\bigg|_{a+b}$$

K - kinematic factor

S_b - spectroscopic factor $A \to B + b$

$\phi(-\vec{p}_B)$ - b-B momentum wave function

$\frac{d\sigma}{d\Omega}\big|_{a+b}$ - half-shell a-b 2-body cross section for $a + b \to c + d$

DWIA

$$|\phi|^2 \to \sum_\Lambda |T^{\alpha L \Lambda}|^2$$

where

$$T^{\alpha L \Lambda} = \frac{1}{(2L+1)^{\frac{1}{2}}} \int \chi_d^{(-)*} \chi_c^{(-)*} \phi_{L\Lambda}^\alpha \chi_a^{(+)} \, d^3r$$

$\chi^{(\pm)}$ - incident/emitted distorted waves

$\phi_{L\Lambda}^\alpha$ - bound state wave function for b calculated in Woods-Saxon well

$$V(r) = \frac{V_0}{1 + e^x} \; ; \quad x = \frac{r - r_0 A^{1/3}}{a}$$

$(+V_{so} + V_{Coulomb})$

Fig. 2

DWIA with Spin-Orbit Distortions

Optical potentials:

$$V + iW \to V + iW + V_{so} \, \vec{\ell} \cdot \vec{s}$$

Distorted waves:

$$\chi^{(\pm)} \to \chi_{\rho\sigma}^{(\pm)} \qquad \rho, \sigma = \pm\tfrac{1}{2}$$

Hence:

$$T^{\alpha L \Lambda} \to T_{\rho_a \sigma_a \rho_c \sigma_c \rho_d \sigma_d}^{\alpha L \Lambda}$$

and (for p,2p)

$$\frac{d^3\sigma}{d\Omega_c \, d\Omega_d \, dE_c} = K' \, S_b \sum_M \left| \sum_{\substack{\Lambda \rho_b \\ \rho_a \sigma_a \rho_c \sigma_c \rho_d \sigma_d}} (L \Lambda \tfrac{1}{2} \rho_b | JM) \right.$$

$$\left. \times (2L+1)^{\frac{1}{2}} \, T_{\rho_a \sigma_a \rho_c \sigma_c \rho_d \sigma_d}^{\alpha L \Lambda} \langle \sigma_c \sigma_d | t | \sigma_a \sigma_b \rangle \right|$$

$\langle \, | \, | \, \rangle$ - p+p $\frac{1}{2}$-shell t-matrix

Notice: factorization of amplitudes, not cross section.

Fig. 3

$d^3\sigma/d\Omega_p \, d\Omega_\alpha \, dE_p$ versus E_p. Sometimes, as an alternative, a single point on an energy sharing distribution is obtained for various detector angle pairs. The motivation for these different choices of geometry and their relative merits has been discussed by Roos.[3]

The recoil momentum of the residual nucleus $\vec{p}_B$ varies smoothly across each kinematic locus and by choosing a so-called quasifree angle pair the point $\vec{p}_B = 0$ can be included. This angle pair will, of course, coincide with free scattering kinematics aside from a small forward folding owing to the α-core binding energy. The quantity $\vec{p}_B$ is seen to be equal and opposite to the struck particle momentum. From the first order diagram we can obtain the plane wave impulse approximation (PWIA) expression shown in Fig. 2. Thus, the predicted cross section involves a product of the free scattering cross section and the square of the struck cluster or nucleon momentum wave function.

Unfortunately, in nuclear physics, modification of this simple diagram due to distortions, that is scattering of the projectile and ejectiles from the core B, is almost always important even in such optimum cases as (e,ep) reactions. The corresponding modification of the PWIA expression is also shown in Fig. 2. This is a factorized distorted wave impulse approximation (DWIA) in which the half-shell 2-body cross section enters as a multiplicative factor.[5] In Fig. 3 we see that when spin dependent distortions are included[6] a more complicated expression results in which there is a factorization of amplitudes. It should be noted that, in some ways, these are inferior calculations to the distorted wave t-matrix approximation developed by McCarthy[7] almost 15 years ago. Our use of such a primitive theory is dictated partially by difficulties in generalizing the earlier work and partially by mere computer usage economics. I shall return to this point later.

As far as the half-shell two-body cross section is concerned in most of the calculations I shall show this is approximated by a nearby on-shell cross section. Calculations employing an extrapolation from on-shell data to the half-shell point have been reported by Miller and Thomas for (p,2p) experiments.[1] For (p,pα) and (α,2α) some estimates have been made by Watson and by Roos and myself using half-shell amplitudes calculated from optical potentials.[8] However, there are no major qualitative changes involved.

RESULTS: NUCLEON KNOCKOUT

The (p,2p) Reaction

Devins will discuss (p,2p) and (p,pn) experiments more extensively in his talk so I will mention only two (p,2p) experiments.

The first, a study of the ^{4}He(p,2p)^{3}H reaction carried out at TRIUMF by Epstein, van Oers and others,[9] is of interest because of sensitivity to relatively high momentum components in the ^{4}He wave function. Data were taken at 250, 350 and 500 MeV incident energy. Results are shown in Fig. 4 for zero recoil momentum and symmetric angles. Also shown are data from SREL,[10] Chicago,[11] Orsay[12] and Maryland.[13] If we omit the new data we note a rather disconcerting discontinuity.[3] Adding the new data suggests a smooth energy dependence supporting the SREL result and suggesting the Chicago data is about 30% low. DWIA calculations are consistent with the general trend of the data. However, the cross sections are somewhat overestimated at both the low and high energies. Some of the 500 MeV data are shown in Fig. 5. Notice that the data extend to momenta ~500 MeV/c which I think is a record for (p,2p) work. The DWIA curves were generated using a code which I wrote in which spin-orbit distortions are included. The optical potentials also included an ℓ-dependent Majorana exchange term and were the result of rather careful global analysis by van Oers. The single particle wave function (or more correctly the projection of ^{4}He onto the ^{3}H ground state) was obtained from fits to electron-^{4}He scattering. The dot-dash curve uses a conventional analysis by Lim[14] in terms of a simple Eckart parametrization. For the continuous curve the electron scattering data were first corrected for meson exchange contributions which, as we see, leads to a wave function with reduced high momentum components. The broken curve uses the meson exchange corrected wave function but the spin orbit terms in the distorting potentials are set to zero. We see that, though negligible below about 250 MeV/c, the spin orbit terms do

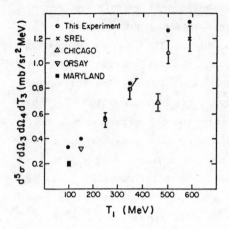

Fig. 4. ^{4}He$(p,2p)^3$H vs. incident energy; Θ TRIUMF data; o DWIA.

Fig. 5. Coplanar symmetric data; —— meson exchange corrected, DWIA with spin-orbit terms; --- same with no spin-orbit terms; -··- Lim Eckart wave function.

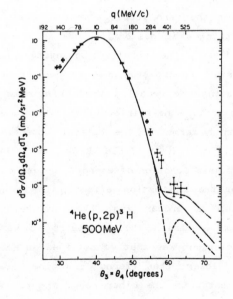

improve agreement with experiment at the largest momenta and are at least as important as getting the wave function right.

The second (p,2p) experiment I should like to discuss is of interest as a reaction mechanism study. It was carried out by my student C. Samanta[15] for the reaction ^{40}Ca(p,2p)^{39}K. The motivation of this work was to study the utility of the DWIA description of the reaction mechanism as distortion effects become progressively more severe. This can be seen in Fig. 6 in which the ratio of distorted wave to plane wave calculations are shown as a function of energy. We see that the ratio ranges from around 1/100 to 46 MeV to ~ 1/4.5 at 150 MeV. At the ends of the range data was already available from Manitoba[16] and IUCF.[17] Samanta chose to fill in intermediate points at 76.1 and 101.3 MeV and to perform a consistent analysis using the IUCF optical potentials[18] and the Elton and Swift analysis of electron scattering for the single particle wave function.[19] Some of Samanta's results are shown in Fig. 7, in which fits to a couple of energy sharing distributions are shown, and in Fig. 8, in which results of a factorization test is shown at each energy. These are data for the zero recoil momentum point at different angle pairs. Since changes in distortion effects are small, and are in any case divided out, the resultant quantity Q should reproduce the p-p scattering cross section. We see that the results are quite encouraging. We recall that the factorization of the cross section is, in fact, a no-spin-orbit result. The modifications on including spin-orbit terms in the optical potentials are seen to be quite small. Finally, in Table I we show the extracted spectroscopic factors as a function of energy. In view of the change of a factor of ~20 in the distortion effects, the consistency, while far from perfect, is encouraging. Even at relatively low energies, our reaction mechanism theory appears to work fairly well. One minor disappointment was that we obtained an absolute spectroscopic factor 30-40% lower than found in other reactions such as (e,ep) experiments. On the other hand, the published analyses of the IUCF data using other optical potentials gives a larger value more consistent with (e,ep) so there is some ambiguity in the analysis.

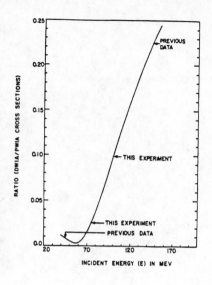

Fig. 6. $^{40}Ca(p,2p)^{39}K$ (2.52 MeV) $L=0, \theta_1 = \theta_2$, $p_B = 0$. Ratio DWIA/PWIA vs. energy indicating energies of Maryland experiment.

Fig. 7.
Energy sharing distributions for $^{40}Ca(p,2p)^{39}K$ $(1/2^+, 2.52$ MeV) $E_p = 101.3$ MeV; —— DWIA without spin orbit terms; -·-·· DWIA with spin orbit terms.

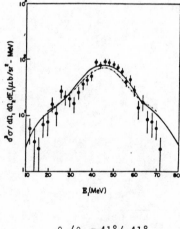

$\theta_1/\theta_2 = 41°/-41°$

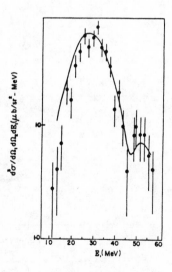

$\theta_1/\theta_2 = 52.2°/-29°$

Fig. 8(a). Factorization Test

Notes:

1) For an L=0 transition take data at various θ_c, θ_d such that $p_B = 0$.

2) Compute

$$Q(\theta^*) = \frac{d^3\sigma^{EXPT}}{d\Omega_c d\Omega_d dE_c} \Bigg/ \left\{ K S_b \sum_\Lambda |T^{\alpha L\Lambda}|^2 \right\}$$

θ^* - two-body c.m. scattering angle.
{ } - varies slowly ($\propto$ K in PWIA)

3) No spin-orbit DWIA predicts $Q(\theta^*) = \frac{d\sigma}{d\Omega}\Big|_{a+b}$

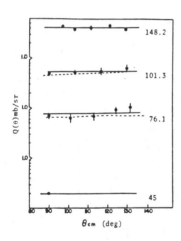

Fig. 8(b). $^{40}Ca(p,2p)^{39}K^*$ (2.52 MeV) L=0 factorization test at E_p = 45, 76.1, 101.3, and 148.2 MeV.

——— DWIA (no spin orbit) i.e., $d\sigma/d\Omega$ (pp)

---- DWIA (spin orbit distortions included).

TABLE I

$^{40}Ca(p,2p)^{39}K$(2.52 MeV $2S_{1/2}$)   L=0

	E_p MeV	Spectroscopic Factor	Ratio DWIA/PWIA
IUCF	148.2	0.98 ± 0.04	0.22
MD	101.3	0.88 ± 0.04	0.10
MD	76.1	1.09 ± 0.07	0.03
MANITOBA	45.0	$1.18 \pm$?	0.01

The $(\alpha,\alpha p)$ and $(\pi,\pi p)$ Reactions

As an alternative to protons, other projectiles are possible in nucleon knockout. I will discuss briefly the $(\alpha,\alpha p)$ and $(\pi,\pi p)$ reactions.

The $(\alpha,\alpha p)$ reaction has been studied at 140 MeV at Maryland by Nadasen and co-workers[20] for ^{2}H, ^{6}Li and ^{19}F targets and, more recently, by Samanta and others[15] for ^{16}O and ^{40}Ca targets. The comparison with (p,2p) reactions is of interest since different radial regions are involved in $(\alpha,\alpha p)$ as can be seen from the histograms of contributions to DWIA cross sections for the two reactions shown in Fig. 9. Results for ^{40}Ca are shown in Fig. 10. The energy sharing distributions are quite well reproduced, although the statistics for the ground state L=2 transition are unimpressive. Notice that, owing to the α-p kinematics, a quasifree geometry obliges us to detect the alpha at a rather forward angle which makes the experiment tough. The extracted spectroscopic factors for the ground and excited state transitions of 10.8 and 2.86, respectively, are more than twice the values found in (p,2p) and, of course, exceed closed-shell sum rule values. A discrepancy of about a factor of 2 was also found for ^{16}O. Samanta investigated various ingredients in the calculations in an attempt to understand the problem. Firstly, she varied the optical potentials and found that the predictions were not surprisingly sensitive mostly to the choice of alpha potential. The spectroscopic factors reported were obtained using so-called shallow alpha potentials which reproduce alpha scattering including the rainbow region. Since there is some precedence from analyses of transfer reactions for ignoring this latter constraint and using a "deep" potential of roughly four times the nucleon optical potential, this was tried. As a result, the discrepancy increased to a factor of 4. Secondly, since $(\alpha,\alpha p)$ is more highly surface localized than (p,2p), the bound proton potential radius parameter was increased in the hope of reducing the ratio of $(\alpha,\alpha p)$ to (p,2p) spectroscopic factors. This did prove possible but only for excessively large radii $(r_0 \sim 2.0$ fm) with excessively small absolute spectroscopic factors

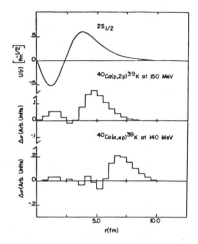

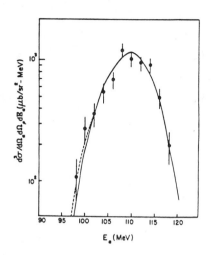

Fig. 9. Radial localization of DWIA for (p,2p) and (α,αp) on ^{40}Ca; U(r) is proton radial wave function; Δσ is obtained by differencing calculations with lower radial cut-offs; L = 0, p_B = 0.

Fig. 10(a). Energy sharing distribution: ^{40}Ca(α,αp)^{39}K(1/2$^+$, 2.5 MeV) θ_α/θ_p = 9.0°/-51.41°, E_α = 1.39.2 MeV; (solid line) deep potential, (dashed line) shallow potential.

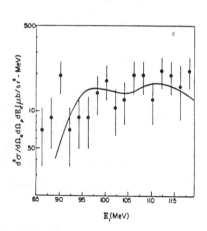

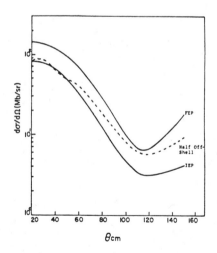

Fig. 10(b). Energy sharing distribution: ^{40}Ca(α,αp)^{39}K(3/2$^+$, 0.0 MeV) θ_α/θ_p = 9.0°/-51.41° at E_α = 139.2 MeV.

Fig. 11. Final energy, initial energy and half-shell cross sections ^{40}Ca(α,αp), 140 MeV.

and was thus rejected. Finally, we speculated that off-shell effects might well be severe for the α-p scattering in (α,αp) at 140 MeV. This is prompted by the observation that only 35 MeV is available in the α-p c.m. system which is to be compared with a binding energy of 10.8 MeV for the $2s_{1/2}$ proton in ^{40}Ca (and more in ^{16}O). The half-shell α-p cross sections were computed using an optical potential (and ignoring Coulomb terms) and are shown in Fig. 11. The half-shell computation is seen to lie lower than the final energy prescription on-shell cross sections used in the analysis and would thus increase the already large spectroscopic factors. Thus, the problem with (α,αp) at present remains unresolved.

Recently, owing to the availability of relatively high intensity pion beams (i.e., about a pico-amp), studies of nucleon knockout by pions have been reported. These experiments are of interest since very little data exists and the reaction mechanism is not well studied. As an example, I will show ^{12}C(π^+,π^+p)^{11}B data obtained by Ziock[21] and co-workers at SIN using the SUSI pion spectrometer in coincidence with a Germanium stack for proton detection. Incident pion energies were between 130 and 200 MeV. In Fig. 12 a missing mass spectrum obtained by Ziock is shown in which a peak corresponding to the residual ^{11}B ground state is very prominent. Also evident are smaller peaks corresponding to the $1/2^-$ state at 2.12 MeV and the $3/2^-$ state at 5.02 MeV. The latter cannot be resolved from the $5/2^-$ state at 4.44 MeV which would not be excited in a single step knockout in most structure models for ^{12}C. However, the experimenters claim that the location of the centroid of the so-called 5.02 MeV peak suggests very little contribution from the $5/2^-$ "multi-step" state. In Fig. 13 we show an energy sharing distribution at an incident energy of 199 MeV for excitation energies < 9.75 MeV. The broken curve is a PWIA calculation. The data shows the characteristic form for an L=1 knockout reaction with a minimum at zero recoil momentum. The continuous line is a DWIA calculation[22] carried out by Roos, Rees and myself at Maryland using Kisslinger type optical potentials for the pion and the IUCF proton potential. We see that the asymmetry is fairly well described by the DWIA calculation. The data yield a

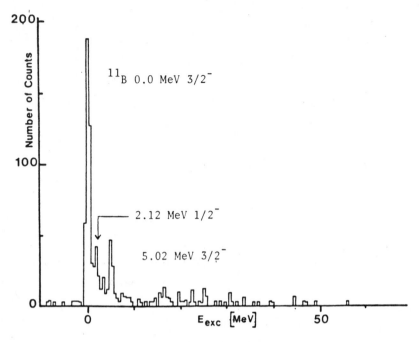

Fig. 12. Missing mass $^{12}C(\pi^+,\pi^+p)^{11}B$, $E_\pi = 130$ MeV, $\theta_\pi/\theta_p = -120°/40°$

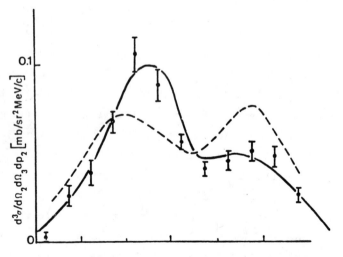

Fig. 13. $^{12}C(\pi^+,\pi^+p)^{11}B$, $E_\pi = 199$ MeV, L=1, —— DWIA; ---PWIA.

Fig. 14. Cluster Knockout

a) Quasi-free reactions

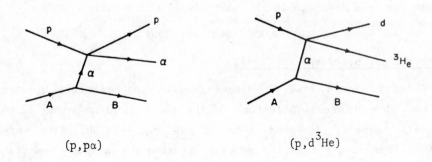

(p,pα)   (p,d³He)

b) "Bound state" wave functions for α-knockout.

$$\langle A | B \boxtimes \alpha \rangle \approx \sqrt{S_b} \ \ \phi_{L\Lambda}^N(\vec{r})$$

where $\phi_{L\Lambda}^N$ is generated in Woods-Saxon well with
parameters r_0 and a, where N is given by

$$2(N-1) + L = \sum_{i=1}^{4} 2(n_i - 1) + \ell_i$$

i.e., for $(1p)^4$ NL = 3S, 2D, 1G

which follows from oscillator shell model provided
we retain only alpha ground state piece of $\langle A | B \rangle$.

spectroscopic factor of ~ 3.1, which is to be compared with a shell model value of 2.85 predicted for the ground state transition alone. While additional studies are in order, these preliminary results suggest that the reaction is reasonably well understood.

CLUSTER KNOCKOUT REACTIONS

The Reactions (p,pα) and (p,d³He)

First I should like to discuss proton induced alpha cluster knockout. The interesting feature of the early Maryland work[8,23] on 1p shell targets, is that both (p,pα) and (p,d³He) data were obtained. Assuming the latter process proceeds via a (p,d) reaction on an alpha cluster as indicated in Fig. 14, we can generalize our DWIA calculations to include this reaction. In the calculations the projection of the target wave function onto the alpha ground state and residual nucleus was approximated by a Woods-Saxon cluster wave function with quantum numbers determined through conservation of oscillator quanta. Typical (p,pα) results are shown in Fig. 15 and (p,d³He) results in Fig. 16. For each target studied, factorization tests were fairly encouraging. In addition, spectroscopic factors extracted from the two reactions were remarkably consistent and in excellent agreement with conventional shell model predictions restricted to a 1p basis. These results, shown in Fig. 17, clearly support the approximation of retaining only alpha particle-like clusters in the reaction description.

Turning now to heavier targets, we can ask whether (p,pα) is a useful spectroscopic tool and whether we can learn anything about cluster momentum wave functions. Results are shown in Fig. 18 for ground state spectroscopic factors for targets from ^{16}O to ^{66}Zn obtained in a 101.5 MeV experiment by Carey and others.[24] The results, normalized to unity at ^{20}Ne, are seen to agree with the Rochester (^{6}Li,d) work[25,26] rather nicely although the latter reaction is somewhat sensitive to the choice of ^{6}Li optical potentials particularly for A > 52. The absolute spectroscopic factors, also indicated in the figure, tend to be somewhat larger than predicted values from nuclear structure theories. For example,

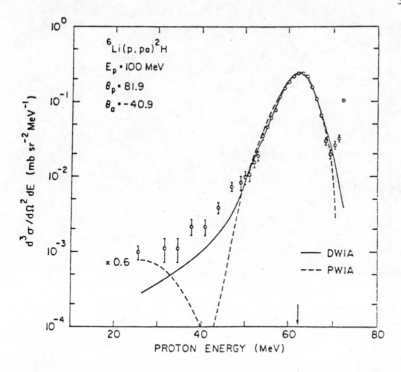

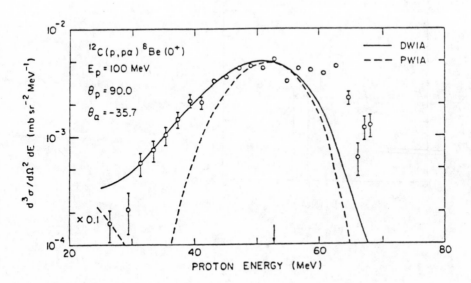

Fig. 15

34

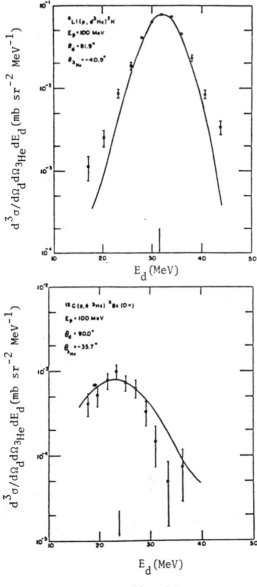

E_d (MeV)

Fig. 16

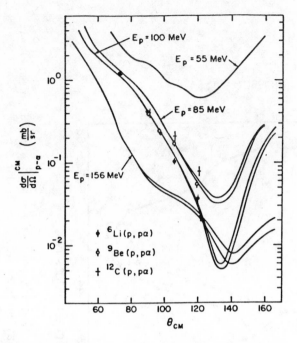

Fig. 17. Factorization test for (p,pα) at 100 MeV; —— free p+α
cross section; data points Q(θ) (see Fig. 8).

TABLE II

Spectroscopic factors: (p,pα) and (p,d^{3}He) at 100 MeV

Target	θ_1/θ_2	(p,pα)	(p,d^{3}He)	Theory
^{6}Li	81.9/-40.9	0.52 ± 0.03	0.59 ± 0.04	
^{7}Li	81.3/-41.0	1.09 ± 0.11	0.94 ± 0.07	1.12[†]
^{9}Be	81.2/-41.0	0.47 ± 0.04	0.43 ± 0.04	0.57[*]
^{12}C	90.0/-35.7	0.56 ± 0.12	0.59 ± 0.04	0.55[*]

[†]LS coupling   [*]Cohen & Kurath

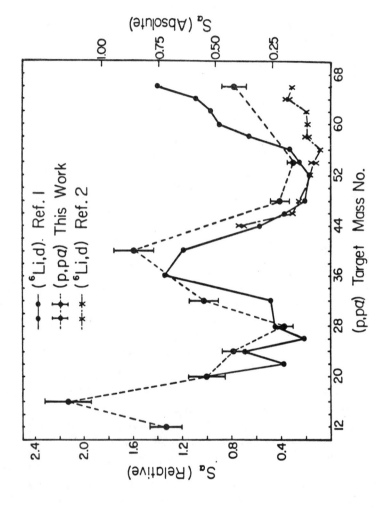

Fig. 18. Alpha cluster spectrocopic factors from (^{6}Li,d) and 100 MeV (p,pα) from Maryland. Ref. 1 and Ref. 2 are different DWBA analyses of (^{6}Li,d) data.[25,26]

the Chung et al. shell model[27] calculations predict values for ^{20}Ne and ^{40}Ca of 0.18 and 0.043, respectively. These are ~ 3 and 20 times smaller than the experimental values. Since we estimate uncertainties in distortions and off-shell effects as no more than about ±30%, the discrepancy for ^{40}Ca is certainly significant. We suspect that it is largely due to the model space truncation in the structure calculation in which ^{40}Ca is a closed (2s1d) shell. This restricts the quantum numbers for the alpha cluster wave function to 5S for ^{40}Ca → ^{36}Ar(g.s.). Including 2p-2h and 4p-4h components in ^{40}Ca could lead to coherent terms with higher principal quantum numbers, e.g., 7S and 9S which would significantly increase the predicted cross section even for quite small admixtures. Clearly, comparisons with improved structure calculations would be of considerable interest.

As far as more modest discrepancies such as for ^{20}Ne are concerned, one should note that the absolute (p,pα) cross section for these medium mass targets is particularly sensitive to the cluster bound state radius. The spectroscopic factors shown were obtained taking r_0=1.3 fm and a=0.65 fm to coincide with the (^{6}Li,d) analysis. However, an increase in r_0 of 0.1 fm increases the predicted cross section about a factor of 2. Whether such a change is reasonable depends upon the guidance available to us from nuclear structure theorists. Since this is generally rather limited (!) we must appeal to the data itself. In Fig. 19 we show calculations for the energy sharing distributions for three of the ground state transitions observed, each normalized at zero recoil momentum. In the data we must ignore regions contaminated by sequential processes in ^{16}O and ^{40}Ca. In the calculations we find that the shapes of the distributions are sensitive to r_0. Thus, the data provide rough limits on the acceptable range of r_0 values ($r_0 \approx 1.2 \rightarrow 1.6$ fm for heavier nuclei) and the experiment does exhibit sensitivity to the cluster momentum wave function. Note that large values of $r_0 \gtrsim 1.9$ fm are unacceptable. Such excessive values have been used in the past in analyses of (α,2α) reactions and various transfer reactions in order to obtain absolute spectroscopic factors in agreement with shell model predictions.

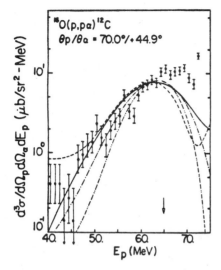

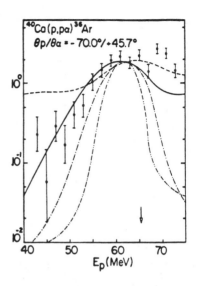

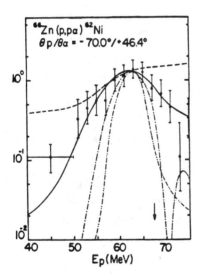

Fig. 19. Energy sharing distributions for 100 MeV (p,pα) ground state transitions. Curves are normalized DWIA calculations for different alpha particle bound state radius parameters: r_0 (0.7,---); (1.3,——); (1.9,-·-·-); (2.5,-··-).

The $(\alpha, 2\alpha)$ Reaction

As in the corresponding nucleon knockout reactions, comparisons of $(\alpha, 2\alpha)$ and $(p, p\alpha)$ are of interest because of the sensitivity of the alpha induced reaction to the extreme nuclear periphery. Among the more extensive studies of the $(\alpha, 2\alpha)$ reaction are the experiments by Sherman and Hendrie[28] at 90 MeV and by Wang and coworkers[29] at 140 MeV. A typical energy sharing distribution from the 140 MeV experiment is shown in Fig. 20. Unlike $(p, p\alpha)$ the distortion effects are sufficiently severe that essentially all sensitivity to the cluster momentum wave function in the target is lost. Thus, bound state potential parameters must be obtained independently. Wang, in his analysis, chose values of r_0 and a which yield a potential similar to the result of folding an α-N potential with the ^{12}C core density distribution. The chosen values were quite similar to those indicated in the $(p, p\alpha)$ study. Nevertheless, the resultant wave function has its nodes at slightly larger radii than a projected 1p shell model wave function. Thus, in view of strong surface localization in $(\alpha, 2\alpha)$ one can argue that the extracted spectroscopic factor will underestimate the shell model value. In fact, the result is ~ 50 times larger than the $(p, p\alpha)$ value which is itself comparable with or somewhat larger than the shell model value. Including reasonable amplitudes of 2p-2h and 4p-4h components does alleviate the problem somewhat, but does not eliminate it. To eliminate the problem requires, for example, a 4p-4h term about 4 times the 0p-0h simple 1p shell model component which is simply not physical.

On the basis of these results, one might suppose that $(\alpha, 2\alpha)$ is not a simple quasifree process. For example, the larger effective alpha clustering probabilities might be induced by the projectile in a multistep process involving inelastic scattering to states with large alpha widths. In fact, most other evidence seems to support a quasifree interpretation. For example, the ratio of spectroscopic factors for the first excited (2+) and ground state transitions in ^{16}O$(\alpha, 2\alpha)$ agree quite closely with the 1p shell model prediction. Similarly, relative spectroscopic factors from one target to another are rather similar to the $(p, p\alpha)$ results. Finally, one can carry out

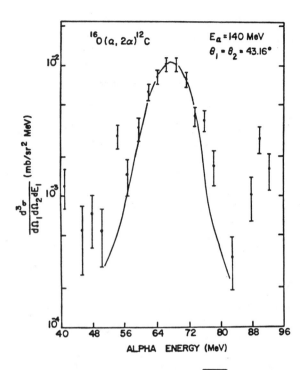

Fig. 20
a) $^{16}O(\alpha,2\alpha)^{12}C$
0.0 MeV L=0 at
140 MeV.

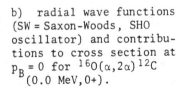

b) radial wave functions
(SW = Saxon-Woods, SHO
oscillator) and contribu-
tions to cross section at
$P_B = 0$ for $^{16}O(\alpha,2\alpha)^{12}C$
(0.0 MeV, 0+).

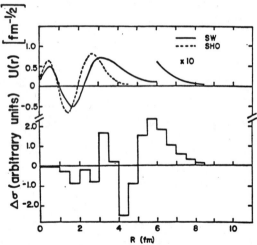

a factorization test to determine whether the reaction shows the angular dependence of the free $\alpha+\alpha$ elastic cross section. In Fig. 21 results for both ^{16}O and $^{20}Ne(\alpha,2\alpha)$ are seen to follow the free cross sections quite well over nearly two order of magnitude. One can conjecture that a multistep process might lead to significant smearing of this distribution.

If we accept a single step quasifree interpretation, then the $(\alpha,2\alpha)$ results imply very large clustering probabilities for ^{16}O and, in fact, for heavier targets. This is in contrast with simple shell model calculations and with $(p,p\alpha)$ studies. This does not necessarily present a problem since the DWIA calculations indicate that $(\alpha,2\alpha)$ is sensitive only to the extreme nuclear periphery where densities are 0.1% or less of central density. This, of course, is not a region we would expect to be well described in most structure calculations. In this region, however, the $(\alpha,2\alpha)$ results suggest that essentially all of the nuclear matter has coalesced into alphas.

Whether this should be taken seriously or not at this point I am uncertain. However, it is interesting to note that we have obtained similar ratios (of $\sim 50 \rightarrow 100$) between alpha induced knockout and proton induced knockout for $(\alpha,\alpha t)/(p,pt)$ and $(\alpha,\alpha d)/(p,pd)$ reactions on ^{16}O whereas, as we pointed out earlier, the corresponding ratio for $(\alpha,\alpha p)/(p,2p)$ is no worse than a factor of 2.

CONCLUSIONS

In summary then, I believe we can feel encouraged. It appears that $(p,2p)$ experiments are beginning to yield good data which should provide constraints on nucleon wave functions. Alpha knockout experiments by protons have started to tell us something about cluster momentum wave functions and the corresponding alpha induced experiments may be telling us something about the nuclear stratosphere. As far as $(\pi,\pi p)$ is concerned, we have to wait and see.

A note of caution is, however, necessary. In general, the experiments I have discussed tend to avoid regions of phase space where the factorized DWIA gives trouble. An example is the forward angle region in coplanar symmetric $(p,2p)$ experiments.[3] This is a region

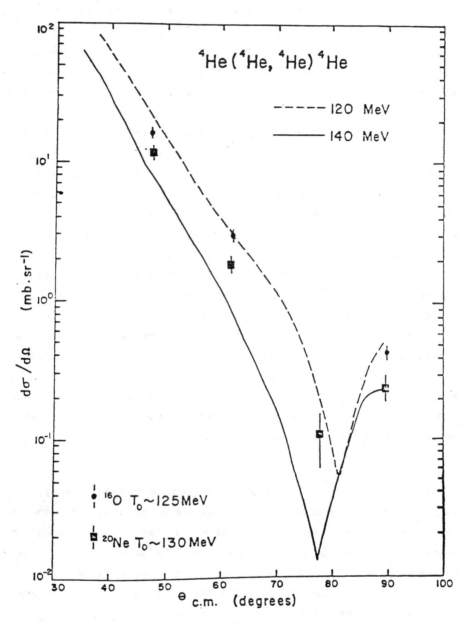

Fig. 21. Factorization test $^{16}O, ^{20}Ne(\alpha,2\alpha)$ at 140 MeV, lines are free cross section data.

where one might hope to learn something about off-shell effects. In order to address this problem and to refine our techniques for determining structure information, improvements in the theory are badly needed.

One useful improvement has been the inclusion of spin-orbit distortions in (p,2p). In addition to the ^{4}He case which we discussed, this has proved useful in analyses of recent (p,2p) experiments using polarized incident protons. However, the important step which is needed is to eliminate the factorization approximation currently used. As I mentioned, this was done in early work by McCarthy[7] for a fairly simple local psuedo-potential. However, studies with more realistic t-matrices would be desirable. A technique for dealing with the problem was described by Koshel[30] at the previous workshop. In addition, Taylor series expansion approximations have been outlined by Redish,[31] by Austern[32] and by Jackson.[33] However, so far rather little numerical work has been reported.

REFERENCES

1. C. A. Miller, Invited paper presented at 9th International Conference on the Few-Body Problem, Eugene, August 1980; C. A. Miller, in Common Problems in Low- and Medium-Energy Nuclear Physics, edited by B. Eastel, B. Goulard, and F. C. Khanna (Plenum Press, New York, 1979), p. 513.
2. W. J. McDonald, Nucl. Phys. A335, 463 (1980).
3. N. S. Chant, in Proceedings of the 3rd International Conference on Nuclear Structure and Nuclear Reactions, edited by W.T.H. van Oers, J. P. Svenne, J.S.C. McKee, and W. R. Falk (AIP, New York, 1978), p. 415.
4. P. G. Roos, in Momentum Wave Functions-1976, edited by D. W. Devins (AIP Conference Proceedings No. 36, New York, 1977), p. 32.
5. N. S. Chant and P. G. Roos, Phys. Rev. C 15, 57 (1977).
6. N. S. Chant et al., Phys. Rev. Lett. 43, 495 (1979).
7. K. L. Lim and I. E. McCarthy, Nucl. Phys. 88, 433 (1966).
8. P. G. Roos et al., Phys. Rev. C 15, 69 (1977); J. Watson et al., Nucl. Phys. A172, 513 (1971).
9. M. B. Epstein et al., Phys. Rev. Lett. 44, 20 (1980); W.T.H. van Oers et al., Phys. Rev. C 25, 390 (1982).
10. C. F. Perdrisat et al., Phys. Rev. 187, 1201 (1969).
11. H. Tyren et al., Nucl. Phys. 79, 321 (1966).
12. R. Frascaria et al., Phys. Rev. C 12, 243 (1975).
13. H. G. Pugh et al., Phys. Lett. 46B, 192 (1973).
14. T. K. Lim, Phys. Lett. 44B, 341 (1973).

15. C. Samanta, Ph.D. Thesis, University of Maryland, 1981.
16. K. H. Bray et al., Phys. Lett. B35, 41 (1971).
17. P. G. Roos et al., Phys. Rev. Lett. 40, 1439 (1978).
18. A. Nadasen et al., Phys. Rev. C 22, 1394 (1980).
19. L.R.B. Elton and A. Swift, Nucl. Phys. A94, 52 (1967).
20. A. Nadasen et al., Phys. Rev. C 19, 2099 (1979).
21. H. J. Ziock et al., Phys. Rev. C 24, 2674 (1981).
22. L. Rees, P. G. Roos, and N. S. Chant, to be published.
23. A. A. Cowley et al., Phys. Rev. C 15, 1650 (1977); A. Nadasen et al., Phys. Rev. C 22, 1394 (1980).
24. T. A. Carey et al., Phys. Rev. C 23, 576 (1981).
25. N. Antaraman et al., Phys. Rev. Lett. 35, 1131 (1975).
26. H. W. Fulbright et al., Nucl. Phys. A284, 329 (1977).
27. W. Chung et al., Phys. Lett. 79B, 381 (1978).
28. J. D. Sherman et al., Phys. Rev. C 13, 20 (1976).
29. N. S. Chant et al., Phys. Rev. C 17, 8 (1978); C. W. Wang et al., Phys. Rev. C 21, 1705 (1980).
30. R. D. Koshel, Nucl. Phys. A260, 401 (1976).
31. E. F. Redish, Phys. Rev. Lett. 31, 617 (1973).
32. N. Austern, Phys. Rev. Lett. 41, 1696 (1978).
33. D. F. Jackson, preprint (unpublished).

ENERGY-MOMENTUM BANDSTRUCTURE OF SOLIDS AS DETERMINED
BY ANGLE RESOLVED PHOTOELECTRON SPECTROSCOPY

R.C.G. Leckey

Research Centre for Electron Spectroscopy and Physics
Department, La Trobe University, Bundoora 3083,
Victoria, Australia.

Following a brief review of basic theory, experimental $E(k)$ band
structures of some third and fifth row transition and noble metals
are presented and discussed in the light of current ideas for the
interpretation of angle resolved photoemission data.   Some aspects
of experimental techniques are discussed and problems involved in
the interpretation of a complex material ($PtTe_2$) in the absence of
a theoretical bandstructure are illustrated.

1.   INTRODUCTION

This paper seeks to illustrate the capabilities of angle resolved
ultraviolet photoelectron spectroscopy (ARUPS) for the determination
of the energy/momentum distribution of electrons in the valence/
conduction bands of solids i.e., the $E(\underline{k})$ band structure of the
material.   The methodology of the technique is discussed in section (2),
which also touches upon the basic theory of the process while section (3)
consists of a brief discussion of recently obtained results from metals
of varying electronic complexity.   The utility of the technique for
determining the band structure of complex solids for which little or
no theoretical information is available is also discussed with particular
reference to $PtTe_2$.

Due to space limitations, this review is necessarily superficial in
many regards; (a) the examples chosen are all conductors – many
interesting studies of semi conducting materials are not covered (b)
no detailed discussion of surface states is included despite the
appearance of such states in certain of the results displayed (c)
little attention is paid to the benefits of controlling the polarization
of the incident photon beam nor of measuring that of the emitted electrons
when the sample is magnetic (d) the possibility of using ARUPS for
structural determinations, particularly of adsorbates, is ignored.
Despite these and other deficiencies, it is hoped that the following will
be of assistance to those working in related fields and will give some
indication of the capabilities of this relatively young technique in
probing the momentum distribution of electrons in solids.

2.   METHODOLOGY

The early historical development of angle resolved studies of the
photoemission process has recently been discussed by Jenkin [1] who
covered the years 1900-1960 .   In its modern form, interest in ARUPS
in relation to the solid state dates from the mid 1970's.   Prior to
that time, photoelectron spectrometers generally accepted a wide
range of electron emission angles and the technique concentrated largely
on the determination of the density of occupied conduction band states.

In what follows it will be assumed that it will be possible to identify
in the observed spectra, the unscattered electrons which have been
photo excited, all other secondary electrons being assumed to constitute
a structureless background of maximum intensity close to zero kinetic
energy.   The interaction of light with the sample, in the dipole
approximation, may be described by a Hamiltonian of the form

$$H = A(\underline{r}) \cdot \underline{p} + \underline{p} \cdot \underline{A}(\underline{r}) \qquad (1)$$

where $\underline{A}$ is the vector potential and $\underline{p}$ is the momentum operator.   The
elastic photo current arising from the interaction of the electromagnetic
field with a solid has been calculated by, for example, Mahan [2] and
may be written in a form similar to the Golden Rule [3].   Thus, the
photo current excited into the direction of a detector $\underline{R}$ by light of energy
$\hbar\omega$ is

$$J(\underline{R}, E, \hbar\omega) = \frac{2\pi e}{\hbar} \left(\frac{e}{mc}\right)^2 \sum_i |M_{fi}|^2 \; \delta(E-E_i - \hbar\omega) \qquad (2)$$

Here E is the kinetic energy of the detected electron, E_i the energy of
the initially occupied state, the sum is taken over all occupied states
and the delta function expresses the Einstein energy conservation condition.
The matrix element M_{fi} concerns initial $|i>$ and final $|f>$ eigenstates of
the Hamiltonian H_o of the solid: $H_o = p^2 + V(\underline{r})$, $V(\underline{r})$ being a self-
consistent potential which should include the presence of the surface.
Thus

$$M_{fi} = <f \, |\underline{A} \cdot \underline{p} + \underline{p} \cdot \underline{A} \, | i > \qquad (3)$$

may be considered as the sum of two contributions; the $\underline{p} \cdot \underline{A}$ term is surface
specific being due to the discontinuity in the dielectric function ε
across the surface. [4]   The photo emitted current due to this effect
is generally much smaller than that due to the first term in M_{fi} and
will not be discussed here.   Although $|i>$ is an eigenfunction of H_o,
it is not a Bloch function since H_o contains the surface which breaks
the translational symmetry normal to the surface plane.   We may
reasonably assume a perfect surface in the sense that periodicity is
maintained parallel to the surface plane and hence $|i>$ may be written
in the two-dimensional form:

$$\psi_i (r) = \exp (i\underline{k}_{//} \cdot \underline{\rho}_{//}) \; \phi_i (\rho, z) \qquad (4)$$

where $\rho = (x, y)$; z being the surface normal.

The final state $|f>$ should describe the free electron nature of the
electron at the detector, its previous transmission to and refraction
at the surface potential step and any damping due to inelastic
scattering in its progress towards the surface.   It has been recognized
for some time that the history of such a photoelectron is the time
reversed history of an incident electron in a low energy electron
diffraction (LEED) experiment and that consequently time reversed LEED
wavefunctions may be used to specify the final state in a photoemission
experiment [2,5].   Such wavefunctions contain terms of the form
$\exp (i \, k_\perp z)$ as well as terms involving the $k_{//}$ periodicity dependance,

where the normal component of $\underline{k}$, $k_\perp$ is complex, the imaginary part describing the inelastic scattering of the electron. $\mathcal{I}$m $k_\perp$ may thus be related to an inelastic scattering mean free path λ via

$$\lambda(E,\underline{k}) = \mathcal{I}m \ (k_\perp \cos \theta)^{-1} \qquad (5)$$

or alternatively to a lifetime Γ_e of the photoexcited state. [6] A central problem in the interpretation of ARUPS concerns the magnitude of $\mathcal{I}$m $k_\perp$ since this in turn controls the extent to which $k_\perp$ is conserved between the original photoexcited state within the solid and the arrival of a free electron at the entrance to the detector.

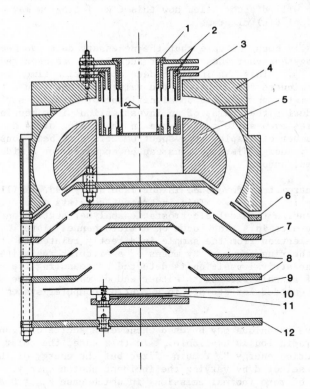

Figure 1. Toroidal energy analyser for angle resolved studies: 1-3 input zoom lens; 4,5 toroidal sectors; 7 analyser focal plane; 8,9 output lens; 10,11 chevron channel plate and position sensitive detector.

In summary, whereas k is conserved in the photoexcitation process, $k_{//}$ is conserved during the transit to the detector aperture but considerable uncertainty generally remains as to the extent to which $k_\perp$ is determined internally. Conservation of $k_{//}$ and the assumption of free electron refraction by the surface potential leads to the result

$$k_{//} = \hbar^{-1} \sqrt{2mE_k} \sin \theta \qquad (6)$$

where E_k is the kinetic energy of the detected electron and θ the polar angle of emission.

An ARUPS experiment therefore consists of determining the locus of peaks in the energy distribution of emitted photoelectrons as a function of (polar) angle of emission. This then enables an energy-momentum diagram $E(k_{//})$ to be constructed for each chosen value of the azimuthal angle ϕ. We will discuss below how this $E(k_{//})$ diagram may be compared with theoretical $E(k)$ diagrams.

It has generally been the case that the necessary data has been recorded using an energy analyser whose angular acceptance has been reduced to $\pm 2^o$ in θ and ϕ. This has involved data acquisition times in the region of 20 minutes for each selected value of θ and ϕ in a typical ARUPS spectrometer and has led to significant problems of surface cleanliness during the course of an investigation since monolayer formation times even at 10^{-10} Torr are in the region of 24 hours. Recently a number of display type spectrometers have been constructed [7,8] whereby energy distributions can be acquired for a wide range of angles simultaneously.

One such spectrometer, developed in our laboratory [9], is illustrated schematically in figure 1. This instrument consists of a cylindrical input zoom lens, a toroidal electrostatic analyser section and an exit lens which results in a ring focus on a microchannel plate. The emission angle of an electron from the sample is directly related to its arrival position at the ring focus. By means of a circular resistive strip delay line the arrival position is detected by measuring the transit time difference of signals flowing to either end of the resistive strip, and is digitally encoded for storage in a control computer system (LeCroy System 3500).

An alternative technique may be used whenever a source of (tunable) synchrotron radiation is available. In this case, the polar angle θ and the detection energy E_k remain fixed but the energy of the initial state may be selected by varying the incident photon energy. Often θ is chosen to be zero (normal emission) in which case $k_{//} = 0$ and all detected electrons must originate along a line in k space. Not all points in the Brillouin zone are generally accesible in this manner because of surface preparation difficulties.

If, in addition to acquiring a set of energy distribution curves for various values of θ from a single crystal face, a similar set of curves is obtained for emission from a second face which is chosen to make a small convenient angle to the first face, and the data are collected without changing the azimuthal angle, then it is possible to locate

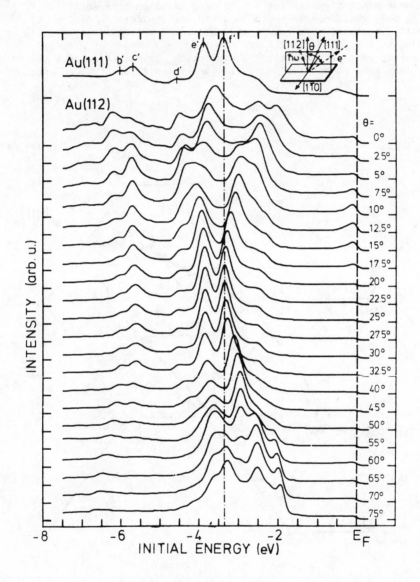

Figure 2. Photoelectron spectra from Au(111) and Au(112) using 21.12 eV
photons.   From reference 24.

precisely the point in k space from which a particular transition
originates.   This technique, originally suggested by Kane[10] is illustrated
in figures 2 and 3.

50

In the example chosen here, taken from reference [24], structure labelled
f^1 in a normal emission spectrum from Au(111) may be seen to be reproduced
at the same binding energy for $\theta \simeq 25°$ for emission from Au(112).
Assuming only the conservation of k_{11} via the application of Eqn. 6,
the initial state $E(\underline{k})$ is completely determined as illustrated in figure 3.
This procedure may be repeated for the (limited) number of clearly
identifiable peaks in the energy distribution curves; additional values
of E, $\underline{k}$ may also be obtained by variation of the incident photon energy.

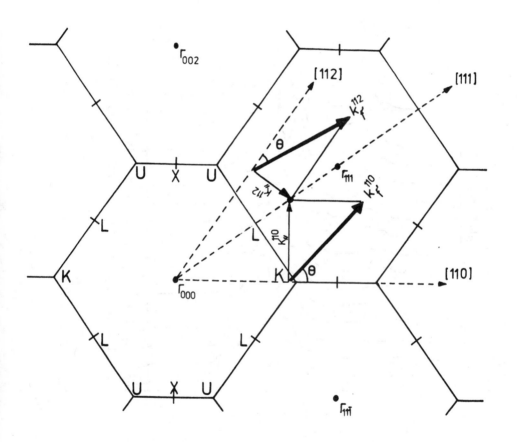

Figure 3.   Section through the Brillouin zone for Au.   The $k_{//}$ values
            for the peak f^1, as calculated using eqn. 6 are shown for the
            (110) and (112) surfaces, thereby identifying the initial
            state location at a point between L and Γ.

It should be noted that, assuming direct (k conserving) transitions, this method also provides information concerning the final state dispersion, although its application to Au has recently been criticized in this regard. [12]   There are also clear difficulties in the preparation of suitable surfaces having the correct angular relationship to each other, for many materials of interest.

3.   A SELECTION OF REPRESENTATIVE RESULTS

Experimental data from a number of laboratories has been combined in figure 4 for dispersion along the threefold axis in Co, Ni, Cu and Zn. Here the lines are largely experimentally determined, but with assistance from published band structure calculations.   Particularly in the case of Zn, the distinction between flat bands of mainly d character and distorted parabolic bands of s,p nature is very obvious.   That this assignment is correct has been verified by observing the variation in relative intensity of structure in the energy distribution curves as the photon energy is varied.   The photoexcitation cross section for d electrons is known to be relatively unchanged in the range $20 < \hbar\omega < 40 eV$

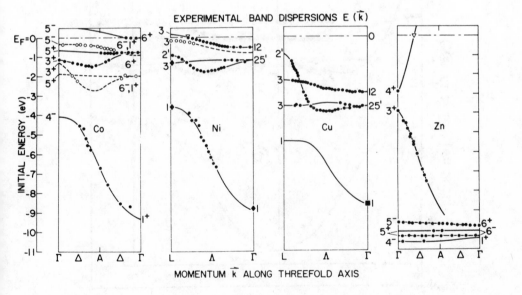

Figure 4.   Experimentally derived E(k) bands for Co, Ni, Cu and Zn. from reference 11.

whereas that for s, p states decreases by a factor of ~10 x as $\hbar\omega$ is increased from 20 eV to 40 eV.   In figure 4 therefore, we may identify the 3d electrons moving from a position straddling the Fermi energy in Co and immediately below the Fermi energy in Ni, to a tightly bound situation in Zn, where they may be regarded as almost localized or core-electron like.   In the case of Ni and Co, majority spin bands have been shown dashed and the ferromagnetic exchange splittings of ~1.25 eV in Co and 0.3 eV in Ni have recently been determined [13].

The data of figure 4 was obtained using the synchrotron radiation method and consequently depends on some prior knowledge of the final state band structure in order to specify $k_\perp$ in each case.   For Cu (111) and other close-packed transition and noble metals, it has been established that a nearly free electron parabolic band of Λ symmetry exists centred at ~4.5 eV above the Fermi energy at the L point of the Brillouin zone and extending across most of the zone towards Γ.   This band has therefore been used to relate the measured energy E_k to the actual electron momentum – using $E_k = \hbar^2 k^2/2m$ – and hence to determine $k_\perp$.

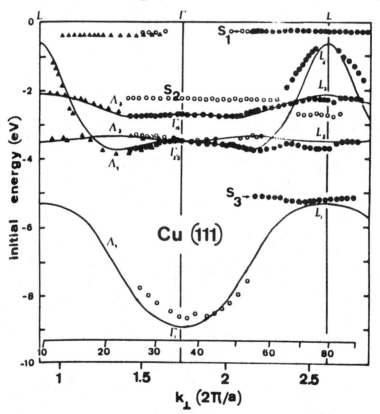

Figure 5.   Experimental data ● compared with theoretical bandstructure – for Cu (111).   From reference 14.

The case of Cu (111) is examined in more detail in figure 5 taken from the synchrotron measurements of Petroff and Thiry [14]. In this figure, the experimentally derived points are compared directly with a band structure calculation due to Burdick [15]. The detailed agreement is generally very good; the experiment also contains three, largely dispersionless structures labelled S_1 - S_3. S_1 and S_3 are found to be excited only with p-polarized light, are sensitive to small amounts of surface contamination and appear in hybridizational band gaps. For these reasons they are taken to be surface states of the Shockley type.[16] S_2, on the other hand, is considered to be a Tamm surface state [17] requiring the existence of a surface layer possessing physical characteristics different (in terms of bond length, potential distribution) from bulk layers. (Shockley surface states result directly from the breakdown of periodicity across the surface and do not necessarily imply a distorted surface layer).

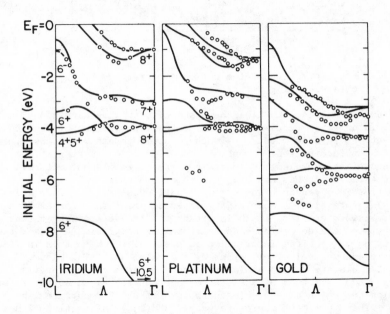

Figure 6. Angle resolved photoemission data for Ir, Pt and Au from reference 11.

If the data for 3d metals shows excellent agreement between experiment and theory, the situation is not quite so good for the 5d metals Ir, Pt and Au, as shown in figure 6. Since a fully relativistic calculation is required for these metals, perfect agreement is scarcely to be expected at this stage and in many ways it is gratifying to see that broad agreement between experiment and theory does exist. The effects of spin-orbit splitting are clearly visible at Γ_7^+ and Γ_8^+ and, although the degeneracy remains at the Γ point, the upper 8^+ band clearly splits along Λ. The d band width is controlled by the extent of spatial overlap of the radial d wavefunctions and in part, by the energy position of the d bands relative to the s,p bands. [11]

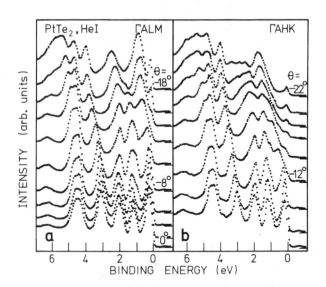

Figure 7.   Representative photoelectron spectra from PtTe$_2$
obtained using HeI radiation.

In the above examples, ARUPS has been shown to provide exceptionally
detailed information concerning the momentum distribution of conduction
band electrons in some metals.   In the examples chosen, the experimental
data has been interpreted, in most cases, with some assistance from
prior theoretical knowledge of certain parts of the bandstructure.   In
this short review it has not been possible to discuss fully the
difficulties involved in transforming experimental $E(k_{//})$ information
into $E(k)$ information.   Such difficulties do exist and, for ARUPS
to be of maximum utility, it is important to discover to what extent
the experiment can determine bandstructure for complex materials for which
little or no theoretical information is available.   With this in mind,
we have recently studied the series NiTe$_2$, PdTe$_2$ and PtTe$_2$ [18] .
These materials are hexagonal layer compounds which cleave readily along
the basal plane which is of assistance in preparing clean surfaces for
inspection but which also precludes the use of the two-surface Kane
method.   The experiment therefore provides $E(k_{//})$ information with the
degree of $k_\perp$ conservation remaining to be established.   The complexity
of the band structure for these materials is foreshadowed by the richness
of the experimental spectra.   Figure 7 shows some illustrative spectra
obtained from the ΓALM surface of PtTe$_2$.   Using such spectra and eqn. 6
it is possible to plot points on an $E(k_{//})$ diagram but, without the
assistance of a theoretical prediction, it is not always clear how these
points should be joined into bands.   Using the information provided by
amplitude modulation as a function of photon energy, it is possible to
assign mainly Pt d character to the rather flat bands with binding energies
close to 3.1 eV and 4.5 eV at Γ, the remaining bands being largely of Te p
character.   In figure 8 the bands shown have been drawn with some
assistance from the results for PdTe$_2$, a material for which a theoretical
band structure has been calculated.

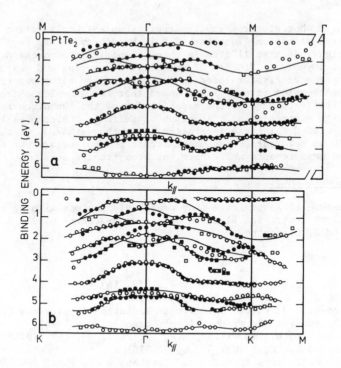

Figure 8.   Experimental energy-momentum dispersion for the
ΓK and ΓM directions in PtTe$_2$ as constructed from data
such as that of Figure 7.

For the case of lighter materials of layer-like composition, it has
proved possible to interpret ARUPS data as if the materials were two-
dimensional.   The weak van der Waals bonding between layers is not
sufficient to cause appreciable dispersion of the bands in the $k_\perp$ direction
for such materials and consequently emission detected at a particular
polar angle can originate from a line in $\underline{k}$ space.   Consequently, although
$k_\perp$ remains indeterminate it is also largely irrelevant.

In the heavy layer compounds studied in our laboratory we have found that
while there is little evidence of dispersion in the $k_\perp$ direction for the
metal d bands, there is significant such dispersion for the p bands
closer to the Fermi energy.   We have consequently found that although
much of the experimental data can be interpreted in terms of a model
which assumes $k_\perp$ is not conserved (an "indirect" model [19]) not all
points on figure 8 can be interpreted fully in this manner.   The
interested reader is referred to reference 18 for further information.

4. SUMMARY

Angle resolved photoelectron spectroscopy is an extremely active and expanding area of topical interest. This brief review has been able merely to highlight a small area in which significant advances have been made in recent years. Studies of non metallic samples and, particularly, of adsorbate and catalyst related investigations are being reported at an ever increasing rate. It is hoped that the examples chosen here served to illustrate some of the capabilities of the technique in the area of energy - momentum distributions of solids. References 20-23 have been included for those interested in further study. Of these, reference 20 is relevant to gas phase atomic and molecular angular distribution measurements which have not been treated here.

5. ACKNOWLEDGEMENT

It is a pleasure to acknowledge many useful discussions with my colleagues in the Research Centre for Electron Spectroscopy, particularly J.D.Riley, J. Jenkin and J. Liesegang.

6. REFERENCES

1   Jenkin, G.J., J. Elec. Spectrosc., 23, 187 (1981)
2   Mahan, G.D. Phys. Rev. B2, 4534 (1970)
3   Ashcroft, N.W. Vacuum Ultraviolet Radiation Physics eds E.E. Koch and R. Haensel (Pergammon-Vieweg, Braunschweig 1974)
4   Feuerbacher, B. and Willis, R.F. J. Phys. C9, 169 (1976)
5   Pendry, J.B., Surf. Sc. 57, 679 (1976)
6   Knapp, J.A. Himpsel, F.J., and Eastman, D.E., Phys. Rev. B19,4952 (1979)
7   Eastman, D.E., Donelon, J.O., Hien, L.C. and Himpsel, F.J., Nucl. Instr. and Meth. 172, 327 (1980)
8   Engelhardt, H.A., Back, W. and Menzel, D., Rev. Sci. Instrum. 52, 835 (1981); 52, 1161 (1981).
9   Riley, J.D., and Leckey, R.C.G., International Patent Application No PCT/AU81/00053
10   Kane, E.O., Phys. Rev. Letts., 12, 97 (1964)
11   Eastman, D.E. and Himpsel, F.J. Inst. Phys. Conf. Ser. 55, 115 (1981)
12   Christensen, N.E., Solid State Comm. 37, 57 (1981)
13   Heimann, P., Himpsel, F.J. and Eastman, D.E., Solid State Comm. 39, 219 (1981)
14   Petroff, Y., and Thiry, P. Appl. Optics 19, 3957 (1980)
15   Burdick, G.A., Phys. Rev. 129, 138 (1963)
16   Shockley, W., Phys. Rev. 56, 317 (1939)
17   Tamm, I., Zeits, Physik 76, 849 (1932)
18   Orders, P.J., Liesegang, J., Leckey, R.C.G., Jenkin, J. and Riley J.D., submitted to J. Phys.C.
19   Ley, L.F., J. Elect. Spectrosc., 15, 329 (1979)
20   Kreile, J. and Schweig, A., J. Elect. Spectrosc. 20, 191 (1980)
21   Himpsel, F.J., Appl. Optics 19, 3964 (1980)
22   Cardona, M. and Ley, L.F., (eds) Photoemission in Solids (Springer Verlag, Berlin, 1978)
23   Feuerbacher, B. and Fitton, B. in Electron Spectroscopy for Surface Analysis ed H. Ibach (Springer Verlag, Berlin, 1977).
24   Heimann, P., Miosga, H. and Neddermeyer, H., Solid State Comm., 29, 463 (1979).

NUCLEAR MOMENTUM DISTRIBUTIONS
FROM (γ,p) and (e,e'p) REACTIONS

J. L. Matthews[*]
Massachusetts Institute of Technology
Cambridge, MA 02139

ABSTRACT

This paper considers the determination of the momentum distribution of nucleons in nuclei from electromagnetic knockout processes such as the (γ,p) and (e,e'p) reactions. Some representative empirical momentum distributions are compared with theoretical predictions. The validity and limitations of the direct single-particle knockout picture are discussed.

INTRODUCTION

In this paper we discuss some aspects of the problem of obtaining information from electromagnetic nucleon-knockout processes on the momentum distribution of nucleons in nuclei. Impulse-approximation diagrams for two such processes - viz. the (γ,p) and (e,e'p) reactions are shown in Fig. 1. In the (γ,p) case, for a given incident photon energy

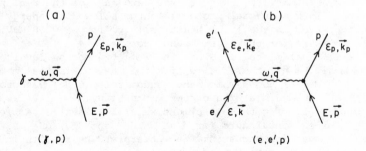

Fig. 1. The (γ,p) and (e,e'p) reactions in impulse approximation.

and proton emission angle one measures the differential cross section $d^2\sigma_\gamma/dE_p d\Omega_p$; in the (e,e'p) case one measures the coincidence cross section $d^4\sigma_e/dE_e d\Omega_e dE_p d\Omega_p$. In either case, the cross section may be written schematically as

$$\sigma \sim C \cdot \sigma_p \cdot S(\vec{p},E), \tag{1}$$

where C contains constants and kinematic factors, σ_p represents an elementary absorption or scattering cross section, and $S(\vec{p},E)$ is the

*Supported in part by the U. S. Department of Energy.

so-called spectral function, i.e. the probability of finding a proton with momentum $\vec{p}$ and energy E in the nuclear ground state. In the independent-particle (shell) model, $S(\vec{p},E)$ becomes

$$S(p,E) = \sum_{\alpha} N_{\alpha} \left| \phi_{\alpha}(p) \right|^2 \delta(E-E_{\alpha}), \tag{2}$$

where α designates a given shell-model orbital; N_{α} is its occupation number ($N_{\alpha} \leq 2j_{\alpha}+1$), E_{α} its energy eigenvalue, and $\phi_{\alpha}(p)$ its momentum distribution, i.e.

$$\phi_{\alpha}(p) = (2\pi)^{-3/2} \int e^{i\vec{p}\cdot\vec{r}} \psi_{\alpha}(r) \, d^3r. \tag{3}$$

We shall show that with certain assumptions $\phi_{\alpha}(p)$, a quantity which can be calculated from various versions of the shell model, can be inferred from experimental data on the (γ,p) and $(e,e'p)$ reactions.

Although superficially similar, the (γ,p) and $(e,e'p)$ processes exhibit important differences which may be categorized as follows:

    i)   ("Dynamic") The elementary amplitudes in Eq. (1) are not the same. For $(e,e'p)$ one may indeed insert the cross section for scattering of an electron by a free proton. For (γ,p) a free proton, of course, cannot absorb a photon (excluding meson production); however, a cross section of the form of Eq. (1) does appear, as will be seen later. Also, we note that $q \neq \omega$ (the photons are virtual) in the electron-induced process.

    ii)   ("Kinematic") Owing to the zero mass of the real photon, there is a minimum initial proton momentum allowed by energy and momentum conservation in the (γ,p) reaction (e.g. for $E_{\gamma} = 60$ MeV, $P_{min} \approx 300$ MeV/c for the least-bound protons).

    iii)   ("Experimental") Since the $(e,e'p)$ process involves a coincidence measurement, there is in effect a maximum initial momentum (also ~ 300 MeV/c) above which the random coincidence background makes measurements impossible with present technology.

In the following three sections we will present a sample of the available experimental data for $d^2\sigma_{\gamma}/dE_p d\Omega_p$ and $d^4\sigma_e/dE_e, d\Omega_e, dE_p d\Omega_p$, discuss the reaction mechanism, and interpret the results in terms of nuclear momentum distributions.

EXPERIMENTAL DATA

The most extensive series of good-resolution (γ,p) and $(e,e'p)$ measurements have been performed by the Glasgow-MIT[1-4] and Saclay[5-7] groups, respectively (see also Ref. 8). Typical proton energy spectra are shown for the ^{6}Li and ^{12}C nuclei in Figs. 2 and 3. In order to extract the quantity $S(p,E)$, and further, $\phi(p)$ from data such as these, one would in principle repeat the measurements at different settings of (ω,θ_p) or (E_e, E_e', θ_p) so as to vary the initial momentum p, and select the initial energy E of interest by examining the appropriate region of the proton spectrum. For example, both spectra in Fig. 3 exhibit a prominent peak $E_s = 16.0$ MeV. In the shell-model picture, these protons have been removed from the $1p_{3/2}$ orbital (with eigenvalue $E = -16.0$ MeV) in the ^{12}C nucleus. The cross sections obtained from the areas under these peaks would be divided by the appropriate values of C [Eq. (1)] in order to determine $S(p,E_{3/2})$, and

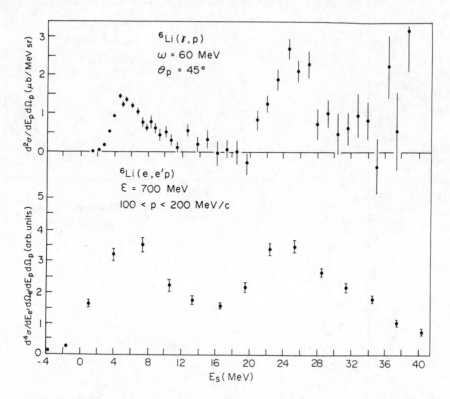

Fig. 2. Differential cross section as a function of proton separation energy $E_s = -E$ in the ^{6}Li(γ,p) [Ref. 1] and $(e,e'p)$ [Ref. 8] reactions. The initial proton momentum in the (γ,p) case is the range $200 \lesssim p \lesssim 300$ MeV/c.

thus, in the shell-model, $\phi_{3/2}(p)$. However, before presenting these results we must discuss the approximations implicit in Fig. 1 and Eq. (1).

DIRECT-KNOCKOUT REACTION MECHANISM

We shall formulate the single-particle direct-knockout reaction mechanism here only for the case of the (γ,p) process. Similar physical arguments hold for the $(e,e'p)$ case, but the theoretical treatment is different in detail and is beyond the scope of this paper [see, e.g., Refs. 5-7].

The differential cross section for the (γ,p_n) reaction [n = 0, 1, 2, . . ., specifying the final state of the residual nucleus] may be written in plane-wave Born approximation [Fig. 1(a)] as[9]

$$\frac{d\sigma}{d\Omega} = C(\omega,\vec{k}_p)\left|\phi_\alpha(p)\right|^2 . \tag{4}$$

We assume that the proton is in an initial state $|\alpha\rangle$ characterized by a wave function ψ_α in coordinate space or $\phi_\alpha(p)$ in momentum space

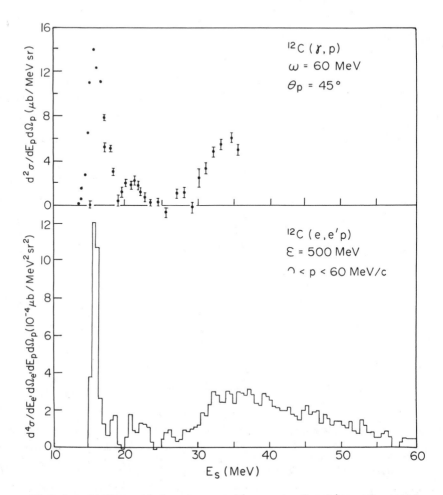

Fig. 3. Differential cross section as a function of proton separation energy $E_s = -E$ in the $^{12}C(\gamma,p)$ [Ref. 1] and (e,e'p) [Ref. 5] reactions. The initial momentum in the (γ,p) case is in the range $200 \lesssim p \lesssim 300$ MeV/c.

[Eq. (3)], and that the final nucleus is left in a hole state |n> corresponding to the removal of a proton from the state |α>. In this approximation $\vec{p} = \vec{k}_p - \vec{q}$ ($\vec{k}_p$ is the measured proton momentum); i.e. the experimental knowledge of (ω,θ_p) uniquely determines $\vec{p}$ and thus each measurement of $d\sigma/d\Omega$ directly yields a measurement of $\phi_\alpha(p)$.

A more realistic picture of the (γ,p) reaction is illustrated by the diagrams in Fig. 4. The dashed lines are intended to signify the interaction of the proton with at least one other nucleon; they could represent a short-range interaction (via the repulsive core of the NN potential), a medium-range interaction (via the one-pion-exchange part of the NN potential), or in the final-state interaction case, the long-range, average effect of the (A-1) system which is generally described by a complex optical potential. (The proton is of course initially

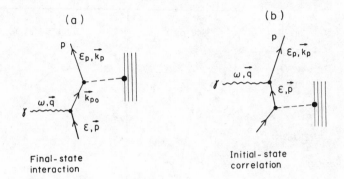

Fig. 4. Final- and initial-state interaction effects in (γ, p) reaction.

bound by the average nuclear potential.) In any case, however one involves the additional particle(s), there is clearly no longer a unique relation between $\vec{p}$, $\vec{q}$, and $\vec{k}_p$, and thus no longer a direct correspondence between cross section and momentum distribution.

If we assume that the most important correction to Fig. 1(a) is the final-state interaction, i.e. Fig. 4(a) with the dashed line representing the proton-nucleus optical potential, the (γ, p) cross section may be written in distorted wave impulse approximation as[10]

$$\frac{d\sigma}{d\Omega} = C \sum_L A_L P_L(\cos\theta_p) . \tag{5}$$

To obtain Eq. (5) we have expanded the electromagnetic operator in multipoles; the coefficients of the Legendre polynomials P_L contain angular-momentum-algebra factors and the integral over the initial and final radial wave functions:

$$A_L \sim \sum_{\ell, \ell', \lambda} B(L, \ell, \ell', \lambda) \int g_{\ell'}(kr) j_\lambda(qr) \frac{\partial}{\partial r} R_{n\ell}(r) r^2 dr . \tag{6}$$

In Eq. (6), $R_{n\ell}$ is the radial part of the initial bound-state wave function [e.g. $\psi_\alpha(\vec{r}) = R_{n\ell}(r) Y_\ell^m(\hat{r})$], and $g_{\ell'}(kr)$ is the radial part of the distorted-wave continuum state, calculated from the complex optical potential. The quantity $\phi_\alpha(p)$ no longer appears explicitly. But is the momentum distribution still present, implicitly, or has the final-state interaction completely "smeared out" the relationship between the initial and final proton momenta? To answer this question, we performed the following calculations: We obtained a representative shell-model momentum space wave function using Eq. (3) and the Woods-Saxon potential parameters given for ^{16}O [$|\alpha\rangle = |1p_{1/2}\rangle$] by Elton and Swift.[11] We then added a "perturbation" to this wave function at $p = 2.7 fm^{-1} = 533$ MeV/c, as shown in Fig. 5(a). Transforming this "perturbed" wave function back to coordinate space and calculating the (γ, p) cross section using Eqs. (5) and (6) for $\omega = 175$ MeV yields the result shown in Fig. 5(b). We see that the effect of the "perturbation" appears in a fairly well-localized region of the angular distribution, broadened somewhat by the final-state interaction. We conclude, qualitatively, that the (γ, p) cross section at $(\omega, \theta_p) =$

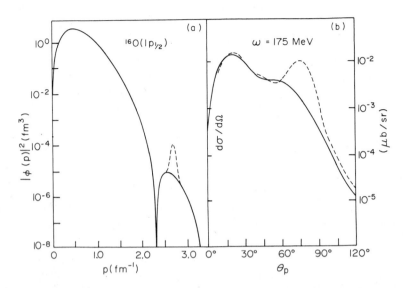

Fig. 5. (a) $1p_{1/2}$ shell-model wave function for ^{16}O (solid curve), with "perturbation" added at $p = 2.7 \text{fm}^{-1}$ (dashed curve). (b) Differential cross section for $^{16}O(\gamma,p_0)^{15}N$ reaction for $\omega = 175$ MeV, calculated in DWIA using shell-model wave function without (solid curve) and with (dashed curve) "perturbation".

(175 MeV, 75°) is <u>mainly determined</u> by a small range of initial proton momenta. (Calculating $\vec{p} = \vec{k}_p - \vec{q}$ for this case, we find $p = 538$ MeV/c.)

This result provides some encouragement for the construction of a model for the (γ,p) reaction which preserves the explicit correspondence between the cross section and initial proton momentum distribution. Referring to Fig. 4(a), we write the cross section in "modified plane wave Born approximation" as follows:[12,13,14,4]

$$\frac{d\sigma}{d\Omega} = D[\omega,\vec{k}_p,V(\varepsilon_p),W(\varepsilon_p)] \, |\phi(p)|^2, \tag{7}$$

where now $\vec{p} = \vec{k}_{p_0} - \vec{q}$ and the measured proton momentum (squared) is

$$k_p^2 = k_{p_0}^2 - (2m/\hbar^2)|V(\varepsilon_p)| . \tag{8}$$

In this picture, $\vec{k}_{p_0}$ is the momentum of the proton immediately after absorbing the photon. The proton must then "climb out of" the potential well $V(\varepsilon)$ due to the rest of the nucleus, with the result that $k_p < k_{p_0}$. The effect of the imaginary part of the potential W is included by means of an absorption factor $\eta(\varepsilon_p) < 1$, to account for some of the protons being lost through inelastic processes. The (energy-dependent) parameters V and W may be obtained from optical-model analyses of proton scattering.[14,15]

We note that the approximate treatment of distortions implied by Eq. (7) maintains the unique relationship between the final and initial proton momenta and the proportionality between the cross section and momentum distribution.

Before applying Eq. (7) to experimental measurements of $d\sigma/d\Omega$ and deducing $\phi(p)$, we first test this approximation by using Eq. (5) with an Elton-Swift shell-model wave function to generate some "pseudo-data". Dividing these calculated cross sections (for $\omega = 60$, 80, 100 MeV; $\theta_p = 30^\circ - 150^\circ$) by the appropriate values of $D(\omega,\vec{k}_p,V,W)$ yields the values for $|\phi(p)|^2$ shown as the solid circles in Fig. 6 (Ref. 13).

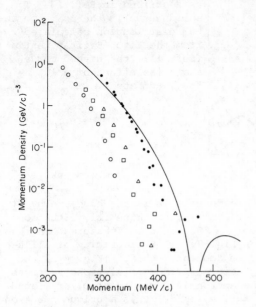

Fig. 6. Test of the approximate treatment of distortions in the (γ,p) reaction. Solid circles: distortions included [Eq. (7)]; open symbols: distortions omitted [Eq. (4)]. (See text and Ref. 13.)

If this treatment of distortions were perfect, these points would lie on the solid curve, which is the momentum distribution corresponding to the wave function from which the pseudodata were obtained. It is possible that the final-state potential parameters could be adjusted to better "fit" the points to the curve; such a procedure, however, is not really justified owing to the approximate nature of Eq. (7). The necessity of including the distortion correction through Eq. (7) is also illustrated in Fig. 6: the open circles, squares, and triangles are the values of $|\phi(p)|^2$ obtained from Eq. (4) [no distortions] for $E_\gamma = 60$, 80, 100 MeV, respectively. It is seen that this prescription fails to produce a unique result for $|\phi(p)|^2$ and that neither set of points agrees at all with the original momentum distribution.

We now apply this procedure to "real" (γ,p) data, obtaining the momentum distributions shown in Figs. 7 and 8, along with some $(e,e'p)$ results.[5,7] The $^{12}C(\gamma,p)$ data[1] represent cross sections in the energy and angular ranges $\omega = 60-100$ MeV, $\theta_p = 30^\circ-150^\circ$, as do the $^{16}O(\gamma,p)$ data,[2] which have in addition been extended to $\omega = 300-380$ MeV at three angles.[16] The (γ,p) results are seen to yield a unique $|\phi(p)|^2$ in both cases for $p \lesssim 500$ MeV/c, and the consistency for $p \gtrsim 500$ MeV (for ^{16}O) is only slightly inferior. (The squares, open circles, and triangles in Fig. 8 denote measurements at $\theta_p = 45^\circ$, 90°, and 135°, respectively.)

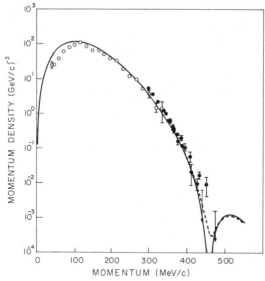

Fig. 7. ^{12}C p-shell proton momentum distribution obtained from (e,e'p) [open circles – Ref. 5] and (γ,p) [solid circles – Ref. 1] measurements as described in Ref. 13. The solid curve is the $1p_{3/2}$ momentum distribution calculated from the Elton-Swift potential (Ref. 11); the dashed curve shows the effect of a 10% $1p_{1/2}$ admixture.

Another way of describing these results is in the high-energy physics language of "scaling". For example, two kinematic situations in which the same initial momentum p = 500 MeV/c is "measured" in the ^{16}O(γ,p$_0$)^{15}N reaction are shown in Fig. 9. (For simplicity of illustration, the final-state interaction "momentum shift" [Fig. 4(a) and Eq. (8)] has been ignored.) The measured differential cross sections in such cases, when divided by the appropriate values of D, are found to yield the same result: the quantity (dσ/dΩ)/D is a unique function of the "scaling variable" p.

INTERPRETATION OF RESULTS

Figs. 7 and 8 illustrate the present state of our knowledge of the $1p_{3/2}$ and $1p_{1/2}$ momentum distribution, in the spirit of Eq. (7), for ^{12}C and ^{16}O, respectively, as obtained from the (e,e'p) and (γ,p) reactions. In order to compare the (e,e'p) and (γ,p) results one must consistently "undistort" both sets of experimental data. This has been done using the same energy-dependent complex optical potential for the outgoing protons in each case, as described in Refs. 13 and 14. (We note that in the published analyses of the (e,e'p) measurements[5,6] "distorted momentum distributions" are extracted from the data and compared with theory. Although this procedure may contain fewer uncertainties it does not allow a comparison between the results of the two types of experiments.)

The solid curves in Figs. 7 and 8 represent the theoretical shell-model momentum distributions calculated using the Woods-Saxon potential parameters given by Elton and Swift.[11] We see that in each case the (e,e'p) points fall below the curve; the same effect has been noted by Mougey et al.[5,6] The (γ,p) data are seen to be in good agreement with the shell-model prediction for ^{12}C and to exceed it by around a factor of two for ^{16}O, for p $\lesssim$ 500 MeV/c. Above this value of the initial momentum the discrepancy increases markedly, becoming $\gtrsim$ 3 orders of

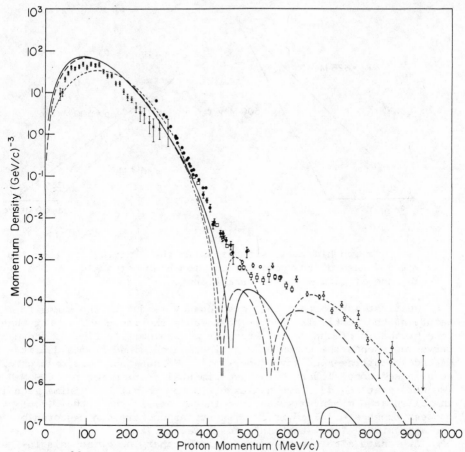

Fig. 8. ^{16}O p-shell momentum distribution obtained from (e,e'p)
[crosses - Ref. 7] and (γ,p) [solid circles - Ref. 2, open symbols -
Ref. 16] measurements, using the method described in Refs. 13 and
14. The solid curve is the $1p_{1/2}$ momentum distribution calculated
from the Elton-Swift potential (Ref. 11). The dashed curve was obtained from the density-dependent Hartree-Fock wave function of
Negele (Ref. 19), and the dotted curve represents the Jastrow
model calculation of Ciofi degli Atti (Ref. 20).

magnitude at p $\gtrsim$ 700 MeV/c.

There are many interesting and important questions in the low-
momentum (p $\lesssim$ 500 MeV/c) region, for both the (e,e'p) and (γ,p)
process. For example, the measured A-dependence of the (γ,p) cross
section in comparison with DWIA theory[17] and the "measured" occupation numbers and sum rules in the (e,e'p) experiments[5,6] have not
been fully explained, and the validity of the cross section factorization hypothesis [Eq. (1)] has not been completely verified.[7,18]
However, in the remainder of this paper we shall concentrate on the
high momentum (p $\gtrsim$ 500 MeV/c) region for ^{16}O, which has so far been
investigated only with the (γ,p) process, and for which the discrepancy
between measurement and shell-model calculation is enormous.

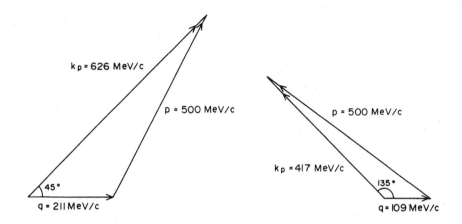

Fig. 9. Two kinematic situations in the $^{16}O(\gamma,p_0)^{15}N$ reaction which "measure" the same initial momentum p ($\vec{p}=\vec{k}_p-\vec{q}$, in the absence of final-state interactions).

We first ask whether this discrepancy is due to the simplified nature of the particular shell-model wave function used, rather than to a breakdown of the direct-knockout mechanism. In Fig. 8. the dashed curve represents the momentum distribution obtained from a density-dependent Hartree-Fock wave function.[19] We see that this prediction does indeed come closer to the experimental results for p $\gtrsim$ 600 MeV/c. (We note that the data are not expected to exhibit the minima seen in the calculated momentum-space wave functions; in the actual (γ,p) process these are filled in by distortions and are not restored by the approximate "undistortion" procedure.)

A perhaps more urgent question is whether the direct single-particle knockout mechanism is valid at all in this region of (very improbable) high momentum components in the single-particle wave function. It is certainly more likely that the photon interacts with a nucleon-nucleon pair which can then share the momentum mismatch. As stated earlier, this involvement of a second particle will in general destroy the unique relationship between $\vec{k}_p$ and $\vec{p}$; such an effort might be responsible for the deterioration of the "scaling" seen in Fig. 8 for p $\gtrsim$ 500 MeV/c. However, one can calculate an "effective" momentum distribution which includes the effect of nucleon-nucleon correlations within the framework of the single-particle picture by employing the Jastrow model. We may write a correlated A-particle wave function $\tilde{\psi}$ in terms of an independent-particle-model wave function ψ as follows:

$$\tilde{\psi}(1,2,\ldots,A) = \psi(1,2,\ldots,A) \prod_{j<k} f(r_{jk}), \qquad (9)$$

where the function $f(r)$ has the properties

$$f(r) \to 0 \quad \text{as} \quad r \to 0$$

$$f(r) \to 1 \quad \text{as} \quad r \to \infty$$

(r is the internucleon spacing). Using the form

$$f(r) = 1 - e^{-r^2/b^2} \tag{10}$$

with b = 0.5fm, and harmonic-oscillator basis functions for ψ, Ciofi degli Atti[20] obtained the "correlated" momentum space wave function shown as the dotted curve in Fig. 8. The "agreement" with the experimental results indicates that nucleon-nucleon correlations of range $\approx$ 0.5fm are almost certainly an important effect in the (γ,p) process, and must be taken into account when attempting to extract information on high momentum components in nuclear wave functions from these data.

CONCLUSIONS

This paper has presented a somewhat superficial discussion of the problem of obtaining nuclear momentum distributions from (γ,p) and (e,e'p) measurements. More (γ,p) and (e,e'p) data exist,[4-8] together with more sophisticated theoretical treatments (e.g. Refs. 7, 21), but the essential features are similar to those illustrated in Figs. 2, 3, 7, and 8. It is clear that in order to extract nuclear momentum distributions, the mechanism(s) of these reactions must be understood in detail. To this end, it is important that both processes be studied, experimentally and theoretically.

We note in Figs. 7 and 8 the meager overlap between (e,e'p) and (γ,p) data. This point is further illustrated by Fig. 10, which shows

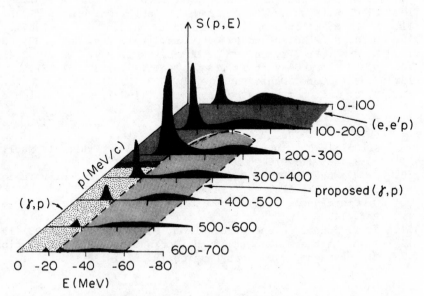

Fig. 10. Schematic spectral function S(p,E) for a 1p-shell nucleus. Kinematic regions in which (γ,p) and (e,e'p) measurements have been performed are indicated, along with a possible extension of the (γ,p) work.

a schematic representation of the spectral function S(p,E). For purely kinematic reasons, as stated earlier, one cannot extend the (γ,p) measurements to lower p. One could in principle extend the (e,e'p) measurements to higher p; in practice, this work is expected to have high priority at future 100%-duty accelerators. With present-day facilities, the (γ,p) experiments could be extended to higher $|E|$, as indicated in Fig. 10, and it is hoped that such work will be carried out in the near future.[22] One will then have mapped out the spectral function over the "entire" (p,E)-plane and will be in possession of a large body of data with which to confront the most sophisticated knockout-reaction theories.

REFERENCES

1. J. L. Matthews, D. J. S. Findlay, S. N. Gardiner, and R. O. Owens, Nucl. Phys. A267, 51 (1976).
2. D. J. S. Findlay and R. O. Owens, Nucl. Phys. A279, 385 (1977).
3. J. L. Matthews et al., Phys. Rev. Lett. 38, 8 (1977).
4. J. L. Matthews, Nuclear Physics with Electromagnetic Interactions (H. Arenhövel and D. Drechsel, eds.), Lecture Notes in Physics Vol. 108 (Springer-Verlag, Berlin, 1979) p. 369.
5. J. Mougey et al., Nucl. Phys. A262, 461 (1976).
6. J. Mougey, Nucl. Phys. A335, 35 (1980).
7. M. Bernheim et al., Nucl. Phys. A375, 381 (1982).
8. K. Nakamura et al., Nucl. Phys. A296, 431 (1978).
9. G. M. Shklyarevskii, JETP (Sov. Phys.) 9, 1057 (1959).
10. D. J. S. Findlay, Ph.D. Thesis, Glasgow University (1975).
11. L. R. B. Elton and A. Swift, Nucl. Phys. A94, 52 (1967).
12. D. J. S. Findlay and R. O. Owens, Phys. Rev. Lett. 37, 674 (1976).
13. D. J. S. Findlay and R. O. Owens, Nucl. Phys. A292, 53 (1977).
14. D. J. S. Findlay et al., Phys. Lett. 74B, 305 (1978).
15. G. Passatore, Nucl. Phys. A248, 509 (1975).
16. M. J. Leitch, Ph.D. Thesis, M.I.T. (1979); M. J. Leitch et al. (to be published). See also Ref. 4.
17. D. J. S. Findlay, D. J. Gibson, R. O. Owens, and J. L. Matthews, Phys. Lett. 79B, 356 (1978).
18. S. Boffi, C. Giusti, F. D. Pacati, and S. Frullani, Nucl. Phys. A319, 461 (1979).
19. J. W. Negele, Phys. Rev. C 1, 1260 (1970), and private communication. See also Ref. 14.
20. C. Ciofi degli Atti, Nuovo Cim. Lett. 1, 590 (1971). See also Ref. 14.
21. J. T. Londergan and G. D. Nixon, Phys. Rev. C 19, 998 (1979).
22. J. L. Matthews, Report of the Workshop on Future Directions in Electromagnetic Nuclear Physics (M.I.T., 1981) p. 337.

CURRENT DEVELOPMENTS IN NUCLEAR STRUCTURE THEORY

Bruce R. Barrett

Department of Physics, University of Arizona, Tucson, Arizona 85721

ABSTRACT

A review of current techniques for computing microscopically nuclear wave functions is given, with particular emphasis given to the exp{S} formalism (or coupled-cluster approximation) and the self-consistent valence-core method. Nuclear structure calculations using wave functions computed by these latter two techniques indicate the importance of two-body and higher-body correlations in the wave functions. A brief introduction is given of the Interacting Boson Model (IBM) for describing nuclear collective motion. The extension of the IBM approach to atoms and molecules is also discussed.

INTRODUCTION

I will split my talk on the subject of Current Developments in Nuclear Structure Theory into two parts. The first part will deal with the Present Status of Microscopic Calculations of Nuclear Wave Functions, a subject clearly of interest to anyone who wants to compute a nuclear momentum density. The second part will cover the Interacting Boson Model of Nuclear Collective Motion. The IBM, as it is called, is the <u>hottest</u> topic in nuclear structure theory today.

I. THE CURRENT STATUS OF MICROSCOPIC CALCULATIONS OF NUCLEAR WAVE FUNCTIONS

In the short time available for this talk, I could not possibly cover all the work that has been carried out in the last few years for calculating nuclear wave functions microscopically. For those interested in more details than I will be able to present here, I would recommend two review articles on the subject. The first is by Svenne[1] and is entitled "Fourteen Years of Self-Consistent Field Calculations: What Has Been Learned." The second I just received in preprint form a couple of weeks ago. It is by Negele[2] and is entitled "The Mean Field Theory of Nuclear Structure and Dynamics." The latter review article is more extensive and covers time-dependent Hartree-Fock theory and the functional integral formalism, as well as the many-body theory of stationary states.

The article by Svenne simply reviews the status of and the results given by Hartree-Fock (HF), Brueckner Hartree-Fock (BHF), renormalized Brueckner Hartree-Fock (RBHF) and density-dependent Hartree-Fock (DDHF) calculations. These approaches contain the well-known Brueckner rearrangement energy (in BHF) and occupation probability effects (in RBHF), but in the end both RBHF and DDHF lead to single-particle energies that are essentially equal to the separation energies, as in Koopmans' theorem.[3]

In the remainder of this part of my talk, I will concentrate on two recent techniques for calculating single-particle wave functions, which include correlations beyond those of the mean field. These are the exp{S} calculations of the Bochum group[4,5] and the self-consistent valence-core calculations of the Iowa State/Ames Laboratory group.[6]

A. I will start with the so-called exp{S} method, which was originated by Coester and Kümmel[7] and developed by the Bochum group.[4,5] It is also referred to as the coupled-cluster approximation. While the methods, mentioned previously in the review article by Svenne, are mainly concerned with determining higher and higher order terms in the expansion for the nuclear ground state energy, the exp{S} method starts by constructing the A-particle wave function, which can be written in the form

$$|\Psi\rangle = \exp\{S\} \; |\Phi\rangle, \tag{1}$$

where $|\Phi\rangle$ is a Slater determinant of single-particle wave functions and S is a sum of many-body operators

$$S = \sum_{n=1}^{A} S_n \tag{2}$$

$$S_n = \frac{1}{(n!)^2} \sum_{\substack{\nu_1 \ldots \nu_n \\ \sigma_1 \ldots \sigma_n}} a^{\dagger}_{\sigma_1} \ldots a^{\dagger}_{\sigma_n} a_{\nu_n} \ldots a_{\nu_1} \langle \sigma_1 \ldots \sigma_n | S_n | \nu_1 \ldots \nu_n \rangle, \tag{3}$$

where the states ν are below the fermi sea (hole states) and the states σ are above the fermi sea (particle states). Hence, the operator S_n is a n-body operator creating n holes in $|\Phi\rangle$ and adding n particles (i.e., np-nh excitations or correlations). The matrix element $\langle \sigma_1 \ldots \sigma_n | S_n | \nu_1 \ldots \nu_n \rangle$ is given by the sum of all <u>linked</u> diagrams with n incoming hole and n outgoing particle lines. Equation (1) describes the most general many-body state $|\Psi\rangle$ for which $\langle \Phi | \Psi \rangle \neq 0$.

When the $|\Psi\rangle$ of Eq. (1) is inserted into the Schrödinger equation and then projected onto n-particle-n-hole states, a set of coupled equations for the S_n is obtained, as indicated by the following equations:

$$H|\Psi\rangle = He^S|\Phi\rangle = Ee^S|\Phi\rangle. \tag{4}$$

Multiplying from the left by exp{-S}, we obtain

$$(e^{-S} He^S - E) |\Phi\rangle = 0. \tag{5}$$

Now project onto the states $\langle\Phi|$, $\langle\Phi|a^{\dagger}_{\nu_1}a_{\sigma_1}$, $\langle\Phi|a^{\dagger}_{\nu_1}a^{\dagger}_{\nu_2}a_{\sigma_1}$, etc. to obtain:

i) $E = \langle\Phi|e^{-S}He^S|\Phi\rangle = \langle\Phi|He^S|\Phi\rangle$

$$= \langle\Phi|H(1 + S_1 + \frac{1}{2} S_1^2 + S_2|\Phi\rangle, \tag{6}$$

since $\langle\Phi|(-S)^m = 0$ for $m>0$.

ii) The equations for the S_n (the np-nh amplitudes) are given by

$$\langle\Phi|a^\dagger_{\nu_1}a_{\sigma_1}e^{-S}He^{S}|\Phi\rangle = 0$$

$$\langle\Phi|a^\dagger_{\nu_1}a^\dagger_{\nu_2}a_{\sigma_2}a_{\sigma_1}e^{-S}He^{S}|\Phi\rangle = 0 \qquad (7)$$

$$\langle\Phi|a^\dagger_{\nu_1}...a^\dagger_{\nu_n}a_{\sigma_n}...a_{\sigma_1}e^{-S}He^{S}|\Phi\rangle = 0.$$

This set of equations is exact and is completely equivalent to the many-body Schrödinger equation. It has the added advantage that it provides a systematic scheme for making approximations to the exact solution by means of truncations of this set of equations. Each equation in this set not only couples to lower-order terms but also to higher order terms. For a given n, the equation for the np-nh projection not only couples S_n to all S_m with $m<n$, but also to the terms S_{n+1} and S_{n+2}.

The reader who is interested in more details regarding the derivation of these equations and their application is referred to the extensive review article on this theory by Kümmel et al.[5]

Instead of discussing detailed calculations of nuclear wave functions using this technique, I want to turn now to the application of wave functions determined by this method to the evaluation of nuclear momentum distributions. These calculations have been performed by Zabolitzky and Ey.[8] They truncate the exp{S} equations after the two-body equation S_2, which is equivalent to the generalized Brueckner-Hartree-Fock (GBHF) approximation. In this way they obtain the maximum overlap single-particle orbitals as well as reliable approximations to the matrix elements of S_2.

Zabolitzky and Ey then use these results for the single-particle orbitals and for S_2 to calculate the momentum distribution $n(q)$, which is related to the one-body density matrix D_1 by

$$n(q) = \sum_{\alpha_1,\alpha_2} \int d\hat{q} \langle q|\alpha_1\rangle \langle\alpha_1|D_1|\alpha_2\rangle \langle\alpha_2|q_2\rangle, \qquad (8)$$

where

$$\langle\alpha_1|D_1|\alpha_2\rangle = \langle\Psi|a^\dagger_{\alpha_2}a_{\alpha_1}|\Psi\rangle/\langle\Psi|\Psi\rangle, \qquad (9)$$

and $\{\langle q|\alpha\rangle\}$ is the complete orthonormal set of Brueckner orbitals in momentum space. The normalization is $\int_0^\infty q^2dq\, n(q) = A$. In the two-body approximation employed by Zabolitzky and Ey, $\langle\alpha_1|D_1|\alpha_2\rangle$ is given diagrammatically by

$$\langle\alpha_1|D_1|\alpha_2\rangle = \qquad\qquad\qquad + \qquad\qquad + \qquad\qquad (10)$$

The first term on the right-hand side is the uncorrelated contribution, while the second and third terms are the correlation contributions. From Eq. (10) we see that by "correlations" Zabolitzky and Ey mean those parts of the full wave function correlations which cannot be expressed in terms of a single Slater determinant. The uncorrelated or single-particle part of the full wave function must be "maximal" and the correlated part "minimal" in some sense. This leads to the maximum-overlap condition $\langle\Psi|\Phi\rangle$ = max. for $\langle x|\nu\rangle$, determining the Brueckner orbitals $\langle x|\nu\rangle$. This condition is equivalent to $S_1 = 0$, i.e., there are no 1p-1h excitations in $|\Psi\rangle$ relative to $|\Phi\rangle$. As a consequence, the single-particle density matrix (8) does not have any particle-hole matrix elements in the two-body approximation.

After calculating n(q) in laboratory coordinates, Zabolitzky and Ey then remove the center-of-mass (cm) motion by inverting the relation

$$n(q) = \frac{1}{(2\pi)^3} \int d^3Q \; n_{cm}(Q) \; n_{int} \; (|\vec{q} - \vec{Q}/A|) \qquad (11)$$

for the internal momentum distribution $n_{int}(q)$. Since $n_{cm}(Q)$ is the cm harmonic oscillator momentum distribution, Eq. (11) can be easily inverted analytically, if n(q) is a gaussian distribution. This is done by expanding n(q) in terms of gaussians.

The results of Zabolitzky and Ey for ^{4}He and ^{16}O are shown in Figs. 1 and 2, respectively, for several different NN interactions. One observes that for momenta less than 2 fm^{-1} the momentum distribution is essentially given by the uncorrelated single-particle motion (Eq. (10), first term) and a 10% correction from the occupation probability term (Eq. (10), third term). The uncorrelated single-particle contribution decreases very rapidly for higher momentum, so that the correlation contribution (Eq. (10), second term) dominates for momenta beyond 2 fm^{-1}, since this term decreases much slower with increasing momentum. The main contributions in the region of momenta between 2 and 4 fm^{-1} comes from the tensor admixtures in the wave function generated by the tensor force present in the NN interaction. For higher momenta, two-body multiple scattering processes from the scalar part of the interaction dominate. No trace can be seen of typical short-range correlations induced primarily by the "repulsive core" of the interaction. At all momenta considered here, the medium-range processes still are of dominating importance.

Zabolitzky and Ey conclude that an experimental determination of only the order of magnitude of n(q) would already be sufficient to see significant influence of nucleon correlations. This is true since the momentum distribution is directly proportional to the square of the Fourier transform of the correlation function, S_2. This is quite different from the results for the form factor F(q) obtained via elastic electron scattering, for which no significant influence of correlations can be found. This latter result occurs since the form factor is proportional to an integral of the correlation functions which smears out all interesting features.

B. I now turn to the self-consistent valence-core technique of Vary et al.[6] This technique consists of first doing a renormalized Brueckner calculation for the nuclear core orbits. The valence single-particle orbits are then renormalized using the Bloch-

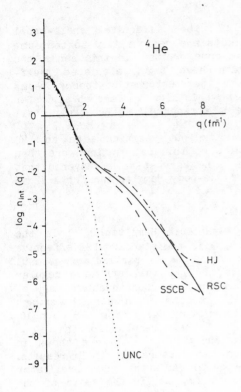

Fig. 1. Momentum distributions for ^{4}He, HJ: Hamada-Johnston potential, RSC: Reid soft core potential, SSCB: de Tourreil-Spring super soft core potential B, UNC: uncorrected, for the RSC potential. The other uncorrelated distributions do not differ appreciably for $q > 2$ fm^{-1}.

Fig. 2. Same as Fig. 1, for ^{16}O.

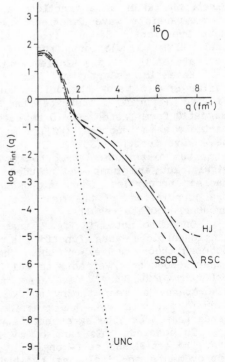

Horowitz-Brandow (BHB) theory of the effective shell-model interaction,[9] so that the valence orbits are renormalized to the same order of perturbation theory as the core orbits. In this sense, the core orbits and the valence orbits have been calculated "self-consistently" to the same order. The perturbation-theory terms included in their calculations are shown in Fig. 3. Vary et al. refer to their valence calculations as being a renormalized Brueckner shell model (RBSM) calculation.

Specific calculations were carried out starting with an ^{16}O doubly closed-shell core. The results of this calculation were then employed in a shell-model study of ^{17}O. Vary et al. discovered that they could approximate the effective one-body Hamiltonian for A=17 in the form

$$H_{eff} = H_o + H_{eff} = T + \bar{U}_{eff}, \tag{12}$$

where H_o is the unperturbed, single-particle Hamiltonian, T is the single-particle kinetic energy, and U_{eff} is the standard BHB effective interaction. The second expression for H_{eff} is particularly useful since the oscillator matrix elements of $\bar{U}_{eff}$ fall off in a regular fashion with increasing radial quantum numbers. Consequently, Vary et al. were able to extrapolate their results in a straightforward manner (using plots of the matrix elements of $\bar{U}_{eff}$) from n=5 to n=10, or to intermediate excitations of $20\hbar\Omega$. Their results for the $d_{5/2}$, $s_{1/2}$, and $d_{3/2}$ single-particle wave functions of ^{17}O are shown in Figs. 4-6. One immediately observes two features of their results. First, the RBSM wave functions are quite different from oscillator single-particle wave functions for r>5 fm, i.e, in the tails of the wave functions. Second, from Fig. 6, we see that the extension of their RBSM calculations from n=5 to n=10 is sufficient to develop accurate tails for the single-particle wave functions.

Recently, McCarthy and Vary[10] have calculated the magnetic form factor of ^{17}O using the self-consistent $d_{5/2}$ wave function given in Fig. 4. Figure 7 illustrates the difference in the results for the magnetic form factor of ^{17}O between using harmonic-oscillator single-particle wave functions (HO) and using the Vary et al. self-consistent $d_{5/2}$ wave function (SCWF). It is clearly seen, as in the case of the momentum distribution, that two-body and higher-body correlations are significant and must be included in any "realistic" calculation of nuclear properties. Harmonic-oscillator wave functions contain only the correlations of the mean field, and these are not sufficient to produce accurate results.

One also notes in Fig. 7 that even the use of a self-consistent single-particle wave function is not enough to produce agreement between theory and experiment. The addition of core-polarization and other second-order effects as well as meson exchange contributions by McCarthy and Vary[10] is also not sufficient to eliminate the discrepancy between theory and experiment. Recent calculations by Coon et al.[11] indicate that the inclusion of three-body-force effects tends to improve the agreement between theory and experiment.

To conclude this part of my talk, I would emphasize once again that the exp{S} and self-consistent valence-core methods are definite improvements for including two-body and higher-body correlations in the nuclear single-particle wave functions. But both techniques could

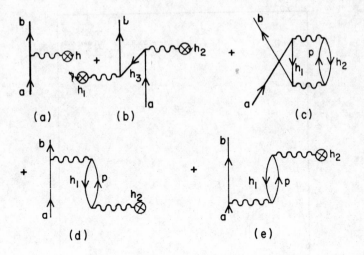

Fig. 3. Lowest order single-particle potential diagrams included in the diagonalization. The X's in the bubbles represent $-U_o$ insertions.

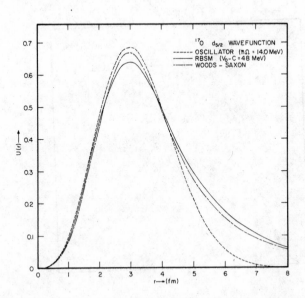

Fig. 4. Wave function for the $d_{5/2}$ state of ^{17}O in the RBSM calculation compared with harmonic-oscillator and Woods-Saxon wave functions.

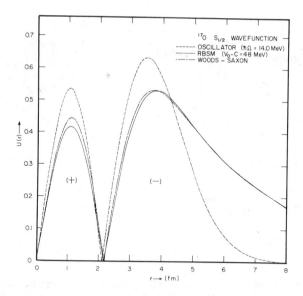

Fig. 5. Wave function for the valence $s_{1/2}$ state of ^{17}O in the RBSM calculation compared with harmonic-oscillator and Woods-Saxon wave functions.

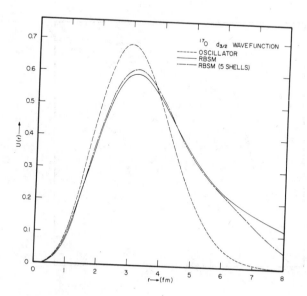

Fig. 6. Wave function for the $d_{5/2}$ state of ^{17}O in the RBSM calculation compared with the harmonic-oscillator wave function and with the RBSM calculation in only 5 oscillator shells. The solid curve represents the full RBSM calculation in 10 oscillator shells.

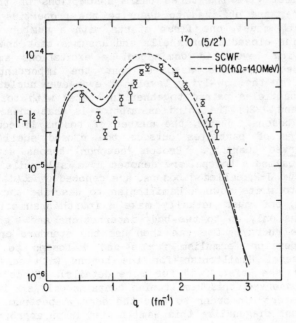

Fig. 7. Transverse magnetic form factor squared, F_T^2, of ^{17}O vs. effective momentum transfer q. The solid line is the sum of the M1, M3, and M5 contributions obtained using the self-consistent $d_{5/2}$ wave function of Ref. 6. The dashed line is the same sum obtained using a simple harmonic-oscillator $d_{5/2}$ wave function. The experimental results are shown for comparison.

be improved themselves. Although the exp{S} method sums the two-body correlations S_2 to all orders, the present calculations do not include three- or four-body correlations (i.e., S_3 or S_4). On the other hand the self-consistent valence-core calculations do include three- and four-body correlations as well as two-body correlations, but only to second order in perturbation theory.

II. THE INTERACTING BOSON MODEL OF NUCLEAR, ATOMIC, AND MOLECULAR COLLECTIVE BEHAVIOR

The Interacting Boson Model (IBM) of Arima and Iachello[12] has been extremely successful in describing the collective properties of medium-to-heavy-mass nuclei. Since Iain Morrison will present a talk later in this Workshop on "Boson-Fermion Models of Nucleon Structure," I will not go into detail concerning the IBM for nuclei, but will instead emphasize its extension to atoms and molecules. For the person interested in more details regarding the IBM for nuclei, two recent review articles on the IBM for nuclei are listed in Refs. 13 and 14.

Let me first give the three basic assumptions of the IBM and explain how they are utilized to describe the properties of nuclei. As in the shell model, one first starts with a number of valence particles outside closed major shells and assumes that the structure of the low-lying levels is dominated by excitations among these particles. Secondly, one assumes that the important particle configurations for the low-lying levels of even-even nuclei are those for identical particles paired together in states with total angular momentum J=0 and J=2. The third assumption is that these pairs can be treated as bosons. Hence, the number of bosons is equal to the number of pairs of particles outside the closed shell and is a strictly conserved quantity. Proton (neutron) bosons with angular momentum J=0, called s bosons, are denoted by $s_\pi(s_\nu)$, while those with angular momentum J=2, called d bosons, are denoted by $d_\pi(d_\nu)$.

In order to write down a Hamiltonian to describe the energy of valence bosons, one must actually make a fourth assumption, namely that interactons only up to two-body interactions are significant in determining the energy. One can then use the standard creation and annihilation operator formalism for s and d bosons to express a completely general Hamiltonian for the bosons with only one- and two-body terms (see Refs. 12-14 for more details). Since the number of bosons is conserved, this Hamiltonian contains only six independent, variable parameters In order to find the energy spectrum of a given nucleus, one must diagonalize this Hamiltonian in an appropriate basis and adjust the parameters to obtain the best possible agreement with the available experimental data.

One of the most intriguing features of the IBM is that it possesses extremely interesting and useful symmetry properties. Since there are six creation operators and six annihilation operators (e.g., $s^\dagger$, $d^\dagger_\mu$, $\mu=0$, ±1, ±2), the group symmetry of the IBM is U(6). Arima and Iachello[12] noted that in the case of U(6) there exist three subgroup chains, when each chain is restricted to contain the angular momentum group, O(3). These three subgroup chains are

I. $U(6) \supset U(5) \supset O(5) \supset O(3) \supset O(2)$
II. $U(6) \supset SU(3) \supset O(3) \supset O(2)$
III. $U(6) \supset O(6) \supset O(5) \supset O(3) \supset O(2)$.

If the Hamiltonian H can be written in terms of only the Casimir operators of a complete chain of subgroups of U(6), then this H is diagonal in the representation of this subgroup chain and possesses what is known as a dynamical symmetry.[12,13] A dynamical symmetry is a systematic breaking of the symmetry of the larger group [e.g., U(6)] by the terms proportional to the Casimir operators of the subgroups. The three dynamical symmetries of the IBM correspond to three well-known limits in nuclear structure theory. The U(5) chain corresponds to the vibrational limit; SU(3), to the rotational limit; and O(6) to the γ-unstable limit. These three limiting cases are tremendously useful in describing and understanding the structure of medium-to-heavy-mass nuclei.

Iachello and collaborators[15-17] have recently extended the IBM technique to explain specific properties of atoms and molecules. In the case of atoms, Iachello and Rau[15] consider the phenomena in atomic physics which appear to be dominated by two-electron correlations. In particular, they investigate the case in which the two electrons are coupled to a total spin of S=0 (i.e., the singlet total spin state). Normally the group structure of this problem would be $U(n^2) \times U_S(2)$, where n is the principal quantum number of the hydrogenic basis and where $U_S(2)$ refers to the spin. Since they consider only S=0, they neglect the spin part and concentrate only on the orbital part $U(n^2)$. They then observe that there is a subgroup chain for $U(n^2)$, which ends with the group O(3), which is of the form

$$U(n^2) \supset O(n^2) \supset \dots \supset O(3).$$

This subgroup chain contains a number of chains, representing all possible ways of breaking $O(n^2)$ down to O(3), but they are only interested in the step from $U(n^2)$ to $O(n^2)$, which represents retaining only two terms in the expansion of the Coulomb interaction, i.e.,

$$\frac{e^2}{r_{12}} = A\, \underline{1} + B\, \underline{P}_{n^2} + \dots , \tag{13}$$

where $\underline{1}$ is the unit operator and $\underline{P}_{n^2}$ is the pairing operator of $O(n^2)$. This latter operator is, apart from a constant, equal to the quadratic Casimir operator of $O(n^2)$. So the step $U(n^2)$ to $O(n^2)$ represents a dynamical symmetry for the phenomena of two-electron correlations in atomic systems. In fact, the structure of Eq. (13), a constant plus a pairing term, appears in many fields of physics and is responsible for the occurence of collective pair states.

As a numerical test, Iachello and Rau compared the actual eigenvalues obtained by diagonalizing the exact Coulomb matrix with those obtained in the $O(n^2)$ scheme, i.e., Eq. (13). The $O(n^2)$ results were in reasonable agreement with the exact results and exhibited all the correct qualitative features of the exact results.

The virtue of the group-theoretic technique of Iachello and Rau is that the expansion of the interaction e^2/r_{12} in terms of invariant operators may be rapidly converging, thus giving rise to major simplifications in the calculations. These simplifications are particularly important when many configurations must be admixed (n large). Although Iachello and Rau have applied their technique only to the study of correlations in two-electron configurations, $(n\ell)^2$, it is clear that the same technique can be used for studying correlations in multielectron configurations. This may lead to considerable simplifications in the calculation of the structure of complex atoms.

Finally, I would like to briefly mention the application of the IBM approach to molecular rotation-vibration spectra.[16,17] The basic idea is to construct a spectrum-generating algebra, or "dynamical algebra," which, within a certain approximation, describes realistic rotation-vibration spectra in three dimensions. The algebra used by Iachello[16] is U(4), the algebra of the unitary group in four dimensions. He introduces four creation ($b_\alpha^\dagger$) and annihilation (b_α) boson operators ($\alpha=1,\dots,4$) and divides these four operators into a scalar (J=0) operator, denoted by $\sigma^\dagger(\sigma)$ and a three-component vector

($J=1$) operator, denoted by $\pi_\mu^\dagger(\pi_\mu)$, $\mu=0$, ±1. Since these operators represent bosons, they satisfy the appropriate __boson__ commutation relations and have parity $(-1)^J$.

As in the case of the IBM for nuclei, one can now use these boson operators to write down a completely general Hamiltonian H for the σ and π bosons involving only one- and two-body terms (see Ref. 16 for more details). In order to find the energy spectrum, one must diagonalize H in an appropriate basis. Since H is constructed with boson operators, spanning the group $U(4)$, the appropriate basis is provided by the totally symmetric irreducible representations of $U(4)$. Since the total number of bosons (i.e., $n_\sigma+n_\pi$) is conserved, H contains only four independent parameters.

The dynamical symmetries connected with the subgroup chains of $U(4)$, including $O(3)$, are

I. $U(4) \supset O(4) \supset O(3) \supset O(2)$
II. $U(4) \supset U(3) \supset O(3) \supset O(2)$.

Using standard group theoretical techniques, it is then possible to construct the most general solution corresponding to both of these group chains.

Iachello has carried out specific calculations for the chain I. His investigations show that a dynamical $O(4)$ symmetry with only linear and quadratic terms corresponds to a Dunham expansion with only three coefficients. In particular, he predicts a value for the ratio of the first two Dunham coefficients, and his predicted value is in reasonable agreement with the experimental value. He also presents a typical spectrum with $O(4)$ symmetry, which describes very well the low-lying rotation-vibration spectrum of diatomic molecules. By including more and more coefficients in his expansion for H, Iachello could improve the agreement with the higher-lying spectrum to any desired accuracy. Thus, the algebraic approach may be viewed as an alternative way to parametrize the observed spectra. The main advantage of the algebraic approach is that it can be used to calculate, in a relatively simple way, other properties besides energy levels, such as excitation probabilities.

Recently Van Roosmalen __et al.__ have extended this algebraic approach to tri- and poly-atomic molecules. In particular, they apply their approach (similar to the one discussed for diatomic molecules) to the study of linear triatomic molecules. As an example, they calculate the low-lying vibrational spectrum of the CO_2 molecule, for which good agreement with experiment is obtained. However, the real advantage of their approach is that it produces __automatically__ the complex vibrational and rotational pattern of triatomic molecules. Furthermore, it is possible to use this same approach to calculate all other properties of triatomic molecules, such as intensities of emission and absorption lines and excitation probabilities from one rotation-vibration level to another. The same approach can be generalized to even more complex molecules.

In conclusion, I would like to emphasize the tremendous usefulness of the algebraic approach of the Interacting Boson Model. It not only accurately describes the experimental data, but also provides a simple physical picture of the dynamics of the system being investigated. In the case of nuclear physics, the IBM for both neutrons and protons offers the exciting possibility of truly understanding nuclear collective motion in terms of a simple microscopic model, the shell model.[13,14] If this effort is successful, then we would at last have a truly unified, microscopic theory of nuclear structure.

ACKNOWLEDGMENTS

I wish to thank J. P. Vary, M. W. Kirson, F. Iachello, S. A. Coon, and R. H. Belehrad for helpful discussions regarding material contained in this talk. I would also like to acknowledge the partial support of this work by the National Science Foundation Grant No.PHY-81-00141.

REFERENCES

1. J. P. Svenne, Advances in Nuclear Physics, edited by J. W. Negele and E. Vogt (Plenum, New York, 1979), Vol. 11, p. 179.
2. J. W. Negele, M.I.T. preprint, CTP#898 (1981), to be published in Rev. Mod. Phys.
3. T. Koopmans, Physica 1, 104 (1933).
4. H. Kümmel, Nucl. Phys. A176, 205 (1971); J. G. Zabolitzky, Nucl. Phys. A228, 272 (1974).
5. H. Kümmel, K. H. Lührmann, and J. G. Zabolitzky, Phys. Rep. 36C, 1 (1978) and references therein.
6. J. P. Vary, R. H. Belehrad, and R. J. McCarthy, Phys. Rev. C 21, 1626 (1980).
7. F. Coester, Nucl. Phys. 7, 421 (1958); F. Coester and H. Kümmel, Nucl. Phys. 17, 477 (1960).
8. J. G. Zabolitzky and W. Ey, Phys. Lett. 76B, 527 (1978).
9. See, for example, B. R. Barrett and M. W. Kirson, Advances in Nuclear Physics, edited by M. Baranger and E. Vogt (Plenum, New York, 1973), Vol. 6, p. 219; and P. J. Ellis and E. Osnes, Rev. Mod. Phys. 49, 777 (1977).
10. R. J. McCarthy and J. P. Vary, Phys. Rev. C 25, 73 (1982).
11. S. A. Coon, R. J. McCarthy, and J. P. Vary, Ames Laboratory preprint, IS-J 643 (1981), to be published.
12. A. Arima and F. Iachello, Ann. Phys. (N.Y.) 99, 253 (1976); 111, 201 (1976); 123, 468 (1979).
13. A. Arima and F. Iachello, Annu. Rev. Nucl. Part. Sci. 31, 75 (1981).
14. B. R. Barrett, Rev. Mex. Fis. 27, 533 (1981).
15. F. Iachello and A. R. P. Rau, Phys. Rev. Lett. 47, 501 (1981).
16. F. Iachello, Chem. Phys. Lett. 78, 581 (1981).
17. O. S. Van Roosmalen, A. E. L. Dieperink, and F. Iachello, Chem. Phys. Lett. 85, 32 (1982).

MOMENTUM DENSITIES IN CHEMISTRY*

Michael A. Coplan
Institute for Physical Science and Technology

John A. Tossell
and
John H. Moore
Chemistry Department
University of Maryland
College Park, MD 20742

A principal interest of the Maryland (e,2e) group is the electronic structure of polyatomic molecules. Potentially, (e,2e) spectroscopy offers great advantages over conventional spectroscopy, photoelectron spectroscopy and even electron and x-ray diffraction. The realization of these advantages depends on the solution of a number of experimental and theoretical problems.

The experimental problems associated with molecular targets in (e,2e) spectroscopy are low signal-to-noise because of the large number of electrons in the targets, closely spaced energy levels, detector contamination, and, in some cases such as the organometallic complex molecules, low vapor pressure. Our spectrometer[1],[2] (Fig. 1) has multiple detectors so that the signal rate is enhanced by a factor of twenty-five over comparable two detector systems. With higher signal rates the signal-to-noise problem becomes tractable. To separate the energy levels the resolution

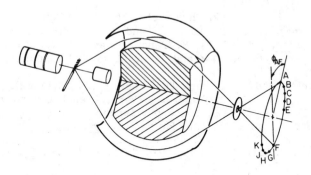

Fig. 1. The multichannel (e,2e) apparatus.

*Work supported by National Science Foundation grant CHE-79-09430

of the spectrometer has been increased by retardation of the
scattered and knocked-out electrons. Currently, at 800eV incident
energy our resolution is 1.0eV. The contamination problem is
particularly serious for a multiple detector system and, though we
continue to use channeltron detectors, we are developing a scinti-
llator/photomultiplier array which should be insensitive to en-
vironmental conditions and have stable gain over long periods of
time. For non-volatile targets we have a heated manifold and
collision chamber which enables us to obtain target densities of
10^{13}-10^{14} cm^{-3} for low vapor pressure substances such as $Cr(CO)_6$.

Once reproducible data of good precision is obtained, the
battle is only half won. To be able to interpret the results in
ways which bring new insight into the electronic structure of
molecules requires that the scattering mechanism portion of the
(e,2e) cross section be separated from the structure portion. We
work at relatively high incident energies and use the noncoplanar
symmetric geometry. In all of our analysis the plane wave impulse
approximation (PWIA) is used, which in conjunction with the non-
coplanar symmetric geometry gives an (e,2e) cross section which
varies only with momentum density of the target electron. The
approximations of the PWIA are certainly open to question, but in-
creasingly the experimental evidence is that at high incident
energies the PWIA gives momentum densities of the correct shape.

Over the last five years the momentum densities for the valence
orbitals of a large number of molecules have been measured by our
group as well as by groups at Flinders, Vancouver and Frascati.
The analysis of the measurements has mostly consisted of comparing
the measurements with suitably averaged momentum density calculations obtained by taking the Fourier transform of calculated wave
functions. A number of qualitative conclusions about the quality
of different calculations have emerged, however the lack of a
quantitative framework within which the measurements and calculations can be discussed is apparent.

We have looked again at how wave functions and densities are
related in position space and momentum space and are now analyzing
our data in terms of functions which allow quantitative comparisons
between theory and experiment.

Figure 2 shows the relations between the relevant functions in
configuration and momentum space. It is important to note that the
Fourier transform of momentum density $\rho(\vec{q})$ gives the function $B(\vec{r})$
which is the autocorrelation function of the wave function in configuration space. Correspondingly, the Fourier transform of the
electron density $\rho(\vec{r})$ is the x-ray structure factor $F(\vec{q})$ which is
the autocorrelation function of the momentum space wave function.

To illustrate the usefulness of this analysis we will take the
example of NO. The momentum densities of the valence electrons of
this molecule have been the subject of three different measurements
on three continents with three different instruments at different
incident energies. The results are in very gratifying agreement
which is one reason for the choice.

Our experimental results for three valence orbitals of NO are

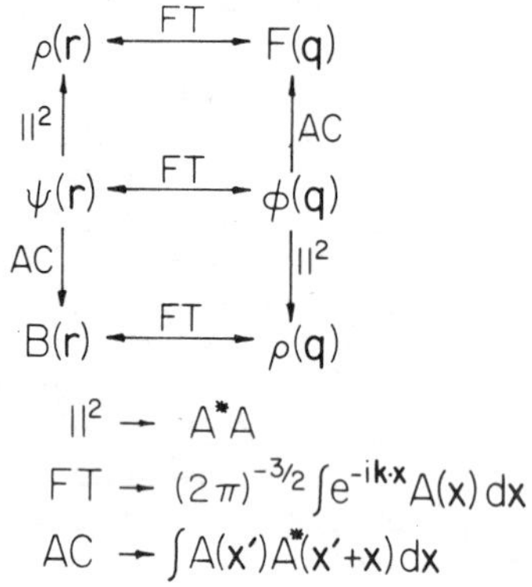

$$\rho(r) \xleftarrow{\quad FT \quad} F(q)$$

$$||^2 \Big\uparrow \qquad\qquad \Big\uparrow AC$$

$$\psi(r) \xleftarrow{\quad FT \quad} \phi(q)$$

$$AC \Big\downarrow \qquad\qquad \Big\downarrow ||^2$$

$$B(r) \xleftarrow{\quad FT \quad} \rho(q)$$

$$||^2 \to A^*A$$

$$FT \to (2\pi)^{-3/2} \int e^{-i\mathbf{k}\cdot\mathbf{x}} A(x)\, dx$$

$$AC \to \int A(x') A^*(x'+x)\, dx$$

Fig. 2. Relation between configuration
space and momentum space
functions (reference 6).

shown in Figure 3. Also shown are spherically averaged momentum distributions calculated from restricted Hartree-Fock (RHF) wave functions[3] and convoluted with our instrument function. These momentum distributions illustrate the two general types: the "s" or symmetric type characteristic of a wave function dominated by a totally symmetric function (such as that for the 5σ in NO or the analogous $\sigma_g 2p$ in O_2), and the "p" or antisymmetric type which is characterized by a momentum space wave function with a minimum or even a node at the origin and maximum amplitude at some intermediate value of momentum. Since the momentum wave function is related to the position wave function by the Fourier transform, the breadth of the momentum distribution is inversely related to the extent of the electron density in position space. Our results imply that the antibonding NO 2π is more localized in position space than the antibonding NO 4σ. The antisymmetric type of momentum distribution can be characterized by the parameter q_m, the value of q for which $\rho(q)$ is a maximum. In general the value of q_m is overestimated by theoretical calculations although the discrepancy decreases as the

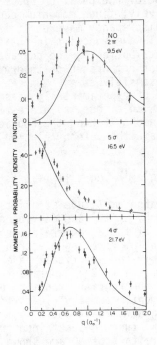

Fig. 3.   Momentum density distribution functions for
valence orbitals of NO
from the (e,2e) experiment
and from RHF wave functions.

accuracy of the calculation improves as shown in Table I in comparing the RHF result to q_m predicted by a spin unrestricted SCF
function employing a limited split valence (SV) basis set.[4]

Table I.   Momentum values q_m for which the
momentum density $\rho(q)$ is a maximum
for orbitals of NO.

	SV (a_o^{-1})	RHF (a_o^{-1})	experiment (a_o^{-1})
4σ	0.73	0.68	0.61
2π	1.04	0.96	0.76

The discrepancy may be the result of an inadequate basis set which underestimates the amplitude of the position space wave function at large distances. Alternatively, since momentum depends on the gradient of distance, the discrepancy may result from too rapid variation of the theoretical wave function in position space. The symmetric type momentum distribution may be characterized by the half width, $q_{\frac{1}{2}}$. However, for comparison of theory and experiment this parameter is rather insensitive.

While direct comparison of experimentally measured momentum densities and theoretical calculations is interesting, a more useful representation can be obtained by taking the Fourier transform of the momentum density to obtain the autocorrelation function of the spatial wave function:

$$B(\vec{r}) = \int \rho_k(\vec{q}) e^{-i\vec{q}\cdot\vec{r}} d\vec{q} = \int \psi_k(\vec{s}) \psi_k(\vec{s}+\vec{r}) d\vec{s}$$

$B(\vec{r})$ is large when ψ_k has large amplitude at points separated by r. This approach was suggested by Smith[5] and developed by Weyrich[6], but of course a similar use of the autocorrelation function has been employed for decades in interpreting electron and optical diffraction experiments. Only the spherically-averaged momentum density can be obtained from the (e,2e) experiment on molecules. However, we have shown that the Fourier transform of a spherically-averaged single-electron momentum density yields the spherical average of the corresponding $B(\vec{r})$ function.[7]

Spherically averaged $B(\vec{r})$ functions obtained from RHF calculations for three valence orbitals of NO are shown in Figure 4. Although all the B(r) functions are qualitatively similar, quantitative differences can be seen by calculating the differences, $\Delta B(r)$, of the autocorrelation functions. For example B(r) for the NO5σ falls off more rapidly than for the NO4σ. This is apparent in the difference function $\Delta B(r) = B(r)_{5\sigma} - B(r)_{4\sigma}$. Similarly, a $\Delta B(r)$ analysis is effective for comparing corresponding orbitals in a pair of molecules or for comparing experimental and theoretical results for the same orbital. As an example we shall consider experimental results and theoretical calculations for NO2π.

The wave function for the NO2π orbital, being antibonding, is polarized toward the less electronegative N atom as illustrated by the contour plot of the RHF NO2π wave function in Figure 5. A less accurate wave function overestimates this polarization as illustrated by the wave function difference map, SV-RHF, in Figure 6. This difference is reflected in a deep minimum in $\Delta B(r) = B(r)_{SV} - B(r)_{RHF}$ (Fig. 7). The minimum occurs at a distance somewhat greater than the bond length owing to the transfer of amplitude from one atomic center to the other. A similar feature appears in the comparison of the theoretical calculation and the experimental result as shown by the $\Delta B(r) = B(r)_{RHF} - B(r)_{exp}$ plot in Figure 8. This suggests that the difference between the actual wave function and the RHF function is similar to that between the RHF and SV functions. Thus the molecular wave function is less polarized than predicted by the RHF function. In this regard it is worth

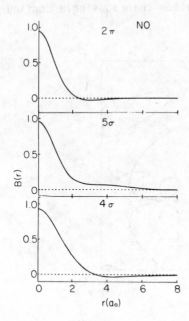

Fig. 4. Spherically averaged
          autocorrelation functions
          from RHF wave functions
          for valence orbitals of NO.

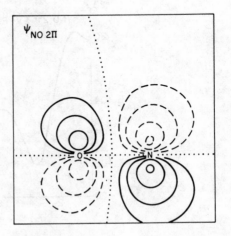

Fig. 5. RHF wave function for 2π
          orbital of NO.

noting that CI or natural orbital calculations generally predict a lesser degree of polarization than do single configuration wave functions.[8]

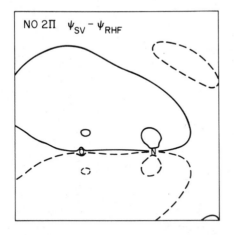

Fig. 6. Difference between SV and RHF
         wave functions for NO2π.

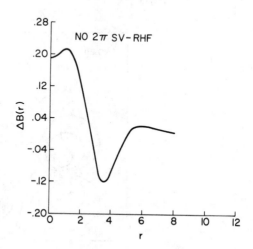

Fig. 7. Difference between B(r) for SV and
         RHF wave functions for NO2π.

In summary, we are applying the (e,2e) technique to molecules of interest to chemists. Employing a number of approximations, many of which deserve further scrutiny, we are attempting to probe the limits of the information content of these experiments in terms of models that are useful and familiar to chemists.

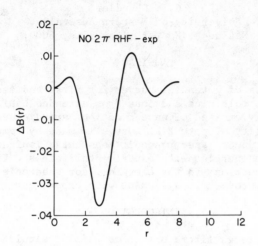

Fig. 8.   Difference between B(r) from
RHF wave function and B(r) from
the (e,2e) experiment for NO2π.

REFERENCES

1. J. H. Moore, M. A. Coplan, T. L. Skillman, Jr., and E. D. Brooks, III, Rev. Sci. Instrum., 49, 463(1978).
2. T. L. Skillman, Jr., E. D. Brooks, III, M. A. Coplan, and J. H. Moore, Nucl. Instrum. and Meth., 155, 267(1978).
3. P. E. Cade and A. C. Wahl, At. Data Nucl. Data Tabl., 15, 1(1975).
4. M. Dupuis, D. Spangler, and J. Wendolowski, Nat. Resour. Comput. Chem. Software Cat., 1, Prog. No. QG01(1980).
5. R. Benesch, S. R. Singh, and V. H. Smith, Jr., Chem. Phys. Lett., 10, 151(1971).
6. W. Weyrich, P. Pattison, and B. G. Williams, Chem. Phys. 41, 271(1979).
7. J. A. Tossell, J. H. Moore, and M. A. Coplan, J. Electron Spectrosc., 22, 61(1981).
8. F. P. Billingsley II and M. Krauss, J. Chem. Phys., 60, 4130(1974).

HARTREE-FOCK AND CI CALCULATIONS ON HOMONUCLEAR
DIATOMIC MOLECULES OF THE FIRST
AND SECOND ROW

A.D. McLean and B. Liu,
IBM Research Laboratory, San Jose, Cal. 95193, U.S.A.

G.S. Chandler,
University of Western Australia,
Nedlands, Western Australia, 6009.

ABSTRACT

Calculations of potential curves for first and second row homonuclear diatomic molecules and ions from extended basis SCF, and singly and doubly excited CI wave functions are discussed. Evaluation of the quality of potential curves is made by comparing equilibrium bond distances, spectroscopic constants, ionisation potentials and dissociation energies with experimental values. It is found that addition of corrections to the energies, for quadruple excitations, substantially improves the agreement with experiment.

INTRODUCTION

There is a large literature on *ab initio* calculations of the electronic structure of homonuclear diatomic molecules[1]. A variety of basis sets, approximations and treatments of correlation have been used in these studies. The motivation for the work reported here was to perform definitive calculations for a series of compounds, producing a set of potential surfaces to be used as reference data.

These computations have been performed with large Slater basis sets, so that the results are near the Hartree-Fock (HF) limit, and configuration interaction (CI) wave functions. The molecules examined are the strongly bound homonuclear diatomic molecules from B_2 to Cl_2 and their singly charged positive ions.

DESCRIPTION OF THE CALCULATIONS

Within the framework of the Born-Oppenheimer approximation, the electronic energy is calculated from the standard electronic Hamiltonian,

$$H_{el} = T_e + V_{eN} + V_{ee} + V_{NN},$$

which takes into account the kinetic energy of the electrons, the electron electron repulsions and the electron nuclear interactions. Thus, all the potential energy surfaces are calculated pointwise for fixed nuclear conformations.

Large Slater basis sets were generated for the calculations. For the first-row atoms, the bases consist of six s-type, five p-type, three d-type and two f-type functions on each atom. For the second row, eight s-type, six p-type, three d-type and two f-type functions were used on each atom. The atomic s- and p- functions differ from the similar sets in Clementi and Roetti's tables[2], by having the outer lobes of the valence s- and p- functions expanded in three rather than two basis functions in order to give additional flexibility in molecular calculations. The d- and f- functions are not used in atomic SCF calculations but are needed to describe polarization effects in the SCF molecular calculations, and are used in both atomic and molecular calculations, for constructing the singly and doubly excited configurations producing correlation effects in the CI calculations.

Potential energy curves were calculated using both the SCF and CI methods. The CI calculations, denoted SDCI, included all single and double excitations from the valence orbitals of the HF configuration. In the case of the first row, the excitations were to the HF virtual orbitals while for the second-row atoms excitations were to a set of functions generated by orthogonalization of basis functions to the HF occupied orbitals. The effect of quadruple excitations on the potential curves was also estimated using both Davidson's relation[3]

$$\Delta E_Q \simeq (1-C_o^2) \Delta E_D$$

and Siegbahn's relation[4]

$$\Delta E_Q \simeq \frac{(1-C_o^2) \Delta E_D}{C_o^2}$$

where C_o is the coefficient of the restricted Hartree-Fock (RHF) configuration in an all-doubles CI wave function and ΔE_D is the all-doubles correlation energy. In fact, C_o and ΔE_D were taken from the SDCI wave function. An overestimate of ΔE_Q results, but it should be small.

Of course, neither the Hartree-Fock, nor the all-singles and doubles calculation arising from it, give a proper description of the whole potential energy surface since, in general, neither lead to the proper dissociation limit. Accordingly, the potential surface calculations were limited to regions around the equilibrium bond distance. Although, dissociation energies have been obtained by obtaining the dissociation limit from separate atomic calculations, in which Hartree-Fock and SDCI level calculations were performed with the same basis sets as used in the molecules.

COMPARISON OF THE POTENTIAL ENERGY
SURFACES WITH EXPERIMENT.

Minima of the calculated potential curves were determined, to
give equilibrium bond distances. Vibrational and rotational levels,
were calculated for these potential curves by numerical intergration
of the radial Schrödinger equation. The calculated vibrational and
rotational levels lying well within the calculated points on the potential energy surface were least-squares fitted to polynomials in
$(v+\frac{1}{2})$ and $J(J+1)$ to yield the vibrational and rotational constants, ω_e,
$\omega_e x_e$, B_e and α_e of (1) and (3). Thus, the vibrational term values
for J=0 can be fitted by,

$$E_{vJ}(J=0) = \omega_e(v+\tfrac{1}{2}) - \omega_e x_e(v+\tfrac{1}{2})^2 + \ldots \tag{1}$$

In addition a purely rotational fit in a vibrational state v is
given by

$$E_{vJ} = B_v J(J+1) - D_v[J(J+1)] + \cdots \tag{2}$$

where the B_v can be fitted to

$$B_v = B_e - \alpha_e(v+\tfrac{1}{2}) + \cdots \tag{3}$$

EXAMPLES OF INDIVIDUAL MOLECULES

Table I shows the second-row molecules and ions examined, their
SCF and SDCI bond distances and energies, the coefficient of the
leading term in the CI expansion, and the number of configuration
state functions (CSF's) in each CI wave function. A number of
states which have not been observed experimentally are listed,
making it worthwhile to briefly examine the electronic structure of
some of these molecules. Al_2 and its ion are straightforward being
built up by filling the $2\pi_u$ orbital in the manner expected from
simple m.o. theory. The ground state of Si_2 however, is interesting
when compared with C_2. In C_2 the extra pair of electrons go into
the π_u orbital to give a $^1\Sigma_g^+$ ground state. Hartree-Fock calculations, however are not able to reproduce this result and predict C_2
to have a triplet ground state[1]. Inclusion of correlation effects
lowers the singlet energy most, and swaps the order of the states[5].
With Si_2 the new electrons go into the $5\sigma_g$ orbital and the actual
ground state is a triplet. Since correlation does not have such a
large differential effect here the HF calculation gives the correct
ordering. Also of interest is the closeness of the Si_2 ($X\,^3\Sigma_g^-$) and
Si_2: $4\sigma_g^2 4\sigma_u^2 5\sigma_g^1 2\pi_u^3; {}^3\Pi_u$ states which at the SDCI level of calculation
are only separated by .27eV, and an inclusion of the correction for
quadruple excitations this is reduced to only .15eV. The remaining
molecules have ground states following the familiar first-row pattern.

Table I: Dominant configurations for the molecular states studied, equilibrium SCF and SDCI energies and coefficients C_o of the leading term in the SDCI wavefunction. All states have a common $1\sigma_g^2 1\sigma_u^2 2\sigma_g^2 2\sigma_u^2 3\sigma_g^2 3\sigma_u^2 1\pi_u^4 1\pi_g^4$ core.

State	Configuration	SCF R_e (Å)	SCF Energy(h)	SDCI R_e (Å)	SDCI Energy(h)	C_o	No. of CSF's
Al_2^+ (X $^2\Pi_u$)	$4\sigma_g^2 4\sigma_u^2 2\pi_u^1$	2.994	-483.564781	2.823	-483.666892	0.938990	2851
Al_2 (X $^3\Sigma_g^-$)	$4\sigma_g^2 4\sigma_u^2 2\pi_u^2$	2.540	-483.770876	2.491	-483.895554	0.931927	2885
Si_2^+ ($^2\Sigma_u^+$)	$4\sigma_g^2 4\sigma_u^1 2\pi_u^4$	1.929	-577.279160	1.964	-577.471254	0.923114	3082
Si_2^+ ($^2\Pi_u$)	$4\sigma_g^2 4\sigma_u^2 2\pi_u^3$	2.147	-577.432597	2.148	-577.618524	0.922847	3435
Si_2^+ ($^4\Sigma_g^-$)	$4\sigma_g^2 4\sigma_u^2 5\sigma_g^1 2\pi_u^2$	2.256	-577.518094	2.261	-577.675761	0.938170	4408
Si_2 ($^1\Sigma_g^+$)	$4\sigma_g^2 4\sigma_u^2 2\pi_u^4$	2.051	-577.692665	2.056	-577.909165	0.917441	3296
Si_2 ($^3\Pi_u$)	$4\sigma_g^2 4\sigma_u^2 5\sigma_g^1 2\pi_u^3$	2.137	-577.755199	2.148	-577.947528	0.930342	5403
Si_2 (X $^3\Sigma_g^-$)	$4\sigma_g^2 4\sigma_u^2 5\sigma_g^2 2\pi_u^2$	2.216	-577.778871	2.238	-577.957103	0.936576	4999
P_2^+ (A $^2\Sigma_g^+$)	$4\sigma_g^2 4\sigma_u^2 5\sigma_g^1 2\pi_u^4$	1.841	-681.119302	1.877	-681.375301	0.946544	5181
P_2^+ (X $^2\Pi_u$)	$4\sigma_g^2 4\sigma_u^2 5\sigma_g^2 2\pi_u^3$	1.926	-681.157054	1.958	-681.395319	0.939986	5517
P_2 (X $^1\Sigma_g^+$)	$4\sigma_g^2 4\sigma_u^2 5\sigma_g^2 2\pi_u^4$	1.850	-681.499323	1.876	-681.767060	0.930336	5311
S_2^+ (X $^2\Pi_g$)	$4\sigma_g^2 4\sigma_u^2 5\sigma_g^2 2\pi_u^4 2\pi_g^1$	1.774	-794.758332	1.804	-795.073765	0.937507	9079
S_2 (X $^3\Sigma_g^-$)	$4\sigma_g^2 4\sigma_u^2 5\sigma_g^2 2\pi_u^4 2\pi_g^2$	1.859	-795.092171	1.879	-795.415518	0.935952	8670

Table I (continued)

$Cl_2^+(X\ ^2\Pi_g)$	$4\sigma_g^2 4\sigma_u^2 5\sigma_g^2 2\pi_u^4 2\pi_g^3$	1.859	-918.599809	1.879	-918.967358	0.938915 9131
$Cl_2(X\ ^1\Sigma_g^+)$	$4\sigma_g^2 4\sigma_u^2 5\sigma_g^2 2\pi_u^4 2\pi_g^4$	1.975	-919.008482	1.984	-919.385102	0.938727 8717

In Tables II and III, spectroscopic constants, equilibrium separations, and dissociation energies for $Si_2(X\ ^3\Sigma_g^-)$ and $Cl_2(X\ ^1\Sigma_g^+)$ are presented and compared with experimental data. Most of the experimental data in this report comes from the compilation of Huber and Herzberg[6]. The patterns of behaviour for the theoretical calculations in Tables II and III are typical of those for the Si species and heavier molecules.

All the theoretical estimates of the dissociation energy are low. The HF values are grossly in error. Inclusion of correlation effects with the SDCI wave functions produces dissociation energies closer to experiment and correction for the quadruple excitations brings them closer again to agreement with experiment. In general, the Siegbahn correction gives the best agreement with experiment

Table II: Spectroscopic constants (cm^{-1}), equilibrium separations, $R_e(\overset{\circ}{A})$ and dissociation energies, $D_e(eV)$, for $Si_2(X\ ^3\Sigma_g^-)$

	ω_e	$\omega_e x_e$	R_e	D_e
SCF	572.4	1.56	2.216	1.97
SDCI	538.7	1.74	2.238	2.55
SDCI + Davidson	519.5	1.88	2.251	2.84
SDCI + Siegbahn	513.9	1.93	2.255	2.90
Experiment[6]	510.98	2.02	2.246	3.24

Consistently low HF predictions of dissociation energies can be understood by picturing bond formation as the pairing up of electrons in localized orbitals, so that additional correlation arises from these pairs. Consequently, there is more correlation in the molecule than in the isolated atoms and the HF dissociation energies must be expected to be low.

Taken as a whole the absolute errors in dissociation energies for the second row diatomics for which experimental data are available range from .97 eV to 3.36 eV for the SCF wavefunctions, .57 eV to 1.57 eV for SDCI wavefunctions .13 eV to .93 eV with the Davidson correction and .06 eV to .79 eV with the Siegbahn correction. The corresponding relative error ranges are 39% to 70%, 21% to 44%, 9% to 24% and 4% to 19%. The average relative errors were 56%, 31%, 16% and 13% respectively,

Turning now to bondlengths it can be seen from the examples in Tables II and III that the HF approximation gives a slightly short bond length. The SDCI wavefunction lengthens the bond and this trend is continued successively with the Davidson and Siegbahn corrections. In some instances this results in the latter corrections giving bond lengths longer than experiment but the errors are invariably small.

Table III: Spectroscopic constants (cm^{-1}) equilibrium separations R_e ($\mathring{A}$) and dissociation energies D_e (eV) for $Cl_2 (X\ ^1\Sigma_g^+)$

	ω_e	$\omega_e x_e$	R_e	D_e
SCF	613.5	1.76	1.975	1.23
SDCI	587.7	1.97	1.984	1.40
SDCI + Davidson	563.5	2.18	1.997	1.92
SDCI + Siegbahn	556.4	2.29	2.000	2.03
Experiment[6]	559.7_2	2.67_5	1.987_9	2.51398_0

Again the best way of summarizing the information is to give error ranges. For the HF calculations errors are of the order of hundredths of an Ångstrom, the range being 0.013 Å to 0.059 Å with an average of 0.039 Å. With the SDCI calculations the range is still of the order of hundredths of an Ångstrom, being 0.004 Å to .027 Å with an average of .015 Å. With the inclusion of the quadruple corrections, errors are in the thousandths of Ångstroms range, with the spread for the Davidson corrections from .001 Å to .009 Å and an average of .004 Å. There is a slightly larger range for Siegbahn's treatment, which shows a tendency to give lengths which are long.

The HF spectroscopic constants, ω_e, are in general too large, meaning that the potential energy surface is too steep near the equilibrium bond distance. Allowance for correlation, corrects this, with improvement increasing successively in the order, SDCI, SDCI + Davidson, and SDCI + Siegbahn. For the HF approximation the range in absolute errors is 53.8 cm^{-1} to 177.3 cm^{-1} with a relative error range of 10% to 22% and an average of 16%. The SDCI ranges are 27.7cm^{-1} to 92.7 cm^{-1}, and 5% to 12%, and an average of 8% while with the addition of the Davidson correction these ranges fall to 3.8 cm^{-1} to 47.5 cm^{-1} and 1% to 6% with an average of 3.2%. The Siegbahn correction results are in slightly better agreement with experiment.

It does not seem worthwhile to examine the much smaller anharmonicity constant $\omega_e x_e$ in detail. It only needs adding that the SDCI wavefunction generally improves on the SCF value and the Davidson and Siegbahn corrections improve on this again.

Table IV: Spectroscopic constants (cm^{-1}), equilibrium separations, R_e(Å) and dissociation energies, D_e(eV) for $Al_2(X\ ^3\Sigma_g^-)$

	ω_e	$\omega_e x_e$	R_e	D_e
SCF	315.7	1.78	2.540	0.53
SDCI	343.6	1.92	2.491	0.93
SDCI + Davidson	344.5	2.03	2.488	1.37
SDCI + Siegbahn	344.2	2.07	2.488	1.44
Experiment[6]	350.01	2.022	2.466	1.57

Table V : Spectroscopic constants (cm^{-1}), equilibrium separations R_e(Å) and dissociation energies, D_e(eV) for Al_2^+: $4\sigma_g^2 4\sigma_u^2 2\pi_u^1; \,^2\Pi_u$

	ω_e	$\omega_e x_e$	R_e	D_e
SCF	160.0	1.76	2.994	0.43
SDCI	208.5	1.93	2.823	0.61
SDCI + Davidson	216.4	1.93	2.797	0.94
SDCI + Siegbahn	217.8	1.93	2.792	0.99

Table VI: Spectroscopic constants (cm^{-1}), equilibrium

separations R$_e$(Å) and dissociation energies,

D$_e$(eV) for B$_2$(X $^3\Sigma_g^-$)

	ω_e	$\omega_e x_e$	R$_e$	D$_e$
SCF	937	10.13	1.636	0.91
SDCI	1028	9.76	1.602	2.05
SDCI + Davidson	1048	9.82	1.590	2.84
SDCI + Siegbahn	1052	9.83	1.586	2.99
Experiment[6]	1051.3	9.35	1.590	3.0$_8$

The discussion till now has been concerned with Si$_2$ and heavier molecules. Al$_2$ and Al$_2^+$ present a different pattern of behaviour. Table IV gives spectroscopic constants equilibrium separations, and dissociation energies for Al$_2$(X $^3\Sigma_g^-$). Nothing new appears with the dissociation energy, which theory gets to be too low. The equilibrium bond distance is unusual though, in that the HF result is too long and correlation serves to shorten it. In line with this, it can be seen that the HF value for ω_e is low and increases when correlation effects are accounted for. After inclusion of estimates for the quadruples the results are not significantly worse than those for the other second-row molecules. Al$_2^+$ shows similar trends. The results are in Table V, but there are no experimental data available for comparison. If, as we might expect from experience with the other diatomics studied, the SDCI plus quadruples calculations give good estimates of bond lengths and spectroscopic constants, then both the HF bond distance and ω_e value are out by a lot more than is conventionally expected. Similar behaviour is also observed for B$_2$ and B$_2^+$. The data is shown in Table VI. B$_2$(X $^3\Sigma_g^-$) shows a similar bond shortening effect to Al$_2$. Again, this effect is greatly increased when an electron is removed to form the positive ion.

In the first-row molecules, it is possible to get some indication of the origin of this effect by examining the SDCI wavefunction. A number of configurations are found to have large coefficients in the B$_2$(X $^3\Sigma_g^-$) wavefunction. Among these is the configuration arising from the $2\sigma_u^2 \rightarrow 3\sigma_g^2$ antibonding to bonding excitation. It would be expected to lead to bond shortening and the effect would be largest for B_2^+: $1\sigma_g^2 1\sigma_u^2 2\sigma_g^2 2\sigma_u^2 1\pi_u^1$; $^2\Pi_u$, where the relative increase in bonding electrons resulting from the excitation is greatest.

The remaining first-row homonuclear diatomics and ions are similar in their behaviour to their second-row counterparts.

IONISATION POTENTIALS

With neutral ground state energies and ionic state energies available, it is possible to calculate adiabatic ionization potentials in which relaxation effects have been accounted for. Table VII compares a selection of the calculated ionisation potentials with the available experimental values.

Table VII: Adiabatic ionization potentials (eV) for selected states

| Ionization | Ionization Potential | | | | |
	SCF	SDCI	SDCI + Davidson	SDCI + Siegbahn	Expt.[6]
$C_2(X\,^1\Sigma_g^+) \rightarrow C_2^+(X\,^2\Pi_u)$	11.06	11.94	12.01	12.00	12.15
$N_2(X\,^1\Sigma_g^+) \rightarrow N_2^+(A\,^2\Pi_u)$	15.31	16.32	16.41	16.42	16.6986
$N_2(X\,^1\Sigma_g^+) \rightarrow N_2^+(X\,^2\Sigma_g^+)$	15.98	15.78	15.62	15.58	15.5808
$O_2(X\,^3\Sigma_g^-) \rightarrow O_2^+(X\,^2\Pi_g)$	11.84	12.03	11.98	11.98	12.071
$F_2(X\,^1\Sigma_g^+) \rightarrow F_2^+(X\,^2\Pi_g)$	15.55	15.72	15.71	15.71	15.686
$P_2(X\,^1\Sigma_g^+) \rightarrow P_2^+(A\,^2\Sigma_g^+)$	10.34	10.66	10.69	10.69	10.81[7]
$P_2(X\,^1\Sigma_g^+) \rightarrow P_2^+(X\,^2\Pi_u)$	9.30	10.11	10.26	10.29	10.53
$S_2(X\,^3\Sigma_g^-) \rightarrow S_2^+(X\,^2\Pi_g)$	9.09	9.31	9.30	9.29	9.36
$Cl_2(X\,^1\Sigma_g^+) \rightarrow Cl_2^+(X\,^2\Pi_g)$	11.13	11.38	11.41	11.40	11.50

The HF wave function results in a fairly acceptable agreement with experiment, the spread in absolute errors being 0.14 eV to 1.39 eV, with a percentage error range of 0.9% to 12%, and an average relative error of 5%. Utilization of the SDCI wave function leads to considerably improved agreement with experiment. Here, the absolute error range is reduced to 0.05 eV to 0.42 eV, with a corresponding relative error range of 0.2% to 2.3%, and an average of 1.4%. When allowance is made for quadruple excitations, the agreement with experiment is further improved, so that, with the Davidson correction, the average relative error has dropped to 1.1%. The Siegbahn correction has a slightly better performance with an average 1.0% error.

CONCLUDING REMARKS

Although the complete set of molecules shown in Table I, and a complementary set of first-row molecules and ions have been studied, the present discussion has necessarily been limited to briefly examining trends in the data, and concentrating on material where comparison with experiment is readily available. This has confined the discussion to bond distances, dissociation energies, ionisation potentials and spectroscopic constants. Other approaches could be suggested. For instance, the completeness of this data lends itself, not only to comparison within periods as discussed, but also to comparison of the electronic structures between periods, which has hardly been touched on here. Also, with the wavefunctions available, properties which are difficult to access experimentally can be readily computed. One important such property is the electric quadrupole moment, and indeed it has been evaluated in the present study. Other properties may be of interest too, and could certainly be obtained.

REFERENCES

1. W.G. Richards, T.E.H. Walker and R.K. Hinkley, Bibliography of *ab initio* Molecular Wave Functions (O.U.P., Oxford, 1971). W.G. Richards, T.E.H. Walker, L. Farnell and P.R. Scott, Bibliography of *ab initio* Molecular Wave Functions. Supplement for 1970-1973 (O.U.P. Oxford, 1974). W.G. Richards, P.R. Scott, E.A. Colbourn and A.F. Marchington, Bibliography of *ab initio* Molecular Wave Functions. Supplement for 1974-1977 (O.U.P. Oxford, 1978).

2. E. Clementi and C. Roetti, At. Data Nucl. Data Tables 14, 177 (1974).

3. E.R. Davidson in The World of Quantum Chemistry edited by R. Daudel and B. Pullman. (Reidel Dordrecht, Holland, 1974).

4. P. Siegbahn, Chem. Phys. Letters 55, 386 (1978).

5. P.F. Fougere and R.K. Nesbet, J. Chem. Phys. 44, 285 (1966).

6. K.P. Huber and G. Herzberg, Molecular Spectra and Molecular Structure IV. Constants of Diatomic Molecules (Van Nostrand Reinhold, New York, 1979).

7. D.K. Bulgin, J.M. Dyke and A. Morris, J. Chem. Soc. Faraday Trans. II 72, 2225 (1976).

MOLECULAR CHARGE DENSITIES AND ELECTROSTATIC INTERACTIONS: A PSEUDOATOM APPROACH

Joel Epstein

Research School of Chemistry, Australian National University

P.O. Box 4, Canberra, A.C.T. 2600. Australia.

ABSTRACT

Using Fourier transform techniques, the electrostatic inter-action between Slater-type charge density functions is evaluated analytically thereby avoiding the use of numerical integration techniques and the bipolar expansion of $|\vec{r}-\vec{r}'|^{-1}$. Electrostatic interactions between atoms of the first three rows of the Periodic Table were accurately described by Born-Mayer expressions of the form $E = -A \exp(-\alpha R)$ for R ranging over several atomic units. Model studies of the N_2-N_2 interaction demonstrated the importance of penetration terms. At large separations when penetration terms can be neglected, pseudoatom point-multipole terms represented the electrostatic interaction more efficiently in general than molecular point-multipole terms.

INTRODUCTION

The electrostatic component is one of several contributions to the change in energy which occurs when two molecular systems are weakly interacting via Coulomb forces. As this component involves only the zeroth order wavefunction $\Psi_o = \psi_a\psi_b$, where ψ_a is the wavefunction of the isolated molecule a, it may be calculated from the charge densities of the isolated molecular species. This paper represents then a departure from the main theme of this workshop in that it discusses an application in chemistry of charge, rather than momentum, density distributions. Nuclear structure is completely ignored - here the nucleus is considered to be a point-charge.

In Rayleigh-Schrödinger perturbation theory, neglecting electron exchange, the electrostatic component is the first order change in the energy of two weakly interacting systems.[1] For some systems, such as two interacting nitrogen molecules used in the model study dis-cussed later in this paper, the electrostatic component is small. For other systems such as those involving hydrogen bonding or drug-receptor interactions, the electrostatic component is larger, and because of cancellation between other components of the interaction energy (polarization, exchange, dispersion and charge transfer contributions), is often a good estimate of the total interaction energy. For example consider the components of the interaction energy for $(H_2O)_2$.[2]

electrostatic	...	-8.0 kcal/mole
exchange	...	9.9 " "
pol. + disp.	...	-0.3 " "
charge transfer	...	-8.2 " "
total		-6.6 " "

094-243X/82/860101-14$3.00 Copyright 1982 American Institute of Physics

Another example is provided by recent *ab initio* calculations of the interaction between para-hydroxyaniline and formamidinium cation used as a model system for drug-receptor interactions.[3] Results using two different basis sets for the configuration of Fig. 1, in which the interplanar spacing is 6.238 a.u., demonstrate the importance of the electrostatic component for this system.[3]

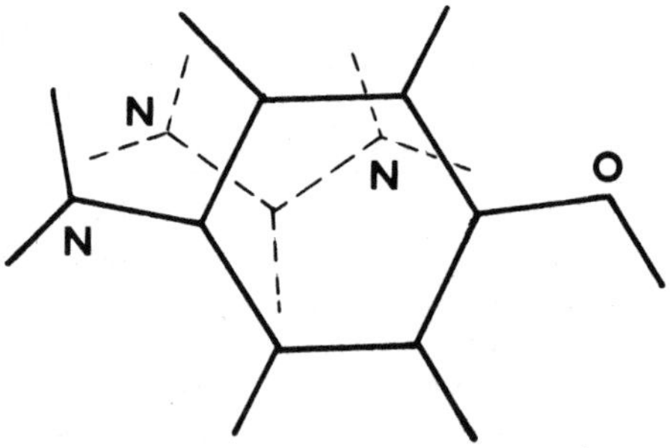

Fig. 1. Configuration for the interaction between para-hydroxyaniline and formamidinium cation.[3]

	STO-3G	4-31G		
electrostatic	-3.99	-7.01	kcal/mole	
polarization	-0.71	-1.62	"	"
exchange	1.36	3.54	"	"
charge transfer + mixing	-0.65	-2.47	"	"
total	-3.99	-7.56	"	"

The purpose of this paper is to present a procedure for evaluating the electrostatic component which is based on a pseudoatom decomposition of the molecular charge densities. Using Fourier transform techniques, the electrostatic interaction between nuclear-centred Slater-type density functions is calculated analytically by a method which avoids the four-region bipolar expansion of Buehler and Hirschfelder.[4] Results are first presented for the interaction between spherical atoms from the first three rows of the Periodic Table. The results of model studies for the N_2-N_2 interaction are then presented, showing the importance of penetration terms and comparing the efficiency of pseudoatom point-multipole terms over molecular point-multipole terms at large intermolecular separations. The method is ideally suited to the study of the electrostatic interaction between molecules using charge densities obtained from X-ray diffraction.[5-7]

THE ELECTROSTATIC COMPONENT

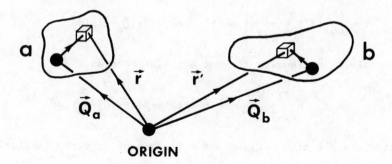

Fig. 2. Co-ordinate system for two weakly interacting
systems a and b.

The electrostatic potential at $\vec{r}'$ due to the electron
distribution of system a is defined by

$$V_a(\vec{r}') = \int \frac{\rho_a(\vec{r}-\vec{Q}_a)}{|\vec{r}-\vec{r}'|} \, d\vec{r} \qquad (1)$$

where $\rho_a(\vec{r}-\vec{Q}_a)$ is the one-electron density of system a. The electrostatic component of the interaction energy between a and b is then
given by

$$E = \sum_{i,j} \frac{Z_{ai}Z_{bj}}{|\vec{R}_{ai}-\vec{R}_{bj}|} - \sum_i Z_{ai} V_b(\vec{R}_{ai}) - \sum_j Z_{bj} V_a(\vec{R}_{bj})$$
$$+ \int \rho_b(\vec{r}'-\vec{Q}_b) \, V_a(\vec{r}') \, d\vec{r}' \qquad (2)$$

where Z_{ai} is the charge of the nucleus located at $\vec{R}_{ai}$. The first
term of equation (2) is the nuclear-nuclear term, the next two terms
are electron-nuclear terms and the last term is the electron-electron
term.

The electron-electron term is the most difficult to calculate
and traditionally involves the bipolar expansion of $|\vec{r}-\vec{r}'|^{-1}$. An
alternative procedure uses Fourier transform techniques. The Fourier
transform of $\rho_a(\vec{r}-\vec{Q}_a)$ is

$$\int \rho_a(\vec{r}-\vec{Q}_a)\exp(i\vec{S}\cdot\vec{r})\,dr = F_a(\vec{S})\exp(i\vec{S}\cdot\vec{Q}_a) \qquad (3)$$

$$\text{where} \quad F_a(\vec{S}) = \int \rho_a(\vec{r}) \exp(i\vec{S}\cdot\vec{r})\, d\vec{r} \tag{4}$$

Using the inverse transform of $\rho_a(\vec{r}-\vec{Q}_a)$,

$$V_a(\vec{r}') = \frac{1}{2\pi^2} \int S^{-2} F_a(\vec{S}) \exp[i\vec{S}\cdot(\vec{Q}_a-\vec{r}')]\, d\vec{S} \tag{5}$$

and the electron-electron term of eqn. (2) becomes

$$E_{ee} = \frac{1}{2\pi^2} \int S^{-2} F_a(\vec{S}) F_b^*(\vec{S}) \exp[i\vec{S}\cdot(\vec{Q}_a-\vec{Q}_b)]\, d\vec{S} \tag{6}$$

Treating the electron-nuclear and nuclear-nuclear terms similarly,

$$E = \frac{1}{2\pi^2} \int S^{-2} \; [F_a(\vec{S}) \exp(i\vec{S}\cdot\vec{Q}_a) - \sum_i Z_{ai} \exp(i\vec{S}\cdot\vec{R}_{ai})]$$

$$\times \; [F_b(\vec{S}) \exp(i\vec{S}\cdot\vec{Q}_b) - \sum_j Z_{bj} \exp(i\vec{S}\cdot\vec{R}_{bj})]^* \; d\vec{S} \tag{7}$$

ELECTROSTATIC INTERACTIONS BETWEEN PSEUDOATOMS

Each molecular charge density may be expanded in terms of nuclear-centred density functions, or pseudoatoms.

$$\text{e.g.} \quad \rho_a(\vec{r}-\vec{Q}_a) = \sum_i \rho_{ai}(\vec{r}-\vec{R}_{ai}) \tag{8}$$

$$\text{Then} \quad F_a(\vec{S}) \exp(i\vec{S}\cdot\vec{Q}_a) = \sum_i f_{ai}(\vec{S}) \exp(i\vec{S}\cdot\vec{R}_{ai}) \tag{9}$$

where the pseudoatom scattering factor $f_{ai}(\vec{S})$ is defined by

$$f_{ai}(\vec{S}) = \int \rho_{ai}(\vec{r}) \exp(i\vec{S}\cdot\vec{r})\, d\vec{r} \tag{10}$$

In terms of pseudoatoms, the electrostatic interaction is

$$E = \frac{1}{2\pi^2} \sum_{i,j} \int S^{-2} [f_{ai}(\vec{S})-z_a][f_{bj}(\vec{S})-z_{bj}]^* \exp[i\vec{S}\cdot(\vec{R}_i-\vec{R}_j)]\, d\vec{S} \tag{11}$$

Each pseudoatom may be represented in terms of multipole density functions i.e. $\rho_\alpha(\vec{r}-\vec{R}_\alpha) = \rho_\alpha(\vec{r}_\alpha) = \frac{1}{4\pi} \sum_\ell \sum_{m=0}^\ell \rho_{\alpha,\ell,m\pm}(r_\alpha) Y_{\ell m\pm}(\hat{\vec{r}}_\alpha)$ (12)

$$f_\alpha(\vec{S}) = \sum_\ell i^\ell \sum_{m=0}^\ell f_{\alpha,\ell,m\pm}(S) Y_{\ell m\pm}(\hat{\vec{S}})$$ (13)

where $f_{\alpha,\ell,m}(\vec{S}) = \int_0^\infty \rho_{\alpha,\ell,m\pm}(r) j_\ell(Sr) r^2 dr$ (14)

and $Y_{\ell m\pm}(\hat{\vec{r}}) = P_\ell^m(\cos\theta_{\vec{r}}) \begin{cases} \cos m\phi_{\vec{r}} \\ \sin m\phi_{\vec{r}} \end{cases}$ (15)

The electron-electron term for the interaction between the multipole component $(\ell,m\pm)$ of pseudoatom α, system a and the multipole component $(u,v\pm)$ of pseudoatom β, system b is (from eqn. (11))

$$E_{ee}(\ell,m\pm,u,v\pm,\vec{R}_{\alpha\beta}) = \frac{1}{2\pi^2} i^\ell (-i)^u \int S^{-2} f_{\alpha,\ell,m\pm}(S) f_{\beta,u,v\pm}(S)$$

$$\times Y_{\ell m\pm}(\hat{\vec{S}}) Y_{uv\pm}(\hat{\vec{S}}) \exp(i\vec{S}\cdot\vec{R}_{\alpha\beta}) d\vec{S}$$ (16)

$$= \sum_{k=|\ell-u|}^{\ell+u}{}' \frac{2}{\pi} \int_0^\infty f_{\alpha,\ell,m\pm}(S) f_{\beta,u,v\pm}(S) j_k(SR_{\alpha\beta}) dS$$

$$\times G(k,\ell,m\pm,u,v\pm,\hat{\vec{R}}_{\alpha\beta})$$ (17)

where $\vec{R}_{\alpha\beta} = \vec{R}_\alpha - \vec{R}_\beta$ and $G(k,\ell,m\pm,u,v\pm,\hat{\vec{R}}_{\alpha\beta})$ is a geometric factor. G and the integral of eqn. (17) are discussed in detail in reference 8. For Slater-type pseudoatom multipole density functions, this integral and the corresponding integrals from the electron-nuclear terms can be expressed in terms of the elementary integral.

$$\frac{2}{\pi} \int_0^\infty \frac{x^{2j-1} \sin(\alpha Rx)}{(1+x^2)^{n+1}} dx = \delta_{j,0} + e^{-\alpha R} \sum_{m=0}^n a_m(\alpha R)^m$$ (18)

where the coefficients a_m are given in reference 8. The electrostatic interaction between the multipole components $(\ell,m\pm)$ of pseudoatom α and $(u,v\pm)$ of pseudoatom β therefore consists of a point-multipole term which is proportional to $R_{\alpha\beta}^{-(\ell+u+1)}$ and penetration terms which decrease exponentially with increasing internuclear distance.

The point-multipole terms involve pseudoatom multipole moments (e.g. pseudoatom charges, dipole moments, quadrupole moments ...). Differencing between the various terms of eqn. (2), which are usually of the same order of magnitude, is avoided by the use of pseudoatom net charges (net charge q = pseudoatom charge - nuclear charge). For example the term $q_\alpha q_\beta/R$ includes the sum of a nuclear-nuclear term and the electron-nuclear and electron-electron terms involving the monopoles of pseudoatoms α and β.

ELECTROSTATIC INTERACTIONS BETWEEN SPHERICAL ATOMS

From eqn. (11), the electrostatic interaction between two spherical atoms separated by the internuclear distance R is given by

$$E = \frac{2}{\pi} \int_0^\infty [f_a(S) - Z_a][f_b(S) - Z_b] j_o(SR) dS \tag{19}$$

where $f_a(S)$ is the form factor of atom a. For neutral atoms, the point-multipole terms, which consist of point-charge terms only, vanish, and the interaction consists of penetration terms only.

A plot of $\ln|E|$ versus R for the interaction between two neon atoms, calculated using Clementi's wavefunction[9] is shown in Fig. 3 for 4.6 a.u. $\leqslant$ R $\leqslant$ 7.0 a.u. For this range of R, $\ln|E|$ is accurately fitted by a straight line, yielding the expression E = -75.08 exp(-2.511R) where all values are in atomic units. The relative root-mean-square error in using this expression for this range of R was 0.23%.

Similar results were found for other atoms of the first three rows of the Periodic Table. For a range of R extending over several atomic units, each electrostatic interaction calculated using Clementi's wavefunctions was well described by the Born-Mayer-type expression E = -Aexp(-αR). Values of A and α for the interactions

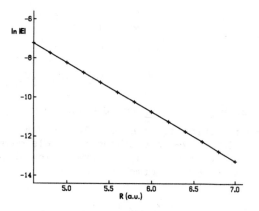

Fig. 3. $\ln|E|$ versus R for the Ne-Ne interaction. The calculated values are indicated by crosses.

H-H, N-N, O-O, F-F, Cℓ-Cℓ, Br-Br, He-He, Ne-Ne, Ar-Ar and Kr-Kr are given in Table I.

Table I. Values of A and α in the expression $E = -A\exp(-\alpha R)$ describing the electrostatic interaction between atoms over the range of R indicated. All values are in atomic units.

H-H ($\zeta=1.0$)	H-H ($\zeta=1.24$)				He-He
0.8994	1.708				8.000
1.588	2.080				2.555

Legend		N-N	O-O	F-F	Ne-Ne
Atom-Atom		43.21	53.74	59.32	75.08
A					
α		1.898	2.129	2.294	2.511

	Range	Error (%)		Cl-Cl	Ar-Ar
H-H	$3.2 \leqslant R \leqslant 5.6$	4.24	($\zeta=1.0$)		
H-H	$3.2 \leqslant R \leqslant 5.6$	3.92	($\zeta=1.24$)	206.1	206.7
He-He	$4.4 \leqslant R \leqslant 6.8$	0.96			
N-N	$4.4 \leqslant R \leqslant 6.8$	0.17		1.832	1.955
O-O	$4.2 \leqslant R \leqslant 6.6$	0.74			
F-F	$3.8 \leqslant R \leqslant 6.2$	1.37			Br-Br / Kr-Kr
Ne-Ne	$4.6 \leqslant R \leqslant 7.0$	0.23		Br-Br	Kr-Kr
Cl-Cl	$5.6 \leqslant R \leqslant 8.0$	0.88			
Ar-Ar	$5.8 \leqslant R \leqslant 8.2$	2.83		420.0	781.5
Br-Br	$6.2 \leqslant R \leqslant 8.6$	1.15			
Kr-Kr	$6.4 \leqslant R \leqslant 8.8$	1.42		1.767	1.944

For the H-H interaction, two density functions were used; one for the isolated atomic wavefunction ($\zeta=1.0$ a.u.$^{-1}$) and one obtained using the standard molecular exponent ($\zeta=1.24$ a.u.$^{-1}$).[10] For the more contracted density function, both A and α were larger. This trend continued across the Periodic Table; A and α both increased as the atomic densities became more contracted. As the atomic densities expanded in going down the Periodic Table, A increased while α decreased.

The accuracy with which the Born-Mayer-type expression describes the electrostatic interaction between spherical atoms is surprising in view of the large number of functions comprising the one-electron densities. The largest error was obtained for the H-H interaction in which the atomic densities were described by single Slater-type functions.

THE ELECTROSTATIC INTERACTION BETWEEN TWO NITROGEN MOLECULES

The N_2 molecular density was decomposed into pseudoatoms by using the generalized scattering factor formalism of Stewart, Bentley and Goodman.[11] At the $[J|K]$ expansion level, the molecular density is represented by the superposition of multipoles to order J on one nucleus and to order K on the other nucleus.

$$\text{i.e.} \quad \rho_M(\vec{r}) = \sum_{j=0}^{J} \rho_{1,j}(r_1) P_j(\cos\theta_{\vec{r}_1}) + \sum_{k=0}^{K} \rho_{2,k}(r_2) P_k(\cos\theta_{\vec{r}_2}) \quad (20)$$

where $\vec{r}_p = \vec{r} - \vec{R}_p$ is the vector from nucleus p at $\vec{R}_p$ to the point at $\vec{r}$. The co-ordinate system is chosen with the z-axis along the internuclear vector $\vec{R}_2 - \vec{R}_1$. The molecular form factor is the Fourier transform of $\rho_M(\vec{r})$ and is given by

$$F_M(\vec{S}) = \exp(i\vec{S}\cdot\vec{R}_1) \sum_{j=0}^{J} i^j f_{1,j}(S) P_j(\cos\theta_{\vec{S}})$$

$$+ \exp(i\vec{S}\cdot\vec{R}_2) \sum_{k=0}^{K} i^k f_{2,k}(S) P_k(\cos\theta_{\vec{S}}) \quad (21)$$

The generalized scattering factors (GSF's) $f_{1,j}(S)$ and $f_{2,k}(S)$ were obtained by a least-squares analysis of the molecular form factor obtained from a calculated wavefunction. The least-squares parameters were the GSF's themselves, resulting in a set of functional equations for the GSF's at each value of S.

An important theorem proved by Stewart, Bentley and Goodman is that at the $[J|K]$ level, the GSF's must necessarily, in superposition, reproduce all the molecular properties of the starting wavefunction of the form $\langle g(r_1) P_j(\cos\theta_{\vec{r}_1})\rangle$ for $0 \leqslant j \leqslant J$ and $\langle h(r_2) P_k(\cos\theta_{\vec{r}_2})\rangle$ for $0 \leqslant k \leqslant K$ where $g(r_1)$ and $h(r_2)$ are arbitrary radial functions. For example at the $[0|0]$ level the GSF's must reproduce the molecular dipole moment; at the $[2|2]$ level the GSF's must reproduce the molecular dipole, quadrupole and octupole moments.[11] This provides a consistent procedure by which the efficiency with which the pseudoatom multipole moments describe the electrostatic interaction at large intermolecular distances (when the penetration terms may be sensibly neglected) may be compared to that of the molecular multipole moments.

GSF's at the $[2|2]$ level for N_2 were obtained from the wavefunction of Cade and Wahl[12] (internuclear distance 2.068 a.u.). The relevant pseudoatom moments were[13]

charges	$c_1 = c_2 = 7.000$ a.u.
dipole moments	$d_1 = -d_2 = 0.04329$ a.u.
quadrupole moments	$q_1 = q_2 = 0.5632$ a.u.

The centre of mass molecular quadrupole moment given by $Q = (4R_1d_1 - 2q_1)$[11] is -0.9474 a.u., in agreement with the value used by Ng, Meath and Allnatt.[14]

One disadvantage of this pseudoatom decomposition is that the pseudoatoms are not unique, but depend upon the level of expansion. For example the GSF's and the pseudoatom properties at the $[1|1]$ or $[3|3]$ level of expansion differ from those at the $[2|2]$ level, even though molecular properties such as the molecular quadrupole moment must necessarily be reproduced for all levels greater than and including $[1|1]$. At the $[1|1]$ level for N_2, $d_1 = -0.2290$ a.u. and the centre of mass quadrupole moment, given by $Q = 4R_1d_1$, is -0.9473 a.u. as required.

Analytical representations of the GSF's were required for the calculation of the N_2-N_2 electrostatic interaction. The GSF's were fitted with linear combinations of 12 Jacobi polynomials, to order 11.

i.e.
$$f_{p,\ell}(S) = 8\gamma^2\sqrt{2\gamma}\frac{(4\gamma S)^\ell}{(4S^2+\gamma^2)^{\ell+2}} \sum_n b_{n,\ell} P_n^{(\ell+3/2,\ell+1/2)}\left(\frac{4S^2-\gamma^2}{4S^2+\gamma^2}\right) \quad (22)$$

The corresponding radial density functions were then obtained in terms of generalized Laguerre polynomials by Fourier inversion.[17]

i.e.
$$\rho_{p,\ell}(r_p) = \frac{1}{4\pi}\gamma\sqrt{2\gamma}(\gamma r_p)^\ell \exp(-\gamma r_p/2)$$

$$\times \sum_n b_{n,\ell}\frac{(2\ell+2n+1)!!}{(2\ell+n+2)!2^n} L_n^{(2\ell+2)}(\gamma r_p) \quad (23)$$

The exponent γ and coefficients $b_{n,\ell}$ were least-squares parameters which minimized

$$\varepsilon = \sum_i W_i[f_{p,\ell}(S_i) - f_{p,\ell}^{GSF}(S_i)]^2 \quad (24)$$

where W_i is a weighting factor.

The analytical representations of the GSF's were constrained to satisfy selected pseudoatom properties, including the pseudoatom multipole moments. For $W_i = S_i^2$ the relative root-mean-square errors of 0.06%, 6% and 1% were obtained for the monopole, dipole and quadrupole GSF components, respectively.

Using these analytical representations the electrostatic component of the N_2-N_2 interaction was calculated as a function of the centre of mass separation, R, for different configurations of the two molecules. The results for three configurations, linear, T and parallel are shown by the dashed curves in Fig. 4(a), 4(b) and 4(c), respectively. Also shown in these Figures are the corresponding electrostatic components calculated from the analytical representations of the GSF's obtained using $W_i = S_i^{-4}$ (continuous curves); the

110

results calculated from the partial wave expansion coefficients
reported by Ng, Meath and Allnatt[14] for the Cade and Wahl wavefunction (crosses); and the results reported by Berns and Van der
Avoird[18] for a wavefunction consisting of Gaussian type functions
(circles).

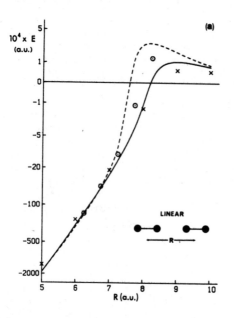
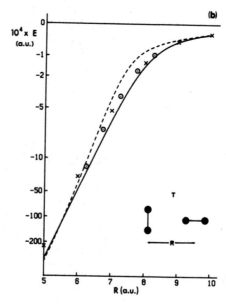

Fig. 4. The electrostatic
component of the N_2-N_2 interaction, calculated from
analytical representations
of the GSF's at the $[2|2]$
level obtained using
$W_i = S_i^2$ (----) and
$W_i = S_i^{-4}$ (———).

X ... results from ref. 14.
⊙ ... results from ref. 18.

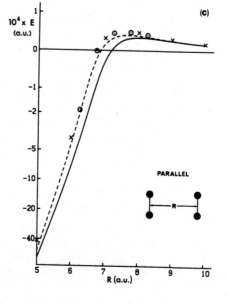

For 7.0 a.u. < R < 10.0 a.u. in the linear configuration, the results from the $W_i = S_i^2$ GSF representations differ markedly from the results of reference 14 which used the same starting wavefunction. Significant differences also exist at R = 5.0 a.u. and R = 6.0 a.u. Note that the arcsinh scale used for the ordinate axis highlights relative differences rather than absolute differences. Similar results were found for the T-configuration. However for the parallel configuration the $W_i = S_i^2$ results are in good agreement with both sets of previous results.

The differences between the present results and those reported previously are principally due to the accuracy of the analytical representations of the GSF's rather than the GSF's themselves. For the linear and T-configurations the discrepancies with the results of reference 14 are largely removed when the electrostatic interactions are calculated from representations of the GSF's obtained with $W_i = S_i^{-4}$. This may be explained by examining differences between the molecular charge densities at the [2|2] level and $\rho_c(\vec{r})$, the charge density calculated from the Cade and Wahl wavefunction. For $W_i = S_i^2$, the difference density itself is uninformative, with the largest features being close to the nuclear positions. The relative difference density, $100 \times [\rho_c(\vec{r}) - \rho_M(\vec{r})]/\rho_c(\vec{r})$ is more revealing and is shown in Fig. 5(a).

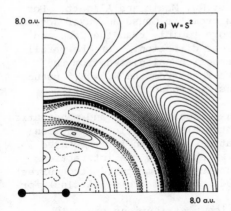 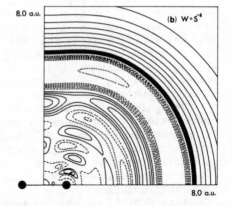

Fig. 5. Relative difference densities for N_2 at the [2|2] level. Contour levels are in steps of 2 for levels less than 10% and in steps of 10 for levels greater than 10%. Continuous lines are positive ($\rho_c(\vec{r}) > \rho_M(\vec{r})$); dashed lines are negative ($\rho_c(\vec{r}) < \rho_M(\vec{r})$).

Within 3.0 a.u. from each nucleus the density from the superposition of pseudoatoms obtained using $W_i = S_i^2$ differs from that of the Cade and Wahl wavefunction by less than 8%. More than 5.5 a.u. away from the centre of mass along the internuclear vector, however, the density from the pseudoatoms decreases more rapidly, to a maximum relative difference of +~340%. For the linear configuration and, to a lesser degree, for the T-configuration, the more contracted density along this direction results in a smaller electron-electron penetration term and the more positive electrostatic interaction shown by the dashed curves of Figs. 4(a) and 4(b). Along the perpendicular bisector of the internuclear vector the density from these pseudoatoms is more similar to the Cade and Wahl density (though again more contracted), producing better agreement for the electrostatic interaction in the parallel configuration.

For R~6-7 a.u., the electrostatic interaction between the two N_2 molecules depends upon the rate of decrease of the N_2 density at large distances from the nuclei. In this region of space the density from the pseudoatoms is strongly influenced by the GSF's at small S. The weighting factor of S^2, however gives smallest weight to these components. A more appropriate weighting scheme is one which gives greatest weight to these components. The relative difference density obtained from analytical representations of the GSF's using $W_i = S_i^{-4}$ is shown in Fig. 5(b). At small distances from the nuclei the differences are similar to those of Fig. 5(a). At large distances, however, the relative differences are dramatically reduced, particularly along the internuclear vector. The resulting electrostatic interaction for the linear and T-configurations, shown by the continuous curves of Figs. 4(a) and 4(b), are in better agreement with the previously reported results of Ng, Meath and Allnatt. For the linear configuration, the results from the Gaussian wavefunction suggest that this density decreases more rapidly than the Cade and Wahl wavefunction along the internuclear direction. For the parallel configuration, the electrostatic interaction calculated from the $W_i = S_i^{-4}$ pseudoatoms is less positive than from the $W_i = S_i^2$ pseudoatoms, indicating that the former density produces the greater electron-electron penetration.

The point-multipole contributions (dashed curves) and penetration contributions (continuous curves) to the electrostatic interaction calculated from the $W_i = S_i^{-4}$ pseudoatoms are shown for the three configurations in Fig. 6(a)-6(c). Also shown are the interactions calculated from the centre of mass quadrupole moments (dotted curves) and from the superposition of isolated spherical atoms (circles) for these configurations.

From Fig. 6. pseudoatom penetration contributions can only sensibly be neglected for R > 8.5 a.u. for the parallel and T-configuration, and for R > 9.5 a.u. for the linear configuration. For all three configurations, the pseudoatom penetration contributions were negative (attractive). The corresponding penetrations from the superposition of isolated spherical atoms were larger in magnitude, illustrating that the pseudoatoms are more contracted. For the linear and parallel configurations, the pseudoatom point-multipole contributions were positive (repulsive). At large R, when the pseudoatom penetration terms may be neglected, the pseudoatom moments described the electrostatic interaction more accurately than the molecular quadrupole moments. For the N_2-N_2 interaction the differences were small. For larger molecules, the efficiency of the

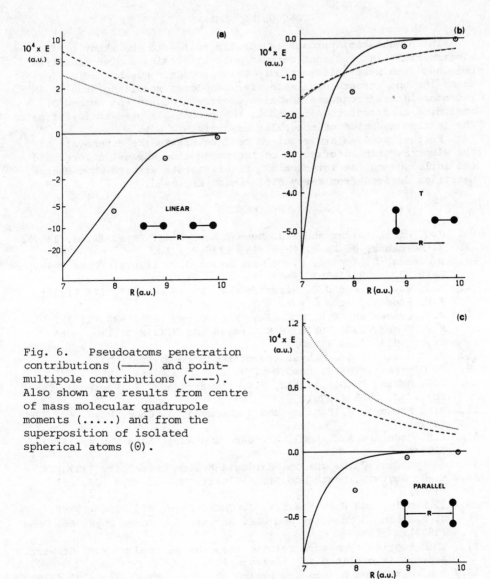

Fig. 6. Pseudoatoms penetration
contributions (———) and point-
multipole contributions (----).
Also shown are results from centre
of mass molecular quadrupole
moments (.....) and from the
superposition of isolated
spherical atoms (⊖).

pseudoatom moments over the molecular moments is expected to be
greater. For the T-configuration, the point-multipole contributions
were negative (attractive). For this configuration, the interaction
calculated from the molecular quadrupoles was negligibly different
from that calculated from the pseudoatom moments. The dashed and
dotted curves of Fig. 6(b) were indistinguishable for R > 8.0 a.u.

CONCLUDING REMARKS

The pseudoatom approach is ideally suited to the study of electrostatic interactions between molecules using charge densities obtained from x-ray diffraction. This procedure may be useful for examining the structure of molecular complexes modelling drug-receptor interactions and may also provide a convenient method of including electrostatic effects in the potentials used to investigate the lattice dynamics of molecular crystals.

Further studies are required to investigate the dependence of the electrostatic interaction on the pseudoatom expansion level and the influence on the interaction of solid state effects when charge densities derived from x-ray diffraction are used.

REFERENCES

1. J.O. Hirschfelder and W.J. Meath, Adv. Chem. Phys. 12, 3 (1967).
2. K. Morokuma, J. Chem. Phys. 55, 1236 (1971).
3. R. Osman, S. Topiol, H. Weinstein and J.E. Eilers, Chem. Phys. Lett. 73, 399 (1980).
4. R.J. Buehler and J.O. Hirschfelder, Phys. Rev. 83, 628 (1951).
5. R.F. Stewart, Acta Cryst. A32, 565 (1976).
6. B.M. Craven and R.K. McMullan, Acta Cryst. B35, 934 (1979).
7. H.P. Weber, J.R. Ruble, B.M. Craven and R.K. McMullan, Acta Cryst. B36, 1121 (1980).
8. J. Epstein, Molec. Phys. (submitted for publication).
9. E. Clementi, IBM J. Res. Dev. 9, 2. Supplement (1965).
10. W.J. Hehre, R. Ditchfield, R.F. Stewart and J.A. Pople, J. Chem. Phys. 52, 2769 (1970).
11. R.F. Stewart, J. Bentley and B. Goodman, J. Chem. Phys. 63, 3786 (1975).
12. P.E. Cade and A.C. Wahl, At. Data and Nucl. Data Tables 13, 339 (1974).
13. J. Epstein, Ph.D. thesis, Carnegie-Mellon University (1978).
14. K.-C. Ng, W.J. Meath and A.R. Allnatt, Molec. Phys. 33, 699 (1977).
15. J. Bentley and R.F. Stewart, J. Chem. Phys. 63, 3794 (1975).
16. J. Epstein, J. Bentley and R.F. Stewart, J. Chem. Phys. 66, 5564 (1977).
17. This Fourier transform relationship was derived by R.F. Stewart. A proof is outlined in reference 13.
18. R.M. Berns and A. Van der Avoird, J. Chem. Phys. 72, 6107 (1980).

BINARY (e,2e) SPECTROSCOPY OF MOLECULES - ELECTRONIC MOMENTUM DISTRIBUTIONS AND MOLECULAR STRUCTURE

Alan Minchinton
School of Physical Sciences, The Flinders University of South Australia,
Bedford Park, South Australia 5042.

ABSTRACT

A survey is presented of some experimental results on small molecules collected in this laboratory using non-coplanar symmetric binary (e,2e) spectroscopy. The (e,2e) technique allows the unique determination of separation energy spectra over a wide energy range, and electronic momentum distributions for individual electron orbitals as well as providing information on spectroscopic factors and electron correlation effects.

INTRODUCTION

The (e,2e) technique can be used to elucidate the electronic structure of molecules principally through the determination of separation (binding) energy spectra and electronic momentum distributions (profiles). Several extensive reviews discussing the (e,2e) technique, mainly as applied to atoms and several molecules, already exist in the literature[1,2] and another review is currently being prepared[3]. This session will be concerned mainly with the non-coplanar binary (e,2e) technique and those molecules studied by the Flinders group. This means that several other important types of electron-electron coincidence experiments that are concerned with other quite different aspects of the ionization process will not be considered in any detail.

These experiments include the technique used by the group at Kaiserlautern headed by Professor Ehrhardt[4,5] which employs low energy incident electrons and asymmetric kinematics. This group reported the first electron coincidence experiment in 1969[6] involving the electron impact ionization of helium. Professor Ehrhardt will be talking about his work later in this workshop.

Experiments using asymmetric kinematics at high incident energies have been carried out by Brion (UBC) and van der Wiel (FOM)[7,8] mainly to measure partial oscillator strengths. The technique is equivalent to the (γ,e) process and shows a quantitative relationship with photoelectron spectroscopy (PES)[9]. These forward-scattering experiments involve distant electron-electron collisions at small momentum transfer and cannot be used to measure momentum distributions. The technique is now commonly called dipole (e,2e) spectroscopy.

The coplanar symmetric geometry in which the outgoing electrons have equal energies and form equal angles with respect to the incident direction was used at high incident energy by the Frascati group in 1972[10] to determine the momentum profile for the 1s state of carbon. They used thin-film targets. The coplanar geometry has also been used by the Flinders group to study the valence states of the inert gases[1,2].

The non-coplanar symmetric binary (e,2e) technique can readily be understood from figure 1. An electron of energy E_o and momentum $\underline{k}_o$

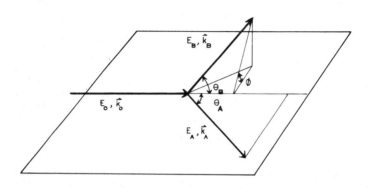

Fig. 1. Schematic diagram of the kinematics of an (e,2e) event.

is incident on a stationary target. Two indistinguishable electrons emerge from the ionizing collision with energies E_A and E_B and momenta $\underline{k}_A$ and $\underline{k}_B$ at angles θ_A and θ_B with relative azimuthal angle $(\phi_A - \phi_B)$ such that $\phi_A = \pi$ and $\phi_B = 0$ as shown in figure 1.

If the kinematics is fully determined, the energy and momentum conservation equations are respectively

$$E_o = (E_A + E_B) + \in_f \tag{1}$$

$$\underline{k}_o = \underline{k}_A + \underline{k}_B + \underline{q} \tag{2}$$

$\in_f$, the separation energy, is the energy difference between the initial state of the target and the final ionic state f; $\underline{q}$ is the recoil momentum of the ion; $\underline{K} = \underline{k}_o - \underline{k}_A$ is the momentum transfer.

In the dipole (e,2e) experiment E_o is large, $\theta_A \approx 0$ and hence the momentum transfer K is small. However, in the binary (e,2e) experiment, to ensure close electron-electron collisions and a knockout mechanism K must be maximized. This, of course, is achieved if $\theta_A = \theta_B = \theta$ and $E_A = E_B$. From (2) then, $-\underline{q}$ is the momentum of the struck electron prior to the collision:

$$q = [(2k_A Cos\theta - k_o)^2 + 4k_A^2 Sin^2\theta Sin^2\tfrac{1}{2}\phi]^{\frac{1}{2}} \qquad (3)$$

In the coplanar symmetric geometry, $\phi = 0$ and momentum profiles are obtained by varying the angle $\theta = \theta_A = \theta_B$ for a given separation energy $\in_f = E_o - E_T$, where E_T is the total energy, $E_T = 2E_A$. Hence, for a stationary target with negligible separation energy the quasi-free scattering angle $\theta_o = 45°$.

Figure 2 shows the results of some coplanar measurements for atom-

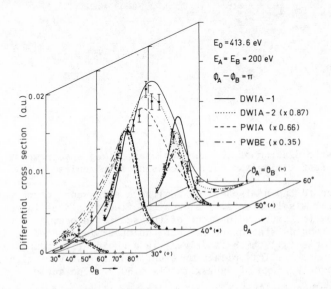

Fig. 2. The coplanar (e,2e) cross-section for the hydrogen atom[11].

ic hydrogen[11] which clearly show the effect of decreasing momentum transfer as θ varies from the quasi-free scattering angle. The differential cross-section peaks when $\theta_A = \theta_B = 45°$, the situation realised in the non-coplanar symmetric technique.

In the non-coplanar symmetric geometry $\phi \neq 0$ and the momentum distribution at a particular separation energy is determined by varying ϕ while keeping $k_A = k_B$ and θ fixed. This technique was first used by the Flinders group in 1973[12] to study the valence orbitals in argon.

EXPERIMENTAL

A schematic diagram of the non-coplanar symmetric spectrometer is shown in figure 3. The identical analysers consist of a three element cylindrical retarding lens followed by a hemispherical dispersing element of 7cm mean radius. The deflection plates Q_3 and Q_4

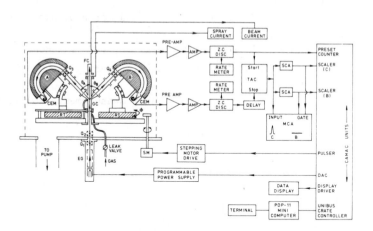

Fig. 3. Schematic diagram of the non-coplanar (e,2e) spectrometer including the fast timing and data processing electronics.

at the entrance to the hemispherical analysers allow correction for minor mechanical misalignments.   Sideways deflector plates in the analysers allow for extra adjustment of the scattered beam.   One of the analysers remains stationary (A) but the other is mounted on a rotating turntable (RT) which is driven via a chain drive connected to a stepping motor.   This planar movement thus allows variation of the angle ϕ_B over a range of approximately $\pm 50°$.   The angles θ_A and θ_B can be mechanically varied a few degrees either side of $\theta = 45°$ but the analysers are generally adjusted to detect electrons at $\theta_A = \theta_B \approx 45°$.

The electrons are produced from a thoriated tungsten cathode (Pelco A.E.I. filaments type 1410) set into a Cliftronics type CE5AH electron gun (EG).   The gun is mounted below the main chamber and is differentially pumped.   Helium gas can be admitted into the electron gun chamber through a leak valve to reduce space charge effects. The electron beam is aligned along the axis of the gas cell (GC) using two sets of quadrupole deflectors Q_1 and Q_2 and is focussed so as to maximize the current collected in the Faraday cup (FC).

The gas being studied is either connected to the system through conventional lectere bottle couplings or if in liquid form is stored in glass flasks connected to the system through glass-stainless steel tubing.   Commercially available gases are used as supplied.   Residual gases are removed from liquid samples using freeze-thaw cycles. The gas passes through a heated leak valve and stainless steel and teflon tubing into the interaction region in the gas cell.

The entire spectrometer is housed in a magnetically shielded stainless steel and aluminium vacuum chamber and is evacuated by an oil diffusion pump. The base pressure is typically 2×10^{-7} torr but the pressure rises to 2×10^{-5}T during the experiments with the pressure in the gas cell being approximately 10^{-3}T. The pressure is monitored using Granville-Phillips Type 274016 ionization gauges employing iridium coated filaments. Magnetic shielding is provided by annealed mu-metal and a set of external Helmholtz coils.

At 1200eV incident electron energy the beam current is typically 30-60μA and the overall energy resolution with 150eV pass energy is 1.8eV. The corresponding momentum resolution is ~ 0.08 a_o^{-1}. At present the electrons passed by the two analysers are detected using channeltrons. However, modifications are being made to replace the channeltrons with position-sensitive detectors (Surface Science Labs. Model 239G) to improve the energy resolution and increase the data acquisition rate. The increased single channel count rates (above the present ~ 400 cps) will also facilitate the implementation of a monochromator in the incident electron beam.

The detection and signal processing electronics are also shown schematically in figure 3. This familiar fast-timing coincidence arrangement has been explained elsewhere[1,13] and will not be discussed further. The data is collected and processed using a PDP-11 computer interfaced with a CAMAC crate. The computer also controls the various spectrometer operations such as incident energy variation and angle variation[1,13]. Under the above conditions a typical experiment on a molecular gas involving the collection of separation energy spectra over a broad energy range at several azimuthal angles plus several spectra over regions of special interest, as well as the momentum distributions (angular correlations) for each orbital, runs for approximately six weeks.

THEORY OF THE (e,2e) REACTION ON MOLECULES

The basic theory for the (e,2e) reaction has already been discussed by Professor McCarthy at this workshop and elsewhere[1,2]. The distorted-wave off-shell impulse approximation for the (e,2e) amplitude is[1]

$$M = \langle \underline{k}_A, \underline{k}_B | T_M | (f|g \rangle \underline{k}_o \rangle \tag{4}$$

where T_M is the antisymmetrized two-body Coulomb t-matrix including exchange (Mott scattering); $|g\rangle$ and $|f\rangle$ are the wavefunctions of the target ground state and final ion state respectively. The amplitude (4) depends on the target and ion structure only through the overlap function $\langle f|g\rangle$ since T_M is independent of the internal ion coordinates.

At sufficiently high energies the distorted waves can be approximated by modified plane waves (eikonal approximation)[1,2] which has the effect of reducing the (e,2e) amplitude to a form that facilitates computation. The (e,2e) cross-section has been compared with calculations using the distorted-wave impulse approximation and the eikonal approximation for all the inert gases including xenon[1]. The results are practically energy independent over the 400eV to 1200eV incident energy range, and in general the DWIA calculations agree with experiment better at 400eV but at 1200eV the plane waves give an excellent fit to the data as well. This is the principal reason for conducting the experiments on molecules at 1200eV incident energy.

We assume that the molecule is in the ground vibrational state $|0>$. The Born-Oppenheimer approximation allows us to write the structure wavefunctions as products of the rotational, vibrational and electronic wavefunctions. Since rotation-vibration states are not resolved in an (e,2e) experiment we can apply closure over the final-state rotation-vibration functions to obtain the differential cross-section for initial angular momentum I[1,2]:

$$\sigma_I \propto |T_M|^2 \sum_{\mu_r} <0| \int d\Omega (2\pi)^{-3} | \int d^3r \, \exp(i\underline{q}.\underline{r})(F|G>|^2 0> \qquad (5)$$

where the sum over μ_r runs over degenerate final states and $|G>$ and $|F)$ are the electronic eigenstates of the molecule and the ion respectively.

The vibrational integral in (5) is an average over the ground-state vibrations; it has been shown for the hydrogen molecule[14] that this average can be replaced by the equilibrium value of the vibrational co-ordinate.

The target ground-state $|G>$ is written as a linear combination of independent-particle basis configurations $|\alpha>$

$$|G> = \sum_{\alpha} a_{\alpha} |\alpha> \qquad (6)$$

The antisymmetrized ion state $|F)$ is written in terms of a basis consisting of a hole in the orbital ψ_j coupled to the target configuration $|\beta>$

$$|F_r) = n_r^{\frac{1}{2}} \sum_{j\beta} t_{j\beta}^{(F)} C_{jr\beta} \psi_j^{\dagger} |\beta> \qquad (7)$$

The Clebsch-Gordan coefficient of the point group of the molecule ensures that the configurations in the sum in (7) all belong to the irreducible representation r of degeneracy n_r.

For closed shell molecules we can make the target Hartree-Fock approximation

$$|G> = |0> \qquad (8)$$

and the target-ion overlap becomes

$$(F|G> = n_r^{\frac{1}{2}} \sum_{j} t_{jo}^{(F)} C_{jro} \psi_j(\underline{r}) |0>$$

$$= \sum_{j=r} t_{jo}^{(F)} \psi_j(\underline{r}) \qquad (9)$$

where here the Clebsch-Gordan coefficient selects configurations for which the orbital representation j is the same as the final state representation r. There is a coherent sum over the different orbitals that contribute to $|F\rangle$. The coefficients $t_{jo}^{(F)}$ are the amplitudes of the configurations $\psi_j^\dagger |0\rangle$ in $|F\rangle$.

We use an LCAO expansion of the orbital $\psi_j(\underline{r})$ in terms of basis functions $u_{\ell m}^{(s)}$

$$\psi_j(\underline{r},R) = \sum_s \psi_s^{(j)}(\underline{r}-\underline{R}_s) \tag{10}$$

$$\psi_s^{(j)}(\underline{r}) = \sum_{\ell m} u_{\ell m}^{(s)}(r) Y_{\ell m}(\underline{\hat{r}}) \tag{11}$$

The calculations are performed at the equilibrium nuclear positions $\underline{R}_s$ with each term having angular properties given by the spherical harmonics $Y_{\ell m}(\underline{\hat{r}})$, so that the differential cross-section for the (e,2e) reaction on a closed shell, rigid molecule in the eikonal and target Hartree-Fock approximation and for the non-coplanar geometry is

$$\sigma \propto S_j^{(F)} n_r \sum_{nm} \int dq \phi_n^*(\underline{q},R_{nm}) \phi_m(\underline{q},R_{nm}) \exp(i\underline{q}.(\underline{R}_m-\underline{R}_n)) \tag{12}$$

where the momentum space wave function is

$$\phi_n(\underline{q},R) = 2\pi^{-3/2} \int d^3r \exp(i\underline{q}.\underline{r}) \psi_n^j(\underline{r}.R) \tag{13}$$

and ψ_n^j is the sum of all the atomic functions centred on the nucleus m in the LCAO for the characteristic orbital. R_{nm} is the set of all equilibrium distances between the nth and m^{th} nucleus and

$$S_j^{(F)} = |t_{jo}^{(F)}|^2 \tag{14}$$

is the spectroscopic factor for the state to contain the orbital j.

We can thus see from (12) that the cross section for any particular state is directly proportional to the square of the momentum-space wavefunction for the characteristic orbital weighted by the spectroscopic factor. In the coplanar geometry the $|T_M|^2$ term is not a constant as it depends on the outgoing momenta, increasing rapidly with decreasing θ. The coplanar experiment thus provides a strict test of the factorised distorted-wave off-shell impulse approximation.[1,2] The spectroscopic factors obey the sum rules

$$\sum_F S_j^{(F)} = 1 \tag{15}$$

which state that the sum of all cross-sections for states of the same symmetry as F is proportional to the number of electrons occupying the single-particle orbital ψ_j of that symmetry. The orbital energy is given by

$$E_j = \sum_F S_j^{(F)} \epsilon_F \tag{16}$$

that is, the weighted centroid of the eigenvalues for all states of the same symmetry as F.

EXPERIMENTAL RESULTS

A list of molecules studied to date in this laboratory is given in table 1. Those molecules up to and including the water molecule were discussed at the previous workshop[15]. Since the present workshop is primarily concerned with the determination and analysis of momentum distributions it is instructive to consider several sets of data in detail for different molecules to gain an understanding of the information derived from the (e,2e) technique.

Table 1 Molecules studied using non-coplanar symmetric binary (e,2e) spectroscopy at Flinders University

	Molecule	Year	Energy (eV)
1.	Methane (CH_4)[16,17]	1973	400
		1976	600, 1200
2,3.	Hydrogen (H_2)[14]	1975	300, 600, 1200
	D_2[14]	1975	600, 1200
	(H_2)[18]	1977	1200
4.	Ethane (C_2H_6)[19]	1976	400, 1200
5.	Nitrogen (N_2)[20]	1977	400, 600, 1200
6.	Carbon Monoxide (CO)[21]	1977	400, 1200
7.	Water (H_2O)[22]	1977	400, 1200
8.	Acetylene (C_2H_2)[23]	1977	400, 1200
9.	Ethylene (C_2H_4)[24,25]	1977	1000
10.	Oxygen (O_2)[26]	1980	400, 1200
11.	Hydrogen Fluoride (HF)[13,27,28]	1979	400, 1200
12.	Hydrogen Chloride (HCl)[13,27,28]	1980	400, 1200
13.	Nitric Oxide (NO)[29]	1980	1200
14.	Nitrous Oxide (N_2O)[30]	1981	1200
15.	Benzene (C_6H_6)[31]	1981	1200
16.	Chloromethane (CH_3Cl)[32]	1981	1200, 1500
17.	Methanol (CH_3OH)[33]	1981	1200
18.	Hydrogen Bromide (HBr)[34]	1981	1200
19.	Hydrogen Iodide (HI)[34]	1981	1200
20.	Bromomethane (CH_3Br)[35]	1981	1200
21.	Iodomethane (CH_3I)[35]	1981	1200
22	Acetonitrile (CH_3CN)[36]	1982	1200

Methanol (CH$_3$OH)[33]

The binary (e,2e) separation energy spectra for methanol are shown in figure 4. The experiment was conducted with an incident

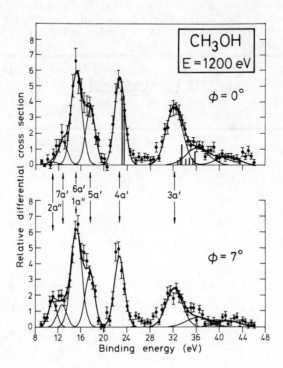

Fig. 4. Separation energy spectra of methanol at 1200eV taken at azimuthal angles of $\phi = 0°$ (q $\approx$ 0.1 a$_o^{-1}$) and $\phi = 7°$ (q $\approx$ 0.6 a$_o^{-1}$)[33].

electron energy of 1200eV plus the separation energy and with an energy resolution of 1.7eV and a momentum resolution of $\lesssim 0.1$ a$_o^{-1}$. The positions and Franck-Condon widths for all the dominant one-hole valence orbital transitions indicated in figure 4 were all derived from photoelectron spectra and were convoluted with the instrumental energy bandwidth to produce the spectral gaussian deconvolutions. The sum of the gaussian peaks agrees well with the experimental points.

Apart from the outer valence orbital (2a'') the intensities of all the transitions are similar in both the spectra, indicating that these orbitals have significant totally symmetric components in the wavefunctions of the corresponding molecular orbitals whereas the 2a'' orbital has nodal symmetry. Indications of the splitting of the spectroscopic strengths of several states is seen in the excess intensity in the region 25-29eV and in the high energy side of the (3a')$^{-1}$ transition extending out to $\approx$ 46eV.

Relative (e,2e) cross-sections were determined from the individual peak areas in figure 4 for comparison with the molecular orbital calculations of Snyder and Basch[37]. It can be seen from Table 2 that the agreement between the measured and calculated cross-sections is

Table 2 Ionization energies and observed relative intensities of the valence orbitals of methanol. The calculated relative intensities using the wavefunctions of Snyder and Basch are given in square brackets.

$(\psi)^{-1}$	Binding energy (eV)	Relative intensities (errors in parentheses)		GF energies[38]	
		$\phi = 0°$ $q \approx 0.1\ a_o^{-1}$	$\phi = 7°$ $q \approx 0.6\ a_o^{-1}$		
2a''	11.0	2(1) [0.4]	20(2) [17]	10.86	
7a'	12.7	21(2) [15]	20(2) [21]	12.42	
6a' } 1a''	15.3	96(4) [76]	100 [100]	15.24 15.64	
5a'	17.6	51(3) [41]	47(3) [46]	17.52	
4a'	22.7	84(4) [131]	67(3) [107]	23.01 23.38	(0.44) (0.44)
3a'	32.2	88(4) [175]	57(2) [93]	32.82 33.08 33.41 33.52 34.14 34.83 35.33 35.58 35.62	(0.050) (0.077) (0.11) (0.11) (0.058) (0.068) (0.062) (0.063) (0.065)
(3a') (4a')	36 38	35(4) 22(4)	19(4) 14(4)		

reasonable for the four outer transitions but that the calculated 4a' and 3a' relative intensities are considerably larger than observed at both angles. The Tamm-Dancoff Green's function results[38] also shown in this table indicate severe splitting of the spectroscopic strengths of both these inner valence orbitals.

The MO calculations assume spectroscopic strengths of unity which suggests that the main $(4a')^{-1}$ and $(3a')^{-1}$ transitions both have spectroscopic strengths of ≈ 0.6 (e.g. 4a' at 0° : 84/131 = 0.64). The Green's function calculations predict that ≈ 88 percent of the 4a' strength is split between two closely lying states and ≈ 70 percent of the 3a' strength is divided among ion states in the range 32.8-35.6eV. The positions and strengths of these states is indicated with the vertical lines in the 0° spectrum in figure 4.

The momentum distributions for the 3a' and 4a' orbitals are shown in figure 5. Also shown are molecular orbital shapes or charge densities in configuration-space that may be more familiar to chemists[39]. The dominant orbital groups for the 3a' MO are σ_{OH} and σ_{CH} and for the

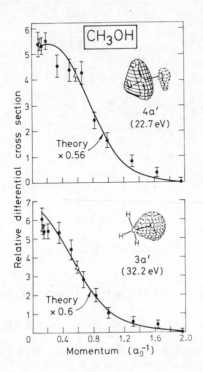

Fig. 5. Experimental and calculated momentum distributions for the 4a' and 3a' orbitals of methanol[33].

4a' is σ_{CH_3}. These symmetries are reflected in the momentum profiles which indicate that these orbitals are dominated by totally symmetric components.

Both the 4a' and 3a' calculated cross-sections have been multiplied by 0.6 and excellent agreement is obtained with experiment. As mentioned earlier this implies that the spectroscopic strength for these orbitals is severely split with the main peaks contributing 60 percent of the total strength. The momentum distribution measured at a separation energy of 36eV (Figure 6) has a shape consistent with ionization from either the 3a' or the 4a' orbital as predicted by the Green's function calculations, further supporting the argument that the inner valence orbitals have their spectroscopic strengths split by electron correlation effects.

A further check of the validity of (e,2e) measurements is provided by the independent results for the cross-sections measured from the momentum distributions and the separation energy spectra. With only one normalization point (6a' + 1a" at ϕ = 7°) the cross-section measurements taken from the energy spectra should coincide with the momentum distribution; these points can be seen as open triangles in all

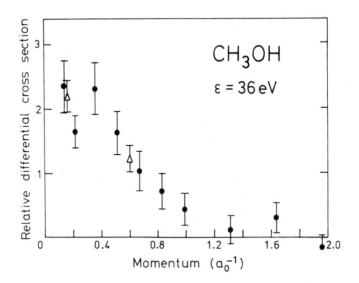

Fig. 6. The momentum distribution for methanol measured at a separation energy of 36eV[33].

the methanol data and show excellent agreement between the different measurements.

Figure 7 shows the momentum distributions measured for the 5a' and (6a' + 1a") orbitals. The 6a' and 1a" peaks could not be resolved with the present energy resolution so that the resultant profile represents the sum of the contributions from both orbitals. Agreement between theory and experiment is excellent. For the 5a' orbital, however, it is evident that the calculations overestimate the contribution of p atomic orbitals, indicated by the reduction in intensity towards low q.

The momentum distributions for the outer valence orbitals (7a' and 2a") shown in figure 8 both show excess intensity relative to the calculated profiles at low q. This is now a common feature observed in (e,2e) experiments and was especially evident in the $1b_1$ orbital of H_2O[22]. The results thus show that more diffuse functions are needed in the wavefunctions to describe the outer spatial regions and hence the low q regions more adequately.

Nitrous Oxide (N_2O)[30]

The binary (e,2e) separation energy spectra for N_2O at ϕ = 0°, 7° and 10° are shown in figure 9. The positions and widths of the four outer valence states were all well known from photoelectron spectroscopy, producing the gaussian curves shown when convoluted with the instrumental energy resolution. The positions of the two inner states were not well defined from previous measurements. We have

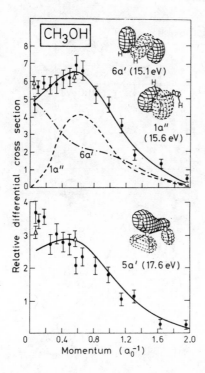

Fig. 7. Experimental and calculated momentum distributions for the 5a' and (6a' + 1a") orbitals of methanol. The 6a' and 1a" calculated cross-sections (dashed lines) have been summed to give the full curve[33].

found the centroids of the best fit gaussians to be at $\approx$ 35.0eV for the $5\sigma^{-1}$ transition and 38.7 $\pm$ 0.5eV for the $4\sigma^{-1}$ transition with Franck-Condon broadened widths of 3.0eV.

The separation energies determined from two separate molecular orbital calculations (SB - Snyder and Basch[37]; G76 - GAUSS-76[40]) are shown in Table 3[30] along with the experimental values. Also listed are the separation energies (and their spectroscopic strengths) for the main transitions obtained using a many-body Green's function calculation[41].

The outer-valence Green's function results agree very well with experiment for the four outer valence states and the 2ph-TDA results for the two spectroscopically split inner states are in excellent agreement with experiment. The MO calculations do not show good overall agreement with experiment especially for the two inner states. As mentioned earlier, in the target-Hartree Fock approximation the orbital separation energy is given by the weighted centroid of the separation energies for all transitions belonging to that orbital

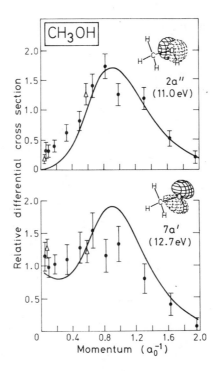

Fig. 8. Experimental and calculated momentum distributions for the outer 2a" and 7a' orbitals of methanol[33].

(equation 16). This means that significant differences between calculated and measured separation energies can be indicative of significant splitting of the spectroscopic strengths of that orbital. Significant correlations are indeed predicted by the Green's function calculation for the 1π, 5σ and 4σ orbitals where disagreement between the MO separation energies and experiment is worst.

Excess intensity at all angles can be seen in the range 22-32eV and above 40eV. From symmetry considerations it is difficult to assign the lower energy data, however the structure above 40eV has definite totally symmetric σ character like either the 4σ or 7σ orbitals. The Green's function calculation predicts the structure to be mainly due to $4\sigma^{-1}$ ionization processes.

If we assume that all the structure above about 36eV is associated purely with the $4\sigma^{-1}$ transition then relative to the total 4σ strength of unity the main peak at 38.7eV has a strength of 0.52 ± 0.05 at $\phi = 0°$, 0.48 ± 0.07 at $\phi = 7°$ and 0.51 ± 0.07 at $\phi = 10°$. This indicates that the 4σ state is severely split with 50 percent of the spectroscopic strength contained in the 38.7eV peak and 50 percent in the structure above 40eV. The GF spectroscopic factors for the 4σ satellites which contribute to the intensity centred around the "main" 4σ peak sum to ~ 0.5.

Table 3 Separation energies (eV) for the main transitions in the valence shell of nitrous oxide. The natural widths are shown in parentheses. The spectroscopic strengths are given in square brackets.

Orbital $(\phi_o)^{-1}$	Expt	SB	G76	GF[a]	GF[b]
2π	12.9	13.25	10.2	12.15 [0.90]	12.72
7σ	16.5	18.94	16.3	15.23 [0.83]	16.38
1π	18.3	21.36	19.4	17.69 [0.69]	18.93
6σ	20.1	22.37	20.0	18.92 [0.84]	20.53
5σ	35.0 (3.0)	39.93	37.1	34.75 [0.19] 34.83 [0.34]	
4σ	38.7 (3.0)	44.85	42.2	38.46 [0.13] 38.63 [0.17]	
E_{tot}^{SCF}(a.u.)		-183.5761	-182.9466		-183.6466

a. 2ph-TDA ionization potentials and spectroscopic strengths[41]

b. Green's function method especially adapted to the outer valence region[41].

The momentum distributions determined from both the MO wavefunctions for the 4σ orbital have been multiplied by 0.5 and are compared with the experimental momentum distribution in figure 10. Agreement between theory and experiment is quite good although the calculations exhibit somewhat sharper peaking at low q than observed.

Overall agreement between theory and experiment for the 5σ orbital is not good especially at low q. However, the results do show that most of the spectroscopic strength is contained in the "main" transition compared with the 63 percent predicted by the GF calculations. Nevertheless, evidence for states of 5σ symmetry is seen in the excess intensity around 31eV, approximately where two 5σ satellites are predicted by the GF calculation. Excess intensity around 0.6 a_o^{-1} in the 6σ momentum profile could possibly be due to contributions from a close-lying transition of π symmetry. The GF calculation predicts a 1π satellite at 19.2eV of approximately the required intensity.

Figure 11 shows the momentum distributions for the $2\pi^{-1}$, $7\sigma^{-1}$ and $1\pi^{-1}$ transitions. The SB curve for the $1\pi^{-1}$ transition has been multiplied by 0.7 (dash-dot curve) and excellent agreement between theory and experiment is obtained. The excess experimental intensity at $q \approx 0$ is due to the finite instrumental resolution. The factor of 0.7 is equivalent to the spectroscopic factor predicted by the GF calculation, thus implying that only 70 percent of the 1π strength is found in the peak at 18.3eV. Evidence for structure of π symmetry is seen around 23eV (a $1\pi^{-1}$ satellite is predicted around 26eV).

Both the MO calculations seriously underestimate the lower momentum components of the cross-section for the $7\sigma^{-1}$ transition and slightly underestimate the higher momentum components. Of course, small errors in normalization can produce misleading results since the cross-section is weighted by the momentum space volume element $4\pi q^2 dq$; a

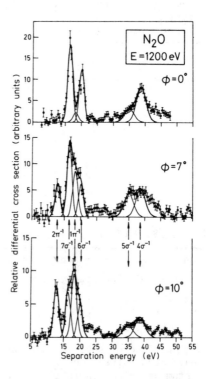

Fig. 9. Separation energy spectra of N_2O at 1200eV taken at the azimuthal angles indicated[30].

small excess of high momentum components will lead to a large reduction in the cross-section at low momentum and vice-versa. Due care is taken to obtain the best overall normalization of theory to experiment by using well defined structures. In this case the MO calculations are normalized to the measured 2π momentum distribution integrated over momentum space.

For the $2\pi^{-1}$ ground state transition both MO calculations overestimate the position of the profile maximum and underestimate the probability of low momentum components. The data thus indicates that both the 7σ and 2π orbitals are more extended in configuration space than predicted by these simple MO calculations.

<u>Halomethanes</u>

As noted, the (e,2e) technique has proven useful as a tool for determining separation energy spectra particularly for inner valence states which are often inaccessible to PES and XPS. As well as determining the separation energies, (e,2e) spectroscopy has the extra advantage of being able to determine the symmetry of all states through the measurement of momentum profiles. The technique has already removed certain controversy over the assignment of states[24,25] and in

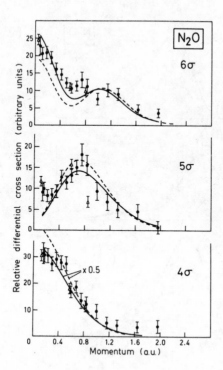

Fig. 10. Measured and calculated momentum distributions for the $6\sigma^{-1}$, $5\sigma^{-1}$ and $4\sigma^{-1}$ transitions in N_2O^{30}. The full curve is due to SB and the dashed curve is due to G76.

the case of the halomethanes is the only technique to have identified the inner valence states. Now that a larger body of information is becoming available through the steady increase in the number of molecules being studied it will be possible to compare trends throughout a specific series of molecules. The change in the nature of the carbon bond in the ethane[19], ethylene[24,25] and acetylene[23] molecules has already been investigated as have trends in the hydrogen halides[13,34].

As part of a series of investigations into the substituted methanes we have recently measured the separation energy spectra for the halomethanes CH_3Cl[32], CH_3Br[35] and CH_3I[35] as well as acetonitrile, CH_3CN[36]. Momentum distributions for all the main transitions have also been measured.

Figure 12 exhibits the separation energy spectra for chloromethane at $\phi = 0°$ and $\phi = 10°$[32]. These spectra represent the first identification of the two inner valence states ($2a_1$ and $1a_1$) at 21.9eV and 25.5 eV respectively with corresponding natural widths of 1.9eV and 2.4eV.

132

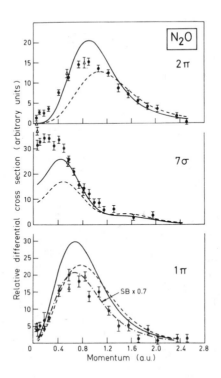

Fig. 11. Measured and calculated momentum distributions for the $2\pi^{-1}$, $7\sigma^{-1}$ and $1\pi^{-1}$ transitions in N_2O[30].

There is considerable intensity above 28eV of the same symmetry as the $1a_1$ orbital (or the $3a_1$ orbital) and which, if assumed to be totally due to $1a_1^{-1}$ ionization processes, constitutes $\approx$ 50 percent of the total $1a_1$ spectroscopic strength at ϕ = 0° and $\approx$ 40 percent at ϕ = 8°.

The weighted mean separation energy of the data above 23eV is 29.7 ± 0.5eV at ϕ = 0° and 30.8 ± 0.5eV at ϕ = 8° which puts the $1a_1$ orbital separation energy at $\sim$ 30eV rather than at 25.5eV as observed for the main transition. This is in reasonable agreement with a MO calculation employing a contracted Gaussian basis set ((9s5p/4s1p/12s9p) /[4s2p/2s1p/6s4p]) which predicts a separation energy of 30.95eV and provides further evidence that the broad correlation effects above 28eV are mainly associated with the $1a_1^{-1}$ transition.

The momentum distribution for the main $1a_1^{-1}$ transition is shown in figure 13 and is compared there with the momentum profile derived from the CGTO wavefunction. Agreement between theory and experiment is excellent if the calculated profile is reduced by 50 percent, the spec-

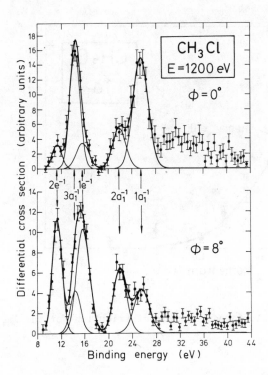

Fig. 12. Separation energy spectra of CH_3Cl at 1200eV taken at $\phi = 0°$ and $\phi = 8°$[32].

troscopic strength of the $1a_1^{-1}$ transition process at 25.5eV. Figure 14 shows the momentum distribution determined at a separation energy of 31.5eV. The dash-dot curve is the calculated $3a_1$ momentum distribution multiplied by 0.43 and the dashed curve is the calculated $1a_1$ momentum distribution multiplied by 0.18. Recent measurements using a Green's function calculation[42] predict that the "main" peaks for the $3a_1^{-1}$ and $1a_1^{-1}$ transitions contribute 70 percent and 80 percent of the total spectroscopic strengths respectively. It thus seems more likely that the high energy structure is principally due to $1a_1^{-1}$ ionization processes.

The separation energy spectra at $\phi = 0°$ and $\phi = 8°$ for bromomethane are shown in figure 15. Similar to the case for CH_3Cl we have positively identified the separation energies and symmetries of the $2a_1$ and $1a_1$ orbitals. The separation energies are respectively 21.70eV and 24.90eV. We have also tentatively determined the separation energies for the corresponding orbitals in iodomethane as 21.4eV and 24.5eV.

134

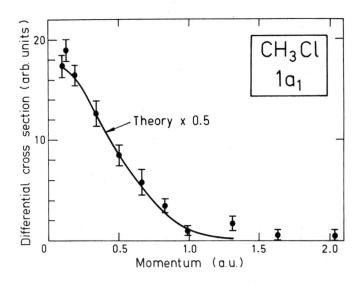

Fig. 13. Measured and calculated momentum distribution for the $1a_1^{-1}$ transition in CH_3Cl.

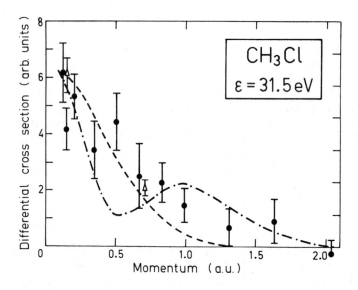

Fig. 14. The momentum distribution for CH_3Cl measured at 31.5eV compared with the calculated $1a_1$ profile (dash) x 0.18 and the calculated $3a_1$ profile (dash-dot) x 0.43.

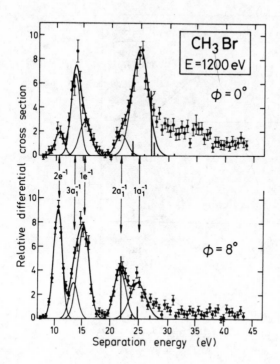

Fig. 15. Separation energy spectra of CH_3Br at 1200eV taken at $\phi = 0°$ and $\phi = 8°$.

For both CH_3Br and CH_3I there is strong electron-electron correlation evidenced by the excess intensity at energies beyond the inner valence state. Recent Green's function calculations[42] also predict that the two inner states have their spectroscopic strengths severely split.

The $1a_1$, $2a_1$ and $3a_1$ momentum distributions for CH_3Br and CH_3I are shown in figures 15 and 17 respectively. In each case the calculated profiles have been normalized to the measured 2e momentum distribution integrated over momentum space. The calculated momentum distributions were derived from CGTO wavefunctions employing quite elaborate basis sets[42], (14s11p5d/9s5p/4s)/[9s6p2d/4s2p/2s] for CH_3Br and (18s4p8d/9s5p/4s)/[13s9p4d/4s2p/2s] for CH_3I. It is very obvious from these results that the MO picture of ionization is not as satisfactory for these larger molecules even if large basis sets containing relatively diffuse functions are employed. It is also apparent that the two inner valence states are more severely split than predicted by the Green's function calculation because the multiplication factors (spectroscopic strengths) needed to accurately relate the calculated

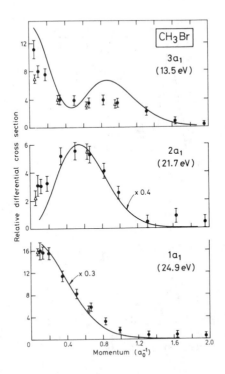

Fig. 16. The measured and calculated momentum distributions for the $3a_1^{-1}$, $2a_1^{-1}$ and $1a_1^{-1}$ transitions in CH_3Br.

and measured momentum profiles are considerably less than those predicted with the GF calculation.

(e,2e) and Chemists

Continuing with the notions expressed in the introductory section for the halomethanes it is fairly obvious that chemists might view the (e,2e) technique with some scepticism as the information gained appears to have little to do with transition rates or heats of vapourization or whatever else it is that concerns chemists. However, it is through the acquisition of large amounts of data that generalized information concerning specific trends is realised. As we have already seen molecular orbital calculations that are used extensively to determine chemical bonding and molecular structure parameters are often seriously in error. The (e,2e) technique which maps the wavefunctions directly (via a fourier transform) should thus provide direct feedback to the quantum chemist or many-body theorist on the state of his basis set!

One of the problems uncovered with the (e,2e) technique is the inability of even quite elaborate MO calculations to correctly predict the low momentum components for outer valence orbitals. This effect can be seen in figures 18 and 19 where the 2e and 1e momentum distribu-

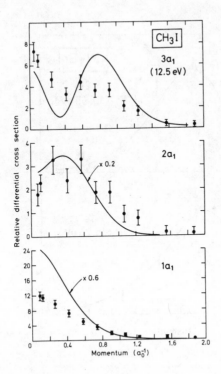

Fig. 17. The measured and calculated momentum distributions for the $3a_1^{-1}$, $2a_1^{-1}$ and $1a_1^{-1}$ transitions in CH_3I.

tions for CH_3Br and CH_3I are shown. The same effect can be seen in the 2a" and 7a' orbitals in CH_3OH (figure 8) and the 2π and 7σ orbitals in N_2O (figure 11). Although at this stage absolute measurements are not possible it appears that moving down the CH_3X (X = Cl, Br, I) series worsens the problem and it thus remains unanswered as to whether adding diffuse functions to already elaborate basis sets is going to cause better agreement between theory and experiment.

The use of the (e,2e) technique for studying trends of interest to the chemist has been well illustrated in a recent series of experiments on the hydrogen halides[13,34]. Separation energy spectra (figure 20) have been measured and comparisons have been made with the isoelectronic rare gases Ne, Ar, Kr and Xe. Similarly the momentum distributions for the valence orbitals have been compared (figure 21). The momentum distributions for HF are very different from the momentum distributions for the other hydrogen halides, reflecting the rather different chemical nature of HF (e.g. HF is significantly hydrogen bonded compared with the others). The results also show that in configuration space the valence orbitals of HBr and HI are quite diffuse whereas the outer orbitals of HF are much more localised at the atomic sites.

138

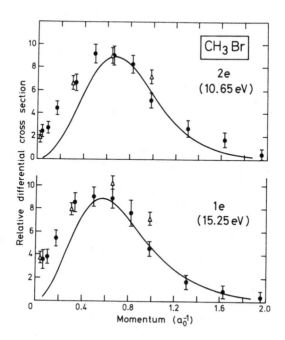

Fig. 18. The measured and calculated momentum distributions for the $2e^{-1}$ and $1e^{-1}$ transitions in CH_3Br. The open triangles were taken from the separation energy spectra (figure 15).

The interpretation of (e,2e) momentum profiles in chemically significant terms is somewhat hampered by the reluctance of chemists to discuss bonding and the like in momentum-space rather than in configuration-space. Dr. John Cook will later discuss the use of the fourier transform of configuration-space charge density maps derived from molecular orbital wavefunctions into momentum-space momentum density maps which can be spherically averaged to produce the (e,2e) momentum distributions. Through the use of such techniques it is possible to accurately predict the shape of momentum distributions and thus possibly provide some direction to the understanding of momentum space wave-functions. The technique has already proved useful in describing the unusual π^*2p momentum profiles for NO and O_2[29].

At present the (e,2e) technique is limited by instrumental reso-lutions and data gathering rates. Even if the resolution was one half to one third the present value individual states for molecules like benzene (figure 22) could be more clearly identified. To this end the use of position sensitive detectors and monochromatic beams on the present apparatus will greatly improve the situation and the further development of high energy Möllenstedt detectors in this lab-

oratory will see the (e,2e) technique become an even more viable analytical tool for both physics and chemistry.

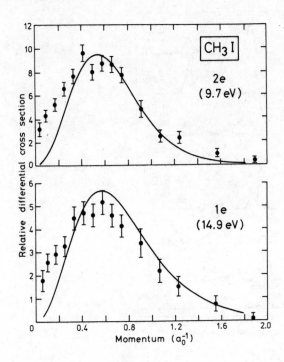

Fig. 19. The measured and calculated momentum distributions for the
$2e^{-1}$ and $1e^{-1}$ transitions in CH_3I.

140

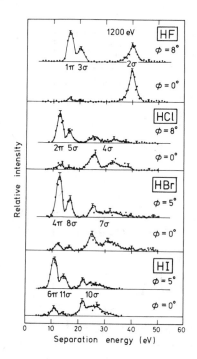

Fig. 20   Separation energy spectra for the hydrogen halides at
1200eV taken at the indicated azimuthal angles[34].

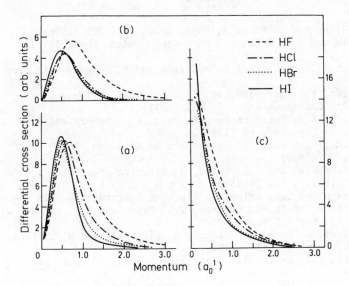

Fig. 21. The measured momentum profiles for the valence orbitals of the hydrogen halides normalized to give approximately the same peak heights for the outer $(np\pi)^{-1}$ transitions[34].

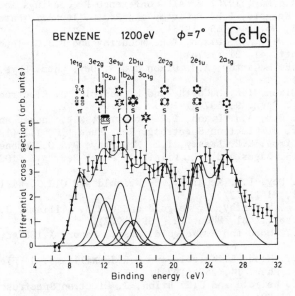

Fig. 22. Separation energy spectrum for benzene at 1200eV and $\phi = 7°$[31].

REFERENCES

1. I.E. McCarthy and E. Weigold, Phys. Reps. 27C, 275 (1976).
2. E. Weigold and I.E. McCarthy, Adv. Atom. Molec. Phys. 14, 127 (1978).
3. C.E. Brion, J.P.D. Cook, I.E. McCarthy and E. Weigold - in preparation.
4. H. Ehrhardt, K.H. Hesselbacher, K. Jung and K. Willmann, Case Studies At. Collision Phys. 2, 159 (1971).
5. H. Ehrhardt, K.H. Hesselbacher, K. Jung, E. Schubert and K. Willmann, J. Phys. B7, 69 (1974).
6. H. Ehrhardt, M. Schulz, T. Tekaat and K. Willmann, Phys. Rev. Lett. 22, 89 (1969).
7. M.J. van der Wiel and C.E. Brion, J. Electron Spectrosc. 1, 309 (1973). Ibid 1, 443 (1973).
8. G.R. Branton and C.E. Brion, J. Electron Spectrosc. 3, 129 (1974).
9. A. Hamnett, W. Stoll, G. Branton, C.E. Brion and M.J. van der Wiel, J. Phys. B9, 945 (1976).
10. R. Camilloni, A. Giardini-Guidoni, R. Tiribelli and G. Stefani, Phys. Rev. Lett. 29, 618 (1972).
11. E. Weigold, C.J. Noble, S.T. Hood and I. Fuss, J. Phys. B12, 291 (1979).
12. E. Weigold, S.T. Hood and P.J.O. Teubner, Phys. Rev. Lett. 30, 475 (1973).
13. C.E. Brion, S.T. Hood, I.H. Suzuki and E. Weigold, J. Electron Spectrosc. Rel. Phen. 21, 71 (1980).
14. S. Dey, I.E. McCarthy, P.J.O. Teubner and E. Weigold, Phys. Rev. Lett. 34, 782 (1975).
15. E. Weigold, Int. Momentum Distribution Workshop, Bloomington, Indiana, U.S.A. (May, 1976). AIP Conference Proceedings No. 36.
16. S.T. Hood, E. Weigold, I.E. McCarthy and P.J.O. Teubner, Nature Phys. Sci. 245, No. 144, 65 (1973).
17. E. Weigold, S. Dey, A.J. Dixon, I.E. McCarthy and P.J.O. Teubner, Chem. Phys. Lett. 41, 21 (1976).
18. E. Weigold, I.E. McCarthy, A.J. Dixon and S. Dey, Chem. Phys. Lett. 47, 209 (1977).
19. S. Dey, A.J. Dixon, I.E. McCarthy and E. Weigold, J. Electron Spectrosc. Rel. Phen. 9, 397 (1976).
20. E. Weigold, S. Dey, A.J. Dixon, I.E. McCarthy, K.R. Lassey and P.J.O. Teubner, J. Electron Spectrosc. Rel. Phen. 10, 177 (1977).
21. S. Dey, A.J. Dixon, K.R. Lassey, I.E. McCarthy, P.J.O. Teubner, E. Weigold, P.S. Bagus and E.K. Viinikka, Phys. Rev. A15, 102 (1977).
22. A.J. Dixon, S. Dey, I.E. McCarthy, E. Weigold and G.R.J. Williams, Chem. Phys. 21, 81 (1977).
23. A.J. Dixon, I.E. McCarthy, E. Weigold and G.R.J. Williams, J. Electron Spectrosc. Rel. Phen. 12, 239 (1977).
24. A.J. Dixon, S.T. Hood and E. Weigold, Abstr. Paper, X ICPEAC, p.384 (1977).
25. A.J. Dixon, S.T. Hood, E. Weigold and G.R.J. Williams, J. Electron Spectrosc. Rel. Phen. 14, 267 (1978).
26. I.H. Suzuki, E. Weigold and C.E. Brion, J. Electron Spectrosc. Rel. Phen. 20, 289 (1980).
27. C.E. Brion, I.E. McCarthy, I.H. Suzuki and E. Weigold, Chem. Phys.

Lett. 67, 115 (1979).

28. I.H. Suzuki, C.E. Brion, E. Weigold and G.R.J. Williams, Int. J. Quant. Chem. 18, 275 (1980).

29. C.E. Brion, J.P.D. Cook, I.G. Fuss and E. Weigold, to be published in J. Electron Spectrosc. Rel. Phen.

30. A. Minchinton, I.G. Fuss and E. Weigold, submitted to J. Electron Spectrosc. Rel. Phen.

31. I. Fuss, I.E. McCarthy, A. Minchinton, E. Weigold and F.P. Larkins, Chem. Phys. 63, 19 (1981).

32. A. Minchinton, A. Giardini-Guidoni, E. Weigold, F.P. Larkins and R.M. Wilson, submitted to J. Electron Spectrosc. Rel. Phen.

33. A. Minchinton, C.E. Brion and E. Weigold, Chem. Phys. 62, 369 (1981).

34. C.E. Brion, I.E. McCarthy, I.H. Suzuki, E. Weigold, G.R.J. Williams, K.L. Bedford, A.B. Kunz and R. Weidman, to be published in Chem. Phys.

35. A. Minchinton, W. von Niessen and E. Weigold. Int. Conf. on Atomic and Molecular Reactions and Structure. Flinders University, South Australia (Feb. 15-17, 1982).

36. A. Minchinton, C.E. Brion and E. Weigold. Int. Conf. on Atomic and Molecular Reactions and Structure. Flinders University of South Australia (Feb. 15-17, 1982).

37. L.C. Snyder and H. Basch, Molecular wavefunctions and properties (Wiley, New York, 1972).

38. W. von Niessen, G. Bieri and L. Åsbrink. J. Electron Spectrosc. Rel. Phenom. 21, 175 (1980).

39. W.L. Jorgensen and L. Salem, The Organic Chemist's Book of Orbitals. (Academic Press, New York, 1973).

40. W.J. Hehre, R.F. Stewart and J.A. Pople, Chem. Phys. 51, 2657 (1969).

41. W. Domcke, L.S. Cederbaum, J. Schirmer, W. von Niessen, C.E. Brion and K.H. Tan, Chem. Phys. 40, 171 (1979).

42. W. von Niessen, private communication.

ELECTRON CORRELATION AND MOLECULAR EFFECTS IN (e, 2e) SPECTROSCOPY

F.P. LARKINS

Department of Chemistry, Monash University, Clayton, Vic., 3168, Australia.

ABSTRACT

The importance of initial and final state correlation effects in (e, 2e) spectroscopy is discussed in general theoretical terms and then the theory is applied to the helium and argon atomic systems. Molecular effects are considered through a discussion of the chloromethane ($CH_{4-n}Cl_n$ n=1 to 4) series of molecules. Correlation effects in molecular systems as manifested by satellite structure are also discussed. The relationship between the intensity of satellite structure in (e, 2e) and photoelectron spectra is also outlined.

1.   INTRODUCTION

In recent years much attention has been directed towards the use of the (e, 2e) technique to elucidate the electronic properties of atoms and molecules. The Flinders University group of McCarthy and Weigold, who are hosting this international workshop, have been foremost in the development of the non-coplanar symmetric binary electron impact technique and in demonstrating its usefulness as a probe of the fundamental electronic structure of various systems. Two review articles by them[1,2] outline the experimental and theoretical developments in the field up to 1978. Photoelectron spectroscopy has also provided much information on the electronic properties of atoms and molecules throughout the past decade. Information from the two techniques is in several respects complementary. As well as an interest in diagram lines representing transitions between the two lowest states of particular symmetry manifolds, satellite structure which arises fundamentally from electron correlation effects and which involves excited states usually of the final symmetry manifold has attracted much attention.[3]

The emphasis in this paper is on the role of electron correlation effects in atomic (e, 2e) spectroscopy with particular reference to the helium and argon systems and on the dependence of orbital momentum distributions on molecular environment. The latter area is discussed with particular reference to the chloromethanes $CH_{4-n}Cl_n$ (n=1 to 4) and to satellite structure in the CH_3Cl system. For atomic systems the relationship between (e, 2e) and photoelectron satellite data is also examined.

2.   ATOMIC SYSTEMS

2.1 Theoretical Considerations

2.1.1   General Formulation

Detailed accounts of (e, 2e) reaction cross-section theory may be found in the literature[1,2,4,5]. In the plane wave impulse approximation for a symmetric geometry the differential cross-section for atoms may be written in atomic units as

$$\frac{d^3\sigma}{d\Omega_A d\Omega_B dE_A} = (2\pi)^4 \frac{k_A k_B}{k_o} |T(q)|^2 \sum_{av} |<\Psi_o^i(N)|\Phi_j^f(N-1)e^{i\underline{q}\cdot\underline{r}}>|^2 \quad (2.1)$$

where k_o, k_A and k_B are the momenta of the incident and scattered electrons respectively. $T(q)$ is the Coulomb T matrix, q is the ion recoil momentum and Σ_{av} represents a summation over final states and an averaging over initial states. $\Psi_o^i(N)$ is the wavefunction for the N-electron initial state of the target and $\Phi_j^f(N-1)$ is the wavefunction for the jth final N-1 electron residual ion state. In this formulation the electron ejected into the continuum by electron impact is decoupled from the residual ion electrons and is represented by a plane wave.

It is convenient to write equation (2.1) for a symmetric geometry with $k_A = k_B$ as [6]

$$\sigma(q) = X(k_o,k_A,q) \; I_j^2(q) \quad (2.2)$$

where the $I_j^2(q)$ term represents the $\sum_{av}$ term in equation (2.1) and is the structure factor, while $X(k_o,k_A,q)$ is the remainder. For the present discussion our interest is principally in the structure factor term.

If correlated wavefunctions of the configuration interaction type are used to represent the initial state and the final ion states[4-6] such that

$$\Psi_o^i(N) = \sum_k b_{ok} \; \chi_{ok}^i(N) \quad (2.3)$$

and

$$\Phi_j^f(N-1) = \sum_\ell c_{j\ell} \chi_{j\ell}^f(N-1) \quad (2.4)$$

where $\chi_{ok}^i(N)$ and $\chi_{j\ell}^f(N-1)$ are the configuration state functions (CSF) expressed as antisymmetrised products of one-electron functions of appropriate symmetry. b_{ok} and $c_{j\ell}$ are the coefficients of the CSF in the initial state and the jth final ion state wavefunction respectively. Now

$$<\Psi_o^i(N)|\Phi_j^f(N-1)\exp(i\underline{q}\cdot\underline{r})> = \sum_{km\ell} b_{ok}c_{j\ell}<\chi_{ok}^{im}(N-1)|\chi_{j\ell}^f(N-1)>\phi_{ok}^{im}(q) \quad (2.5)$$

where $<\chi_{ok}^{im}(N-1)|\chi_{j\ell}^f(n-1)>$ is the overlap matrix between the residual (N-1) CSF, $\chi_{ok}^{im}(N-1)$, after elimination of the Nth electron associated with orbital ϕ_m and the ℓth CSF $\chi_{j\ell}^f(N-1)$ from the jth final ion state. $\phi_{ok}^{im}(q)$ is the mth momentum space wavefunction associated with ϕ_m for the N_{th} electron in the CSF, $\chi_{ok}^i(N)$. Furthermore,[4] since

$$I_j^2(q) = \sum_{av} |<\psi_o^i(N)|\Phi_j^f(N-1)>\exp(i\underline{q}\cdot\underline{r})>|^2 \tag{2.6}$$

from (2.5)

$$I_j^2(q) = \sum_{av} |\sum_k \sum_m \sum_\ell b_{ok}c_{j\ell}<\chi_{ok}^{im}(N-1)|\chi_{j\ell}^f(N-1)>\phi_{ok}^{im}(q)|^2 \tag{2.7}$$

$$= N \sum_k \sum_{k'} \sum_m \sum_{m'} \sum_\ell \sum_{\ell'} n_{km}^{\frac{1}{2}} n_{k'm'}^{\frac{1}{2}} b_{ok}b_{ok'}c_{j\ell}c_{j\ell'} \tag{2.8}$$

$$x <\chi_{ok}^{im}(N-1)|\chi_{j\ell}^f(N-1)><\chi_{ok'}^{im'}(N-1)|\chi_{j\ell'}^f(N-1)>\phi_{ok}^{im}(q)\phi_{ok'}^{im'}(q)$$

n_{km} and $n_{k'm'}$ are the number of equivalent target electrons with the same nl and exponent values in the state being ionised. N is a constant.

This is the result for the most general case when correlated wave-functions are used to describe both the initial and final states of the target.

It is instructive to consider three limiting cases.

2.1.2 Uncorrelated Initial and Final States

If the initial and final states of the system may be adequately represented by single configurations such that from equations (2.3) and (2.4) we have

$$\psi_o^i(N) = \chi_{o1}^i(N) \tag{2.9}$$

and

$$\Phi_j^f(N-1) = \chi_{j1}^f(N-1) \tag{2.10}$$

that is, k,ℓ,b_{o1} and c_{o1} are all equal to 1 then from equation (2.7) and the above

$$I_j^2(q) = N \sum_m \sum_{m'} n_{1m}^{\frac{1}{2}} n_{1m'}^{\frac{1}{2}} <\chi_{o1}^{im}(N-1)|\chi_{j1}^f(N-1)>$$

$$x <\chi_{o1}^{im'}(N-1)|\chi_{j1}^f(N-1)>\phi_{o1}^{im}(q)\phi_{o1}^{im'}(q) \tag{2.11}$$

The orbitals $\phi_m(q)$ and $\phi_{m'}(q)$ from the initial state must have the same symmetry. More than one term results when relaxation effects from the N electron initial state to the (N-1) final ion state are included. For example, if we consider ionisation from the 3s orbital of the argon atom and include relaxation effects for the ion then equation (2.9) will include contributions from ϕ_{1s_i}, ϕ_{2s_i} as well as ϕ_{3s_i}.

However, if the assumption is made that only contributions from a single ground state target orbital are important then only a single term remains such that

$$I_j^2(q) = nN|<\chi_{o1}^{im}(N-1)|\chi_{j1}^f(N-1)>|^2|\phi_{o1}^{im}(q)|^2 \tag{2.12}$$

where m represents the one-electron orbital ionized in the target. Furthermore, with a frozen orbital approximation the overlap determinant is unity for j corresponding to the ion ground state and zero otherwise.

The (e, 2e) cross-section formula has been most commonly used in this form. It is known as the target Hartree Fock approximation, although this is not strictly correct.

2.1.3 Initial State Correlation

If the initial state is correlated, but the final ion state is uncorrelated, then from equations (2.3), (2.8) and (2.10) we have

$$I_j^2(q) = N \sum_k \sum_{k'm} \sum_{m'} n_{km}^{\frac{1}{2}} n_{k'm'}^{\frac{1}{2}} b_{ok} b_{ok'} \tag{2.13}$$

$$\times \; <\chi_{ok}^{im}(N-1)|\chi_j^f(N-1)> <\chi_{ok'}^{im'}(N-1)|\chi_j^f(N-1)> \phi_{ok}^{im}(q) \phi_{ok'}^{im'}(q)$$

This is the formula which is applicable to, for example, the helium system to be discussed in detail in section 3. The initial two electron state is correlated while the final one-electron ion states are represented by exact hydrogenic type states.

If the cross-section is dominated by the contribution from a single target orbital, as in the Hartree Fock case then m = m'.

$$I_j^2(q) \simeq nN \sum_k \sum_{k'} b_{ok} b_{ok'} <\chi_{ok}^{im}(N-1)|\chi_j^f(N-1)>$$

$$<\chi_{ok'}^{im}(N-1)|\chi_j^f(N-1)> |\phi_{ok}^{im}(q)|^2 \tag{2.14}$$

Furthermore, in many cases because of orthogonality and symmetry constraints only one term corresponding to k = k' remains in equation (2.14) such that

$$I_j^2(q) \simeq Nn \; b_{ok}^2 |<\chi_{ok}^{im}(N-1)|\chi_j^f(N-1)>|^2 |\phi_{o1}^{im}(q)|^2 \tag{2.15}$$

This is simply the Hartree Fock result (equation 2.12) weighted by b_{ok}^2, the square of the coefficient of the leading configuration in the ground state CI wavefunction.

2.1.4. Final State Correlation

If the initial state is uncorrelated but the final ion state is correlated then from equations (2.4), (2.9) and (2.8) we have

$$I_j^2(q) = Nn \sum_m \sum_{m'} \sum_\ell \sum_{\ell'} n_{1m}^{\frac{1}{2}} n_{1m'}^{\frac{1}{2}} c_{j\ell} c_{j\ell'} \tag{2.16}$$

$$\times \; <\chi_{o1}^{im}(N-1)|\chi_{j\ell}^f(N-1)> <\chi_{o1}^{im'}(N-1)|\chi_{j\ell}^f(N-1)> \phi_{o1}^{im}(q) \phi_{o1}^{im'}(q)$$

Now, as for the previous case, if $m = m'$ and $\ell = \ell'$ then

$$I_j^2(q) \simeq N n_{j\ell} c^2 |<\chi_{ol}^{im}(N-1)|\chi_{j\ell}^f(N-1)>|^2 |\phi_{ol}^{im}(q)|^2 \tag{2.17}$$

This expression is therefore essentially the Hartree Fock expression (equation 2.12) weighted by the square of the mixing coefficient, $c_{j\ell}^2$, for the leading configuration in the jth final ion state CI wavefunction.

2.2 Satellite Lines

The ionisation of a system which results from electron impact may in principle leave the ionized system in one of many states of appropriate symmetry in a particular manifold. The relative population of the atomic states of appropriate symmetry in the manifold will depend upon the relative cross-sections for the various transition processes.

For a symmetric system with $E_A = E_B$ from equations (2.1) and (2.2)

$$\frac{\sigma_{sat}(q_s)}{\sigma_{main}(q_m)} = \frac{k_o^{main}}{k_o^{sat}} \frac{|T_{sat}(q_s)|^2}{|T_{main}(q_m)|^2} \frac{I_{sat}^2(q_s)}{I_{main}^2(q_m)} \tag{2.18}$$

In general terms, from equation (2.7), for final states j and p

$$\frac{I_{sat}^2(q_s)}{I_{main}^2(q_m)} = \frac{\sum_{av} | \sum_k \sum_m \sum_\ell b_{ok} c_{j\ell} <\chi_{ok}^{im}(N-1)|\chi_{j\ell}^f(N-1)>\phi_{ok}^{im}(q_s)|^2}{\sum_{av} | \sum_k \sum_m \sum_\ell b_{ok} c_{p\ell} <\chi_{ok}^{im}(N-1)|\chi_{p\ell}^f(N-1)>\phi_{ok}^{im}(q_m)|^2} \tag{2.19}$$

If one assumes a single target orbital is dominant for the satellite and the main line then for $q_m = q_s$, equation (2.19) becomes

$$\frac{I_{sat}^2(q)}{I_{main}^2(q)} = \frac{\sum_k \sum_\ell b_{ok}^2 c_{j\ell}^2 |<\chi_{ok}^{im}(N-1)|\chi_{j\ell}^f(N-1)>|^2}{\sum_k \sum_\ell b_{ok}^2 c_{p\ell}^2 |<\chi_{ok}^{im}(n-1)|\chi_{p\ell}^f(N-1)>|^2} \tag{2.20}$$

It is evident that within this model the ratio of the structure factors for the two lines is independent of q. However, even when the satellite line intensity is measured at the same q value as the main line intensity the kinematic factors in equation (2.18) do not cancel.[6]

Equation (2.20) may be further reduced for the special cases discussed previously:

a)   For the Hartree Fock case, from equation (2.12)

$$\frac{I_{sat}^2(q)}{I_{main}^2(q)} = \frac{|<\chi_o^i(N-1)|\chi_{sat}^f(N-1)>|^2}{|<\chi_o^i(N-1)|\chi_{main}^f(N-1)>|^2} \equiv R_{HF} \tag{2.21}$$

b)   For initial state correlation only, from equation (2.15)

$$\frac{I_{sat}^2(q)}{I_{main}^2(q)} = \frac{b_{sat}^2}{b_{main}^2} \times R_{HF} \tag{2.22}$$

c)   For final state correlation only, from equation (2.17)

$$\frac{I_{sat}^2(q)}{I_{main}^2(q)} = \frac{c_{sat}^2}{c_{main}^2} \times R_{HF} \tag{2.23}$$

Such formulae will apply in selected cases, but not in others. For example, equation (2.23) is generally considered to be applicable for the determination of the relative intensity of ^2S final ion state satellites associated with the ionisation of the 3s orbital of argon, with R_{HF} equal to unity however equation (2.22) is not applicable to the helium satellite spectrum over a wide range of q.

2.3   Relationship between (e, 2e) Satellite Intensities and Photoionisation Satellite Intensities

In photoelectron spectroscopy, following a similar notation to that developed in previous sections the ratio of the intensity of two lines in the spectrum is given by [7]

$$\frac{P_{sat}(E)}{P_{main}(E)} = \frac{\sum_{if} | \sum_k \sum_\ell b_{ok} c_{j\ell} <A\chi_{j\ell}^f(N-1)\phi_{\varepsilon_j}^f(1)|\Sigma d_u|\chi_{ok}^i(N)>|^2}{\sum_{if} | \sum_k \sum_\ell b_{ok} c_{p\ell} <A\chi_{p\ell}^f(N-1)\phi_{\varepsilon_p}^f(1)|\Sigma d_u|\chi_{ok}^i(N)>|^2} \tag{2.24}$$

when d_u is the dipole operator and $\phi_{\varepsilon_j}^f$ is the continuum orbital.

This expression should be compared with equation (2.19). Now, in the high photon limit when only the leading terms in the expression which involves the dipole matrix element between the orbital from which the primary electron is ejected and the continuum electron are retained and when the dipole matrix elements involving different continuum orbitals, but the same target orbital, are assumed to be approximately equal equation (2.25) reduces to

$$\frac{P_{sat}}{P_{main}} = \frac{\left| \sum_{k} \sum_{\ell} b_{ok} c_{j\ell} < \chi_{ok}^{im}(N-1) | \chi_{j\ell}^{f}(N-1) > \right|^{2}}{\left| \sum_{k} \sum_{\ell} b_{ok} c_{p\ell} < \chi_{ok}^{im}(N-1) | \chi_{p\ell}^{f}(N-1) > \right|^{2}} \qquad (2.25)$$

This expression is the multiconfiguration analogy of shake theory.[8] It is applicable only in the region where the intensity ratio is independent of incident photon energy.

This expression is identical with the limiting form for (e, 2e) processes, namely equation (2,20) only when cross terms in equation (2.25) are ignored. Experimental cases in the high q limit and the high photon limit are not available to compare with the predictions.

3. THE HELIUM SATELLITE SPECTRUM

The spectrum which results from ionisation of the ground state of helium has been studied experimentally in considerable detail by electron impact and photoionisation spectroscopy[9-14]. The ionisation may be represented as

$$\text{He } (n=1) \xrightarrow[h\nu]{e} \text{He}^{+} (n=1,2,3...) + e'$$

where the final ion states correspond to 1s, 2s, 2p, 3s, 3p, 3d hydrogenic type states.

The experimental (e, 2e) studies[9-10] have probed the momentum distribution (q dependence) of the n = 1 main line and of the n = 2 satellite lines. The q dependence of the $\sigma(n = 2)/\sigma(n = 1)$ cross-section ratio has then been determined. The results obtained for this ratio along with some theoretical results to be discussed shortly are shown in figure 1.

Photon induced ionisation studies have provided information on the photon dependence of the main (n = 1) and satellite (n = 2) line cross-sections within 100 eV of threshold[11-12] as well as information on the $\sigma(2p)/\sigma(2s)$ cross-section ratio[13-14]. The latter ratio has not yet been determined by (e, 2e) spectroscopy. Photoelectron results for the $\sigma(n = 2)/\sigma(n = 1)$ cross-section ratio is shown in figure 2[11] along with some recent relaxed Hartree Fock calculations by Larkins and Richards using equation (2.24) and the velocity form of the dipole operator.

The photon induced $\sigma(n = 2)/\sigma(n = 1)$ ratio is in general higher than the electron induced result, but the results from figures 1 and 2 are not directly comparable. The limiting relationships, equations (2,20) and (2.25), where photon one electron induced results may be compared require that the cross-section ratio be independent of q and hν. This regime has clearly not been reached with the experimental results presented in figures 1 and 2.

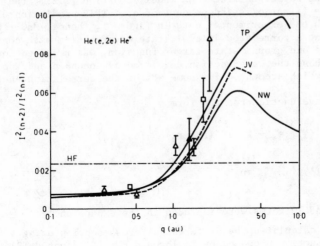

Figure 1

The incident energy independent curve for the ratio of the
intensities of the n = 2 to n = 1 helium (e, 2e) lines as a
function of the ion recoil momentum q. See text for
explanation of theoretical curves.

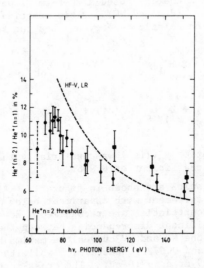

Figure 2

The ratio of the intensities of the n = 2 to n = 1 helium
photoelectron lines as a function of incident energy.
Theoretical values are by Larkins and Richards (unpublished).

Within the target Hartree-Fock model, equation (2.21), the $\sigma(n = 2)/\sigma(n = 1)$ ratio is independent of q. This prediction clearly does not agree with the experimental data in figure 1; hence, McCarthy et al[9] recognized that a correlated initial state wavefunction with contributions from more than one ground state target function, was required for the helium atom. They chose the wavefunction developed by Joachain and Vanderpoorten[16] (denoted JV) which accounted for some 98% of the correlation energy.

The JV wavefunction has the form

$$\psi_o^i(N) = \frac{1}{4\pi} \sum_{\ell=0}^{\lambda} F_\ell(r_1, r_2) \, P_\ell(\cos\theta_{12}) \tag{3.1}$$

$$F_\ell(r_1, r_2) = \sum_{m,n} B_{mn}^{(\ell)} \, r_>^{\ell+m} \, r_<^{\ell+n} \exp^{-\beta(r_>+r_<)} \tag{3.2}$$

A total of 70 terms were included in the expansion.

Larkins[6] calculated the differential cross-section using a simple four-term CI wavefunction proposed by Taylor and Parr[17] (denoted TP) to describe the ground state of the helium atom. This wavefunction accounted for some 85% of the correlation energy and had the form

$$\psi_o^i(N) = b_{o1} \, \chi_{o1}^i(1s,1s') + b_{o2} \, \chi_{o2}^i(2p^2) + b_{o3} \, \chi_{o3}^i(3d^2) + b_{o4}\chi_{o4}^i(4f^2) \tag{3.3}$$

The leading configuration $\chi_{o1}^i(1s,1s')$ had a coefficient of 0.998. Recently, Larkins and Richards[15] used a 20-term CI wavefunction proposed by Nesbet and Watson[18] (denoted NW). This function accounted for some 97.7 per cent of the correlation energy and had the form

$$\psi_o^i(N) = \sum_{k=1}^{4} b_{ok} \, \chi_{ok}^i(s^2) + \sum_{k=5}^{10} b_{ok} \, \chi_{ok}^i(s,s') + \sum_{k=11}^{13} b_{ok} \, \chi_{ok}^i(p^2)$$

$$+ \sum_{k=14}^{16} b_{ok} \, \chi_{ok}^i(pp') + \sum_{k=17}^{18} b_{ok} \, \chi_{ok}^i(d^2) + b_{o,19}\chi_{o,19}^i(dd') + b_{o,20}\chi_{o,20}^i(f^2) \tag{3.4}$$

The leading configuration $\chi_{o1}^i(1s^2)$ was essentially the Hartree Fock type configuration with a coefficient of 0.996.

The theoretical results are shown in figure 1. All three wavefunctions provide satisfactory agreement with experiment below q = 2.0au, when the large experimental uncertainties are recognized. However, at larger q values there are significant discrepancies between the predicted cross-section ratios. For the NW and JV curves the maxima both occur at q = 3.0 au with values of 0.061 and 0.072 respectively, while the TP curve has a maximum value of 0.099 at q = 7.0 au. The cross-section ratio with the NW function decreases to 0.04 at q = 10.0 au. The photoionisation cross-section ratio has been determined experimentally to be 0.05 ± 0.01 for a photon energy of 1487 ev.[19]

When the $\sigma(n=2)/\sigma(n=1)$ ratio for helium is accurately determined from (e, 2e) experiments in the range q=1 to 5 au a sensitive test will be available to discriminate between the quality of various correlated wavefunctions. It is interesting to note that even though the NW wavefunction is dominated by a single configuration it is the small admixture of other configurations which is responsible for the structure in the q dependence of the cross-section ratio.

Another feature of the problem is the contribution of the $\sigma(2s)$ and $\sigma(2p)$ components to the total $\sigma(n=2)$ cross-section. Table 1 shows the values for the $\sigma(2p)/\sigma(2s)$ (e, 2e) cross-section ratio under various kinematic conditions.

TABLE 1

Differential (e, 2e) cross-section ratio for populating the He^+ (2p) to He^+(2s) final ion states (%)

θ(deg)	TP	JV	NW
45	0.26	1.8	0.36
49	1.40	14.1	2.04
53	2.74	23.2	3.07

Kinematic conditions: symmetric coplanar geometry

$$\phi = 0, \quad \theta_A = \theta_B \quad E = 800 \text{ eV}$$

TP : Taylor and Parr wavefunction[17]

JV : Joachain and Vanderpoorten wavefunction[16]

NW : Nesbet and Watson wavefunction[18]

The ratio is predicted to be very small with the TP and NW correlated functions; however, with the JV function the ratio is an order of magnitude larger. While there is no experimental (e, 2e) data on this ratio, experimental photoelectron data for the $\sigma(2p)/\sigma(2s)$ photoionisation cross-section ratio has been obtained by direct[14] and indirect methods[13]. These results along with three calculations[20-22] are shown in Figure 3. There remains to be resolved a significant discrepancy between the two sets of experimental values within 20 eV of threshold. Furthermore, there is a large deviation between the theoretical values near threshold. These photoelectron results will be discussed in detail elsewhere[22]. It is interesting to note that the $\sigma(2p)/\sigma(2s)$ ratio is of order 0.5–2.5 in the photon energy range 70–150eV, while the corresponding predicted (e, 2e) ratio with the TP and NW functions is only .003 to 0.03 under the conditions investigated, namely q = 0.1 to 2.0 au. These results cannot of course be directly compared.

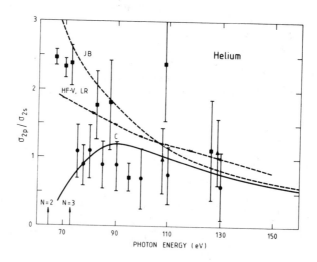

Figure 3.

    The ratio of the 2p to 2s cross-section in the helium
photoelectron spectrum as a function of incident photon
energy. Experimental points Woodruff and Samson Ref. 14, (■);
Bizau et al. Ref. 13, (●).

    Weigold and McCarthy[1] have also measured the q dependence of the
relative cross-section for leaving the helium system in the 1s final ion
state. They have shown that the shape dependence is well represented by
the momentum distribution of the 1s Hartree Fock orbital. It has
been shown that the momentum distributions of the TP[6] and the NW wavefunctions[15] are also in good agreement with experiment. Hence, it was
concluded that the q dependence of the relative 1s differential cross-
section is not a sensitive test of the quality of the wavefunction beyond
the Hartree Fock limit. However, for wavefunctions of lesser quality there
may be real differences in the q dependent momentum shape profiles. It is
in the q dependence of the 2s and 3s final helium ion states that there are
clear difference between the theoretical predictions of the TP and NW
correlated wavefunctions. Consequently, it seems likely that it will be
the satellite and not the main lines which provide the most stringent
test of the quality of correlated wavefunctions.

4. THE ARGON VALENCE SHELL SATELLITE SPECTRUM

The valence shell photoionisation and (e, 2e) spectra for the noble gases have been extensively studied[1,2,3,23,24]. In addition to the s and p diagram lines considerable satellite structure has been observed at higher binding energies. In contrast to the helium case the satellite structure is dominated by final state correlation effects although there may be some contribution to the satellite intensity resulting from initial state correlation effects at low photon energy or low q values.

The initial state wavefunctions for the closed shell noble gas systems in their ground state which accounts for valence shell correlation may be reasonably represented by a CI expression of the general form

$$\Psi_o^i(^1S) = b_{o1}|ns^2np^6> + \Sigma b_{o\ell}|ns^2np^4n'd^2>$$

$$+ \Sigma b_{ok}|ns^2np^4n''s^2> + \Sigma b_{om}|ns^2np^4n''p^2> \qquad (4.1)$$

The final ion states have ^2S, ^2P and ^2D symmetry[3,24]. The ^2D states result from the ionisation of nd orbitals in configurations mixed in the initial state. Wavefunctions which account for correlation in the valence shell have the form

$$\Phi_j^f[ns\,^2S] = c_{jo}|ns\,np^6> + \Sigma c_{j\ell}|ns^2np^4(^1D)n'd>$$

$$+ \Sigma c_{jk}|ns^2np^4(^1S)n''s> \qquad (4.2)$$

$$\Phi_k^f[np\,^2P^o] = c_{ko}|ns^2np^5> + \Sigma c_{kj}|ns^2np^4 \begin{pmatrix} ^3P \\ ^1D \\ ^1S \end{pmatrix} n''p> \qquad (4.3)$$

$$\Phi_\ell^f[nd\,^2D] = \Sigma c_{\ell k}|ns^2np^4 \begin{pmatrix} ^3P \\ ^1D \\ ^1S \end{pmatrix} n'd> + \Sigma c_{\ell j}|ns^2np^4\ (^1D)n''s> \qquad (4.4)$$

The argon n=3, for krypton n=4 and for xenon n=5. Furthermore, n' has values $\geq$ n and n'' values $\geq$ n+1. Dyall[25] has calculated the intensity of various ^2S, ^2P and ^2D satellite lines in the argon, krypton and xenon (e, 2e) spectra using wavefunctions of the form given above with Rydberg orbitals as the correlating functions.

The (e, 2e) results for argon with E = 400 eV, $\phi = 0^o$ and $\theta = 45^o$, 42.3^o or 39.5^o are summarized in Table 2. The values were determined using equation (2.1) with the structure factor $I_j^2(q)$ given by equation (2.8).

The important finding is that ^2D satellites, which result from initial state correlation effects through the $3s^23p^4nd^2$ configurations are significant, under certain kinematic conditions, along with the ^2S satellites.

TABLE 2

Theoretical Intensities of the ^{2}S, ^{2}P and ^{2}D lines in the (e, 2e)

Valence Spectrum of Argon for E = 400 eV, $\phi = 0$[a]

Final State	Binding Energy	θ		
	eV	45°	42.3°	39.5
$3s^2 3p^5$ $^2P^o$	15.8	11.4	18.7	110.4
$3s\ 3p^6$ ^2S	29.2	100	100	100
$3s^2 3p^4(^1S)4s$ ^2S	36.5	2.3	3.1	2.6
$(^1D)3d$ ^2D	37.2	4.9	0.0	5.4
$(^1S)3d$ ^2D	38.0	1.5	0.0	1.8
$(^1D)3d$ ^2S	38.6	11.1	11.5	12.6
$(^1D)4d$ ^2D	40.5	2.7	0.0	3.3
$(^1D)4d$ ^2S	41.2	7.2	7.6	8.4
$(^1S)5s$ ^2S	42.4	1.9	1.5	2.1
$(^1D)5d$ ^2S	~42.5	1.9	2.7	2.4
$(^1S)4d$ ^2S	~42.8	2.8	0.1	5.2
Total ^{2}S Satellite Intensity		24.4	26.4	28.1
Total ^{2}D Satellite Intensity		11.9	0.1	15.7

a   Only lines with predicted intensity greater than 1 per cent of ^{2}S diagram line are shown. No $^2P^o$ satellite lines make a significant contribution under these kinematic conditions.

b   Data from ref. 25.

The latter result from final state correlation effects in the ^2S manifold. In particular, while for $\theta = 42.3°$ the ^2D satellite intensity is negligible, for $\theta = 45°$ it is almost 50 per cent of the ^2S satellite intensity and for $\theta = 39.5°$ it is 56 per cent of the ^2S satellite intensity. The final state $^2P°$ satellites are predicted to be negligible under all the kinematic conditions ivestigated. These relative intensity values would seem to be an overestimate and warrant further investigation, but they do correspond to very low q values (q < 0.3). The principle of the potential effect of $2D$ satellites is undoubtedly correct.

In the experimental analysis of the argon (e, 2e) spectrum[1,2,23] it was assumed that the satellites were exclusively due to ^2S final ion state correlation effects and then the 'spectroscopic factors' were calculated using equations (2.17) and (2.23). By adopting a frozen orbital model such that $<\chi^{im}_{o\ell}(N-1)|\chi^f_{i\ell}(N-1)>$ in equation (2.17) is unity for $\ell = 1$ and zero otherwise the coefficients c^2_{j1} (spectroscopic factors) can be related directly to the intensities of the main and satellite lines since $\sum_j c^2_{j1} = 1$. c_{j1} is the coefficient of the $|3s\ 3p^6,^2S>$ configuration in the CI expansion for the jth ^2S state. The summation is over all ^2S states. The assumptions embodied in this approach will not be equally valid for all the noble gas systems under all kinematic conditions.

The previous workers[1,2] justified their conclusion from ϕ angular distribution studies of the q dependence of main and satellite lines in argon at a few binding energies. They showed that the shape profiles are similar for the main and satellite lines. However, the energy resolution at which the experiments were carried out prevented any distinction between satellites of ^2D and ^2S symmetry. Consideration of the momentum space wavefunctions show that differences between the 3s and 3d profiles may be very difficult to detect except at q values < 0.5 au. The n=4 momentum space Rydberg wavefunctions amplitudes are also appreciably larger than the corresponding n=3 functions for q < 0.5 au. Furthermore from a consideration of the published data (ref 1, p346, fig 10.3) it is evident that for q > 1.0 the ratio of the satellite to main line intensity increases in a manner like that reported in figure 1 for helium.

The full interpretation of the argon and other noble gas systems is more complex than the simplified approach which has thus far been adopted. More experimental data at higher resolution is required to establish the importance of initial and final state correlation effects including relaxation effects as well as the extent of contributions from satellites of different symmetry.

5.   MOLECULAR EFFECTS

Theoretical work on the momentum distribution of molecular wavefunctions can reasonably be said to originate from the pioneering studies of Coulson and Duncanson who published a series of five papers in 1941[26-30]. However, despite this early theoretical work there was little interest in the field of molecular momentum wavefunctions until the extensive (e, 2e) studies on molecules over the past decade. At present much of the work consists of relating the experimental cross-section data to theoretical data derived from molecular orbital calculations with Gaussian basis sets of double zeta quality. In particular, the wavefunctions of Snyder and Basch[31] have been extensively used to match the shape of the momentum distribution. It has been shown that they, in general, provide distributions in good qualitative agreement with many of the well resolved diagram lines.

Minchinton et al in this volume discuss recent experimental studies in molecular (e, 2e) spectroscopy.

The theory of the (e, 2e) reaction for molecules has been developed by McCarthy[32] for the plane wave and the distorted plane wave impulse approximations. A discussion may be found in this volume and in previous reviews[1-2]. To date the effort has concentrated on evaluating the structure factor with Hartree-Fock type molecular orbitals and a frozen orbital approximation such that following equation (2.12) the molecular structure factor $I_j^2(q)$ is given by

$$I_j^2(q) = \text{constant } x |\psi_{ol}^i(q)|^2 \tag{5.1}$$

where $\psi_{ol}^i(q)$ is the momentum transform of the approximate Hartree Fock one-electron molecular orbital. The M.O. wavefunction is usually expressed as a linear combination of the basis set functions.

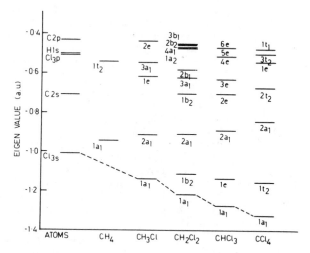

<u>Figure 4</u>
        The calculated valence eigenvalues for the chloromethanes
        and their constituent atoms using contracted Gaussian basis sets

To enhance our understanding of the field we have considered in detail the chloromethane series of molecules, $CH_{4-n}Cl_n$ n=1,...4 using the same Gaussian basis sets for the molecules and the isolated atoms. Ab initio m.o. calculations were performed with the following basis sets:[33,34]

        Carbon   (9s, 5p) contracted to [4s, 2p]

        Hydrogen   (4s, 1p) contracted to [2s, 1p]

        Chlorine   (12s, 9p) contracted to [6s, 4p]

Full details of the basis set may be found in reference 34.

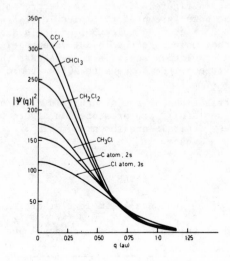

Figure 5.

The momentum distributions of the $1a_1$ valence orbitals of the chloromethanes and of the valence s orbitals of the constituent carbon and chlorine atoms.

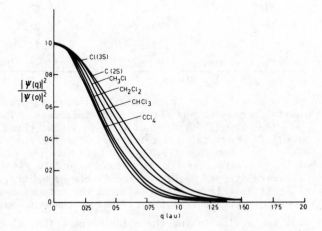

Figure 6.

The momentum distributions of the $1a_1$ valence orbitals of the chloromethanes relative to $|\psi(0)|^2$ value.

The calculated valence orbital eigenvalues for the series of molecules and the atoms are shown in figure 4. The one-electron m.o. notation varies according to the point group of the molecule (T_d to C_{2v}). The momentum distributions for selected molecular orbitals are reported here. For all the chloromethane molecules the lowest valence eigenvalue is of a_1 symmetry. A Mulliken population analysis reveals that this m.o. is predominantly an admixture of the C 2s and the Cl 3s atomic orbitals suitably symmetry adapted.

The momentum distributions of the a_1 orbitals are given in figure 5 and the profiles relative to $|\psi(o)|^2$ are presented in figure 6. The principal characteristics of these orbitals are shown in table 3. All these profiles have the classic 's-type' shape.

TABLE 3

Characteristics of the valence $1a_1$ molecular orbitals in the

chloromethanes and of the chlorine and carbon valence s orbitals

Species	ε eV	$\|\Psi_{1a_1}(o)\|^2$	FWHM Profile q au
Cl 3s	29.2	114.4	1.13
C 2s	19.2	153.5	1.04
CH_3Cl	31.0	179.8	0.96
CH_2Cl_2	33.1	246.5	0.85
$CHCl_3$	34.7	289.0	0.81
CCl_4	35.9	327.1	0.77

For the chloromethane series the $1a_1$ eigenvalue increases from CH_3Cl to CCl_4 and the differential cross-section for q=o also increases. The cross-section value at q = o for the $1a_1$ orbital of CCl_4 is almost three times the value for the atomic Cl 3s orbital and twice the value for the atomic C 2s orbital. The C 2s orbital eigenvalue (19.2 eV) is less than the Cl 3s eigenvalue (29.2 eV), however, the value for $|\Psi(o)|^2$ from Table 3 is significantly larger reflecting the more diffuse nature of the C valence orbital. The half width of the a_1 profiles (Figure 6) depend upon the molecular environment and show a systematic decrease from the Cl atom to the CCl_4 molecule.

Experimental data are available for the chloromethane (CH_3Cl) molecule. The theoretical and experimental differential cross-sections for the $1a_1$ valence orbital of CH_3Cl are shown in figure 7. There is good qualitative agreement between the observed and calculated q dependence of the profiles.

This factor provides support for the conclusion that the basis set chosen is adequate to obtain near Hartree-Fock molecular orbitals, and correct shape profiles. However, as for the atomic helium system discussed in section 2, the complete picture is more complex. As a consequence of normalizing the theoretical cross-section values to the experimental $1e_1$ values there is approximately a factor of 2 difference between the calculated and experimental cross-section values for the $1a_1$ orbital. This factor reflects the point that the $1a_1$ intensity is spread through the $1a_1$ satellites lines as well as the $1a_1$ main line[34]. In the calculation electron correlation effects and relaxation effects have been ignored. The effect is in principle analogous to the argon valence spectrum where there is ^2S intensity splitting with an estimated 56 per cent of the ^2S intensity[23] in the main line with the remainder spread throughout the satellites.

Another set of molecular orbitals of interest is the outermost valence orbitals of dichloromethane (CH_2Cl_2) where there is a near degenerate set of four orbitals (Figure 4). The $1a_2$, $4a_1$, $2b_1$ and $3b_1$ orbitals have calculated eigenvalues of 13.0, 12.9, 12.4 and 12.3 eV respectively. From a Mulliken population analysis the dominant contribution to these molecular orbitals comes from chlorine 3p atomic orbitals. The differential cross-sections calculated within a plane wave impulse approximation with E = 1200 eV, $\vartheta = 45°$ and variable ϕ are shown in figure 8. The values are plotted relative to the maximum in the $1a_2$ curve set equal to 1.0. All four orbitals have a different q dependence reflecting the various different symmetries of the orbitals. In the range q = 0.5 to 1.0 all the orbitals have differential cross-sections of the same order of magnitude. Because of the near degeneracy of these orbitals (within 1 eV) it is unlikely that they could be studied independently in any experimental investigation.

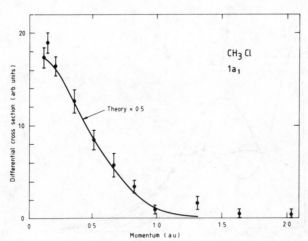

Figure 7.

Experimental differential cross-section of the $1a_1$ valence orbital of CH_3Cl. The full line is the q dependence predicted with the ab initio molecular orbital wavefunction discussed in the text. The weighting factor of 0.5 accounts for correlation and relaxation effects.

162

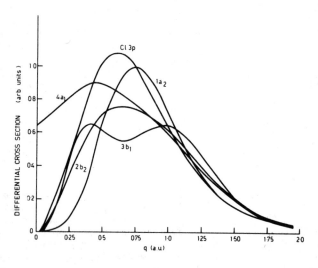

Figure 8.

The differential cross-section values for the near degenerate $1a_2$, $4a_1$, $2b_2$ and $3b_1$ molecular orbitals of dichloromethane CH_2Cl_2 as a function of the ion recoil momentum.

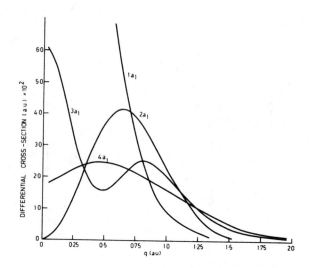

Figure 9.

The differential cross-section values for the four a_1 valence orbitals of dichloromethane CH_2Cl_2 as a function of the ion recoil momentum.

The dichloromethane molecule also has four valence orbitals of a_1 symmetry. Their orbital energies are as follows $1a_1$ 33.1 eV, $2a_1$ 24.8 eV, $3a_1$ 16.9 eV and $4a_1$ 12.9 eV. The $1a_1$ orbital is principally an admixture of Cs and Cℓs atomic orbitals, the $2a_1$ is Cs, Cℓs and Hs, the $3a_1$ Cp. Cℓs and p, while the $4a_1$ is dominated by Cℓp contributions. The differential cross-sections for the four orbitals calculated with E = 1200 eV, ϕ = 45° and variable ϕ are shown in figure 9. The momentum distributions for the orbitals are distinctly different reflecting the different atomic orbital character of the molecular orbitals even though they are all of a_1 symmetry. Further analyses of the data for the chloromethanes are in progress.

6. SATELLITE STRUCTURE IN MOLECULAR (e, 2e) SPECTRA

The experimental (e, 2e) spectrum for many molecules reveals evidence of satellite structure at higher binding energies than the diagram lines also was the case for several atomic systems including helium and argon discussed earlier. This structure again provides evidence for the importance of electron correlation effects in the initial or final states of the molecule.

The (e, 2e) spectra of CH_3Cl is shown in Figure 10[34]. Under the kinematic conditions used E = 1200 eV, ϕ = 0, θ = 45° there is strong satellite structure beyond the $1a_1$ peak. The q dependence at ε = 31.5 eV is also shown along with the calculated differential cross-section for the $1a_1$ and the $3a_1$ valence orbitals multiplied by 0.18 and 0.43 respectively. It was concluded that the structure above 28 eV largely belongs to correlation effects associated with the $1a_1$ orbital, which is strongly chlorine 3s in character. This case is again similar to the argon 3s system.

Hence, it is apparent that for a detailed account of the q dependent (e, 2e) spectra of many molecules it will be necessary to have correlated molecular orbital wavefunctions for the ground state and the relaxed final ion states of the molecule. Furthermore, as for the atomic case, it is probable that it will be the q dependence of the satellite lines rather than the main lines which will provide the most sensitive test of the quality of the ab initio molecular orbital wave functions.

ACKNOWLEDGEMENTS

I am indebted to K.G. Dyall, R.M. Wilson and J.A. Richards for assistance with the calculations reported herein. Valuable discussions with E. Weigold, I.E. McCarthy and A. Minchinton, as well as the use of their molecular momentum transform program, are very gratefully acknowledged.

Financial support from the Australian Research Grants Committee is also much appreciated.

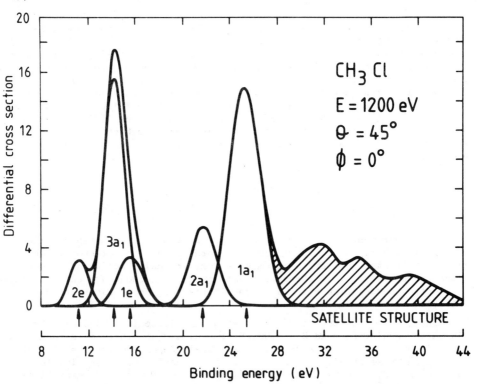

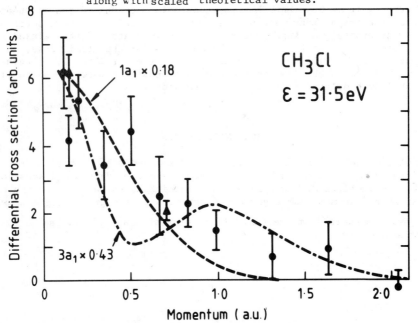

Figure 10A. The (e, 2e) spectrum of chloromethane for $E = 1200\,eV$, $\theta = 45°$, $\phi = 0$.
Figure 10B. The momentum distribution of the satellite structure at 31.5 eV along with scaled theoretical values.

References

1. McCarthy, I.E. and Weigold, E., (e,2e) Spectroscopy,
   Phys.Rep. 27C, 275 (1976).

2. Weigold, E. and McCarthy, I.E., (e,2e) Collisions, Adv. At. Mol Phys.
   14, 127 (1978).

3. Dyall, K.G., and Larkins, F.P., Satellite Structure in Atomic Spectra,
   I J.Phys. B. At. Mol. Phys. 15, 203 (1982)
   II ibid 15, 219 (1982)

4. Levin, V.G., Structure of Wavefunctions of Atoms in the (e,2e) reaction,
   Phys. Lett. 39A,125 (1972).

5. Levin, V.G., Neudatchin, V.G., Pavlitchenkov, A.V., and Smirnov, Yu. F.,
   On the display of basic properties of the molecular electronic wave-
   functions in the (e,2e) quasielastic knockout experiments,
   J.Chem.Phys. 63, 1541 (1975).

6. Larkins, F.P., The (e,2e) satellite spectrum of helium,
   J.Phys.B.At. Mol. Phys. 14, 1477 (1981).

7. Larkins, F.P., Electron Correlation Effects in atomic photoelectron
   spectra,in Molecular Physics and Quantum Chemistry into the 80's.
   ed. P.G.Burton (Wollongong University, 1980).

8. Aberg, T., Asymptotic Double Photoexcitation Cross Sections of the
   Helium Atom Phys.Rev. A2, 1726 (1970); Ann.Acad.Sci.Fenn. A6,308 (1969).

9. McCarthy, I.E., Ugbabe, A., Weigold, E., and Teubner, P.J.O., (e,2e)
   Reaction as a Probe for Details of the Helium Wave Function,
   Phys.Rev.Lett. 33, 459 (1974).

10. Dixon, A.J., McCarthy, I.E., and Weigold, E., Excitation of the n=2 states
    of He$^+$ in the ionisation of helium, J.Phys.B. At.Mol.Phys. 9, L195 (1976).

11. Wuilleumier, F., Adam, M.Y., Sandner, N., and Schmidt, V.,
    Photoionisation of helium above the n=2 photoionisation threshold,
    J.Physique Lett. 41, L373 (1980).

12. Woodruff, P.R., and Samson, J.A.R., Measured Cross Sections for
    photoionisation of Ground state He to He$^+$(n=2),
    Phys.Rev.Lett. 45, 110 (1980).

13. Bizau, J.M., Wuilleumier, F., Dhez, P., Ederer, D.L., Chang, T.N.,
    Krummacher, S., and Schmidt, V., The final state symmetry for the N=2
    states in photoionized helium determined by theory and experiment.
    Phys.Rev.Lett. to be published (1982).

14. Woodruff, P.R., and Samson, J.A.R., Measurements of Partial Cross-
    Sections and Autoionisation in the Photoionisation of helium to He$^+$(n=2),
    Phys.Rev.A. (1982).

15. Larkins, F.P., and Richards,J.A., The helium (e,2e) satellite spectrum,
    J.Phys.B. At.Mol.Phys. 15, L (1982).

16. Joachain, C.J., and Vanderpoorten, R., Configuration interaction
    wavefunctions for two-electron systems, Physica,46, 333 (1970).

17. Taylor, G.R., and Parr, R.G., Superposition of Configurations: The
    Helium Atom, Proc.Natl.Acad.Sci.USA 38, 154 (1952)

18. Nesbet, R.K., and Watson, R.E., Approximate wave functions for the ground state of helium, Phys.Rev. $\underline{110}$, 1073 (1958).

19. Carlson, T.A., Krause, M.O., and Moddeman, W.E., Excitations accompanying photoionisation in atoms and molecules and its relationship to electron correlation, J.Physique, $\underline{10}$, C4, 76 (1971).

20. Jacobs, V.L., and Bourke, P.G., Photoionisation of Helium above the n=2 threshold, J.Phys.B. At.Mol.Phys. $\underline{5}$, L67 (1972).

21. Chang, T.N., Photoionisation of helium to $He^+(n=2)$, J.Phys.B.At.Mol. Phys. $\underline{13}$, L551 (1980).

22. Richards, J.A., and Larkins, F.P., to be published.

23. Williams, J.F., High resolution energy and angular correlations of the scattered and ejected electrons in electron impact ionisation of argon atoms, J.Phys.B.At.Mol.Phys. $\underline{11}$, 2015 (1978).

24. Dyall, K.G., and Larkins, F.P., A configuration interaction study of the valence shell photoelectron spectrum of argon, J.Electron Spectroscopy $\underline{15}$, 165 (1979).

25. Dyall, K.G., Electron correlation effects in atomic satellite spectra. PhD. Thesis, Monash University (1980); Dyall, K.G., and Larkins, F.P., (unpublished).

26. Coulson, C.A., Momentum Distribution in Molecular Systems, Part I The single bond, Proc.Camb.Phil.Soc. $\underline{37}$, 55 (1941).

27. Coulson C.A., and Duncanson, W.E., Momentum distribution in molecular systems. Part II Carbon and the C-H bond. Proc.Camb.Phil.Soc. $\underline{37}$, 68 (1941)

28. Coulson, C.A., Momentum distribution in molecular systems Part III Bonds of higher order, Proc.Camb.Phil.Soc. $\underline{37}$, 74 (1941).

29. Duncanson, W.E., Momentum distribution in molecular systems Part IV The Hydrogen molecular Ion H_2^+. Proc.Camb.Phil.Soc. $\underline{37}$, 397 (1941).

30. Coulson, C.A., and Duncanson, W.E., Momentum distribution in molecular systems Part V. Momentum distribution and the shape of the compton line for CH_4, C_2H_6, C_2H_4 and C_2H_2 Proc.Camb.Phil.Soc. $\underline{37}$, 407 (1941).

31. Snyder, L.C., and Basch, H., Molecular wavefunctions and properties (Wiley, New York) (1972).

32. McCarthy, I.E., Theory of the (e,2e) reaction on molecules, J.Phys.B.At.Mol.Phys. $\underline{6}$, 2358 (1973).

33. Larkins, F.P., and Wilson, R.M., unpublished.

34. Minchenton, A., Giardini-Guidoni, Weigold, E., Larkins, F.P., and Wilson, R.M., Momentum distributions and separation energies for the valence orbitals of chloromethane. J.Chem.Phys. to be published (1982).

PION PRODUCTION IN NUCLEON-NUCLEON SCATTERING

I.R. Afnan
School of Physical Sciences, The Flinders University of South Australia,
Bedford Park S.A. 5042 Australia

I. INTRODUCTION

Over the past ten years a major effort has gone into the use of pions as a tool for the study of momentum distribution of nucleons in nuclei. This has mainly concentrated on the (p,π) reaction and its time reversed process (i.e. (π,p) reactions). The advantage that pions have over other probes such as electrons, photons, and protons is that in pion production (absorption) we have a minimum momentum transfer of $Q = \sqrt{2m_\pi m_N} \simeq 500$ MeV/c. Thus, pion production (absorption), probes only the high momentum component of the nucleon wave function. Although we now have considerable experimental differential cross section and polarization asymmetry data, there has been little information gained about the momentum distribution of nucleons in nuclei[1]. This is mainly due to our lack of understanding of the reaction mechanism. In particular, does the momentum transfer in the reaction go to one nucleon or to a cluster of two or more nucleons? In the first instance we gain information about single particle wave function, while in the second case we are studying the two or more particle correlations. The early experiments[2] on $(p,\pi^\pm)$ reactions favoured the momentum transfer taking place on a single nucleon in π^+ production making (p,π^+) a potentially useful tool to study single nucleon momentum distribution in nuclei. This was most clearly illustrated by the facts that[2] (i) the (p,π^+) cross section was $\sim$ 50 times larger than the (p,π^-) cross section. (ii) the (p,π^+) angular distribution was forward peaked indicating a direct knockout, which in this case would correspond to the proton emitting a π^+ and the final neutron being captured by the nucleus Fig. 1a. On the other hand, the (p,π^-) cross section is isotropic favouring a higher order process in which the pion gets emitted as a result of the incident proton, interacting with one or more nucleon in the target (see e.g. Fig. 1b).

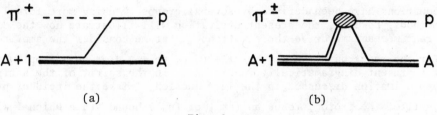

(a)   (b)

Fig. 1

The striking similarity of (d,p) stripping with the mechanism for (p,π^+) reactions (Fig. 1a), where the proton is a n-π^+ bound state,

0940243X/82/860167-16$3.00 Copyright 1982 American Institute of Physics

suggested that pion production can be used to determine the high momentum component of the nucleon wave function and thus complement the information gained from (d,p) stripping. Unfortunately, detailed numerical results were not very successful, because unlike stripping, (p,π^+) is a high momentum transfer reaction and the final results were very sensitive to such things as the input initial and final optical potentials, and the choice of $N \rightarrow \pi N$ vertex[3].

The fact that (p,π^-) is forbidden in the single nucleon mechanism, Fig. 1a, yet the cross section for (p,π^+) and (p,π^-) are comparable at large angle suggests that the two nucleon mechanism Fig. 1b does play a role in (p,π^+) reaction, and should be included in the description of pion production. However, to include both mechanisms requires a certain degree of caution in avoiding problems of double counting when adding the two mechanisms and at the same time including distortion in the initial and final state.

In the present talk, I would like to present a model that will allow me to describe pion production in proton-proton scattering which includes both the single nucleon and the two nucleon mechanism. Furthermore, distortion will be included in both the initial and final state selfconsistently. In this way I avoid the problems of double counting. The final equations, which have been derived in a more rigorous manner[4,5], will describe the reactions

$$p + p \rightarrow p + p \qquad \text{(1-a)}$$

$$p + p \leftrightarrow \pi + d \qquad \text{(1-b)}$$

$$\pi + d \leftrightarrow \pi + d \qquad \text{(1-c)}$$

Although I am predominantly interested in the reaction (1-b), the results for the other two reactions will give a measure of how good the distortion in the initial and final states are. Furthermore, the amplitudes will satisfy two- and three-body unitarity which implies that the effect of the three-body final states is included.

II. THEORY

In the present section I would like to present a simple derivation of a set of equations that will describe the reactions in Eq. (1) starting from the two basic reaction mechanisms discussed above. The final equations have been derived by several groups[4,5] using more formal methods, however, the present derivation will illustrate how the different mechanisms drive the equations. Let me consider the amplitude for the reaction $p + A \rightarrow \pi + (A+1)$ (e.g. $pp \rightarrow \pi^+ d$ or $pn \rightarrow \pi^- pp$) which I represent diagramatically in Fig. 2. In the spirit of the stripping approximation discussed in the introduction, I take the incident projectile, the proton, to be an $(n\pi^+)$ or $(p\pi^0)$ bound state which I will call N'. Then the diagram in Fig. 2 can be approximated by the diagram in Fig. 3a, which in lowest order is given by the diagram in Fig. 3b. This mechanism which corresponds to a single nucleon exchange can in principle describe π^+ production, but gives zero cross

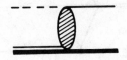

Fig. 2

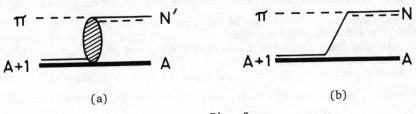

(a)   (b)

Fig. 3

section for π^- production. To get a description of π^- production I
need to go to the next higher order process, which is commonly referred
to as the two nucleon mechanism, and is given by the diagram in Fig. 1b.
However, if I am to include both the one- and two-nucleon mechanism in
my description of (p,π) reactions, I need to replace Fig. 1b by a microscopic model for this mechanism to avoid over counting. In Fig. 4,
I present such a microscopic description in which either the incident
nucleon or one of the target nucleons gets into the Δ state (a P_{33} π-N
resonance) and that decays by pion emission. Note that in the case

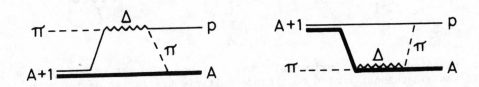

Fig. 4

when A is a nucleon the two diagrams in Fig. 4 are identical and can
be represented by a single diagram in Fig. 5. In this diagram I have
maintained the stripping convention that the nucleon that emits the
pion is a π-N bound state. We will find that this convention of
treating one of the nucleons differently will lead to an undercounting
problem[6] that I will have to resolve if I am to get the correct cross
section. Although the Δ in Figs. 5 and 6 usually refers to the (3,3)
π-N resonance, it can in general correspond to the π-N amplitude in all
channels. In particular, it can include the P_{11} channel for which the

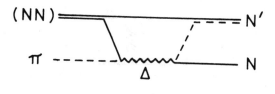

Fig. 5

amplitude has a pole corresponding to the nucleon. This microscopic
two-nucleon mechanism which incorporates one pion-exchange followed by
nucleon-exchange contributes to both the π^+ and π^- production amplitude.

The above two mechanisms combined should describe pion production.
However, in the spirit of the distorted Wave Born Approximation (DWBA)
I should include distortion in both the initial and final states. If
I include distortion in the initial state, I get basically the Koltun
Reitan model[7], whose extension is often referred to as the standard
model for pp → π^+d and pp → πNN, and corresponds to the evaluation of
the diagrams in Fig. 6[8]. The fact that I have not included ρ-exchange
is partly compensated for by having form factors for the πNN and πNΔ
vertices. Furthermore my final results without ρ-exchange are in good

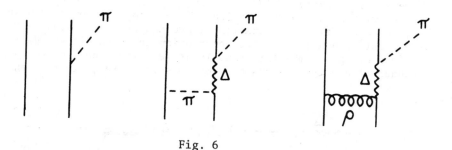

Fig. 6

agreement with the experimental cross section. To get the distortion
in the π-d channel I need to include multiple scattering of the pion
off the deuteron to all orders. Thus the pp → πd reaction in the
DWBA is given by the diagrams in Fig. 7, where in Fig. 7b I have

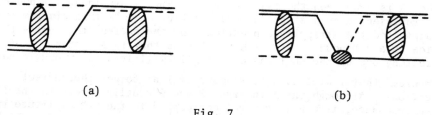

(a)   (b)

Fig. 7

included the full πN amplitude in the intermediate state. In the P_{11}

channel, the π-N amplitude has the nucleon pole allowing me to write
the amplitude as the sum of two terms, a nucleon pole term and a background term as illustrated in Fig. 8, and written more explicitly as

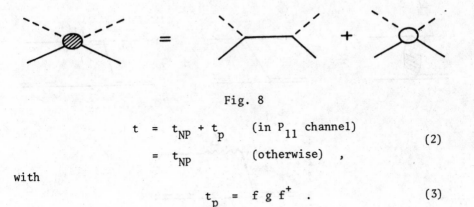

Fig. 8

$$t = t_{NP} + t_p \quad \text{(in } P_{11} \text{ channel)}$$

$$= t_{NP} \quad \text{(otherwise)} ,$$

(2)

with

$$t_p = f \, g \, f^+ .$$

(3)

Here f is the πNN vertex, while g is the nucleon propagator. The above
splitting of the π-N amplitude allows the diagram in Fig. 7b to be
written as a sum of two diagrams (Fig. 9). However, the first diagram

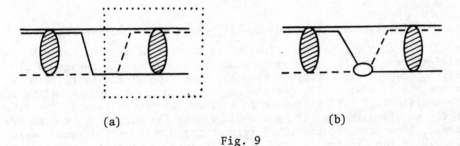

(a)   (b)

Fig. 9

(Fig. 9a) corresponds to N-N distortion followed by one pion exchange
which gives more N-N distortion. In other words, the quantity in the
dotted square in Fig. 9 is nothing more than N-N distortion, and in
that case the diagram in Fig. 9a is identical to that in Fig. 7a.
Thus putting distortion in, as in Fig. 7, leads to double counting,
which should be avoided.

There are two ways of overcoming this double counting problem.
Use the DWBA but for the π-N amplitude employ only t_{NP} i.e. you don't

have intermediate N-N states. This would be the logical path for
(p,π) reactions on heavier nuclei where it is not possible to solve
the many body problem. Alternatively one can try to generate the
distortion in a self consistent manner by deriving an integral equation for the amplitude. In the present case I will follow the second
approach, for the simplest pion production reaction (i.e. pp → πd), and
generate the distortion in both the initial and final states. I will
first consider the amplitude for pp → πd as the sum of the one- and
two-nucleon mechanism with no distortion, as illustrated by the diagrams

172

in Fig. 10. Here one observes that the diagram in the dotted rectangle in the second and third term on the right hand side (r.h.s.) of
the equation in Fig. 10, corresponds to the amplitudes for NN' → πd

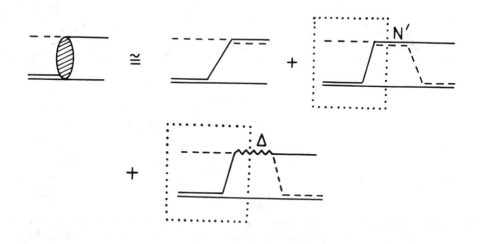

Fig. 10

and NΔ → πd respectively. From this point, the Δ corresponds to t_{NP}
in all π-N channels. The iteration of the second term on the r.h.s.
will generate the NN' → NN' amplitude due to one pion exchange (OPE)
only. However, if I include also the iteration of the third term
on the r.h.s. I will also get the contribution of NN' → NΔ → NN' to the
NN' amplitude. This allows me to replace the second term on the r.h.s.
of the equation in Fig. 10 by the diagram in Fig. 11a. In a similar
manner, I can replace the third term on the r.h.s. of the equation in
Fig. 10 by the diagram in Fig. 11b. In this way I have generated the

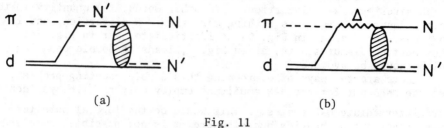

(a)   (b)

Fig. 11

distortion in the N-N channel and I know the mechanism that produces
this distortion is OPE plus the second order potential for NN' → NΔ →
NN'. Furthermore, there is no ρ-exchange in this model which will
result in a weak spin-orbit interaction. I will come back to this
point later when I present the results for this model. I now can
write the equation in Fig. 10 including the effect of distortion in

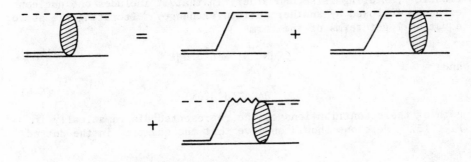

Fig. 12

the nucleon-nucleon channel diagramatically in Fig. 12, and algebraically as

$$X_{dN'} = Z_{dN'} + Z_{dN'}\tau_{N'}X_{N'N'} + Z_{d\Delta}\tau_{\Delta}X_{\Delta N'} \quad . \quad (4)$$

Here $Z_{dN'}$ and $Z_{d\Delta}$ are the potential due to single nucleon exchange, while $\tau_{N'}$ and τ_{Δ} are the propagators for NN' and NΔ intermediate states. This equation basically gives the amplitude for pp $\to$ πd, in terms of the amplitude, for NN' $\to$ NN' and NN' $\to$ NΔ. To make the model completely self consistent, I need to get a similar set of equations for $X_{NN'}$ and $X_{\Delta N'}$. This can be achieved by going through the same procedure developed above for $X_{dN'}$, or alternatively by realising that the above equation is one of the set of three-body equations for the πNN system. The connection with the three-body equations is the result of the assumption that the incident proton can, in the spirit of the stripping approximation, be considered as (π^+-n) bound state. Furthermore, the stripping approximation is based on the observation that (p,π^+) reactions in heavier nuclei are forward peaked. The complete set of equations are of the form

$$X_{dN'} = Z_{dN'} + Z_{dN'}\tau_{N'}X_{N'N'} + Z_{d\Delta}\tau_{\Delta}X_{\Delta N'}$$

$$X_{N'N'} = Z_{N'N'} + Z_{N'N'}\tau_{N'}X_{N'N'} + Z_{N'\Delta}\tau_{\Delta}X_{\Delta N'} + Z_{N'd}\tau_d X_{dN'} \quad (5)$$

$$X_{\Delta N'} = Z_{\Delta N'} + Z_{\Delta N'}\tau_{N'}X_{N'N'} + Z_{\Delta\Delta}\tau_{\Delta}X_{\Delta N'} + Z_{\Delta d}\tau_d X_{dN'}$$

Here $Z_{N'N'}$, $Z_{N'\Delta}$, $Z_{\Delta N'}$ and $Z_{\Delta\Delta}$ are potentials that arise from single pion exchange, $Z_{dN'}$ and $Z_{d\Delta}$ are the potential due to nucleon exchange, while τ_d, τ_{Δ} and $\tau_{N'}$ are the πd, NΔ and NN' propagators in intermediate states.

    As stated before, in the spirit of DWBA, one should include distortion in both the initial and final state. I have already demonstrated how the N-N distortion is included. To demonstrate that the π-d distortions are also included, I need to iterate the above equations. The reason for this is that the lowest order contribution to π-d elastic scattering is second order, in that it includes one nucleon exchange followed by another nucleon exchange. Iterating the above equation I get terms of the form

$$Z_{dN'} \tau_{N'} Z_{N'd} \tau_d X_{dN'} \qquad \text{(6-a)}$$

and

$$Z_{d\Delta} \tau_\Delta Z_{\Delta d} \tau_d X_{dN'} \qquad . \qquad \text{(6-b)}$$

Both of these contributions can be represented diagramatically in Fig. 13. Here one should observe that the quantity in the dotted

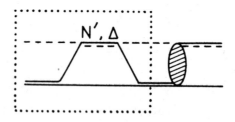

Fig. 13

rectangle is nothing more than the lowest order contribution to π-d elastic scattering, and can be considered as the optical potential in the π-d channel. In solving the full set of integral equations I not only include the full distortion in the π-d channel due to elastic scattering but the effect of real absorption is included because of the coupling to the N-N channel.

The above set of coupled equations, which in principle describes the reactions in Eq. (1)[9], are based on: (i) Including the one- and two-nucleon mechanism on equal footing with no double counting. (ii) Introducing distortion in the initial and final states in a self consistent manner. (iii) Assuming the incident proton is a (π⁺-n) bound state. Although this last assumption allowed me to write a three-body equation for the pp → πd reaction it does lead to several serious under-counting problems which can reduce the magnitude of the calculated cross section by a significant amount. These problems are most clearly ill-ustrated by examining the three-body equations for the πNN system in their most general form i.e.

$$U_{\alpha\beta} = \bar{\delta}_{\alpha\beta} G_o^{-1} + \sum_{\gamma \neq \alpha} t_\gamma G_o U_{\gamma\beta} \qquad \text{(7)}$$

where $\bar{\delta}_{\alpha\beta} = 1 - \delta_{\alpha\beta}$, G_o is the three-body free Green's function while t_γ are the π-N and N-N subsystem two-body amplitude. Let me define t_i, i = 1,2 to be the amplitude for the interaction of the pion with the ith nucleon, while t_3 is the N-N amplitude in the πNN space. Equation (7) is represented diagramatically in Fig. 14. To illustrate the problem with this equation I divide the π-N amplitude into a pole and non-pole part as illustrated in Fig. 8. This allows me to write the second term on the r.h.s. of the equation in Fig. 14 as shown in Fig.

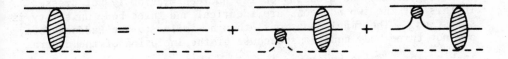

Fig. 14

15.   In the first term on the r.h.s. of the equation in Fig. 15, I
have the pion absorbed on one nucleon and then remitted by the same

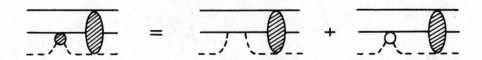

Fig. 15

nucleon.   Furthermore, the nucleon that absorbs the pion gets dressed,
and is labeled as N'.   The problems with the equations as they stand
are:
(a) In intermediate states we have N-N' configurations which should
     be identical nucleons.   However, since N' is dressed we don't
     have identical nucleons and the Pauli exclusion principle is not
     automatically satisfied.   In previous applications of these
     equations[9], the Pauli principle was imposed by proper selection
     of channels in intermediate states.
(b) Both nucleons in intermediate states should be able to emit a
     pion.   This in general is not possible within a three-body model
     and is only overcome by including on the r.h.s. of the equation
     in Fig. 15 the diagram in Fig. 16.

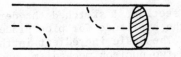

Fig. 16

(c) The fact that N' is dressed while N is not leads to a partial
     violation of three-body unitarity.
All three of these problems can be resolved by introducing the diagram
in Fig. 16 into the equation in Fig. 14.   The contribution of this
diagram is two fold.   It not only allows either nucleon to emit the
pion irrespective of its previous history, but by iterating the equa-

tions I generate the dressing on the nucleon previously not dressed[10]. In this way the two nucleons are identical and three-body unitarity is satisfied. The final equations, though similar to the Faddeev-form, are not three-body equations because of the inclusion of the diagram in Fig. 16. These equations are given by[11]

$$U_{\alpha\beta} = \bar{\delta}_{\alpha\beta}G_0^{-1} + \sum_{\gamma\rho} \bar{\delta}_{\alpha\gamma}B_{\gamma\rho}G_0 U_{\rho\beta} \tag{8}$$

where

$$B = \begin{pmatrix} t_1 & V_{12} & 0 \\ V_{21} & t_2 & 0 \\ 0 & 0 & t_3 \end{pmatrix} \tag{9}$$

with

$$V_{ij} = \bar{\delta}_{ij}f_i g f_j^+ \tag{10}$$

and given diagramatically in Fig. 17.

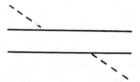

Fig. 17

By restoring the equivalence of the two nucleons I have overcome the approximation that the incident proton only is a $(\pi^+ n)$ bound state which is a nonrelativistic concept that should not strictly be applied to the πNN vertex.

III. RESULTS

In the previous section, I sketched the derivation for a set of equations that give the amplitude for pion production in nucleon-nucleon scattering. The basic ingredient in the derivation was the inclusion of one- and two-nucleon mechanism, and the requirements that the distortion in both the initial and final states be included self-consistently. In this way, I avoided the obvious double counting problem, and at the same time got the amplitude for π-d and N-N elastic scattering. The final equations are one set of coupled integral equations for all the amplitudes for the reactions in Eq. (1). The input to these calculations are the π-N amplitude and the N-N interaction in the πNN space. The dominant π-N amplitudes are the P_{11} and P_{33} partial waves. In particular, the P_{11} plays a central role for pion production in that it determines the πNN vertex. The deuteron (i.e. the $^3S_1$-$^3D_1$ channel) plays a similar role to the P_{11} in the N-N sub system.

It is now an accepted fact that pp → πd is sensitive to the short range behavior of the deuteron wave function because of the high momentum transfer in the reaction[7]. It is therefore essential that the deuteron wave function have the correct short range behavior. What is often not realised is the fact that the πNN vertex plays the role of the wave function for a π-N bound state corresponding to the nucleon, and the above argument for the importance of the short range behavior of the deuteron wave function can be carried over to the πNN vertex. In other words, pp → πd might be a tool to study the off-shell behavior of the πNN vertex. However, before I demonstrate the sensitivity of pion production to the πNN vertex and the success of this model in describing this reaction I should examine the distortion in the initial and final states.

In Fig. 18 I present the results of this model[12] for the singlet

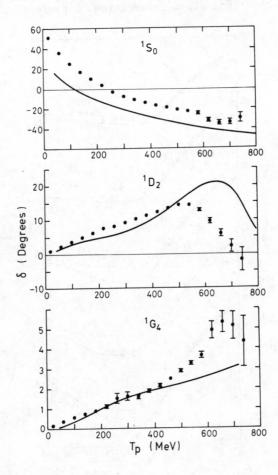

Fig. 18 Singlet p-p phase shifts as predicted by theory. The experimental results are from Ref. 19.

proton-proton phase shifts.   Considering the lack of parameter adjustment, the agreement is good, particularly in the $^1D_2$, which is the dominant amplitude for pp → πd.   Furthermore, a detailed analysis of this
amplitude in terms of argand plot indicates a dibaryon resonance at
E_R = 2160 MeV with a width of Γ/2 = 100 MeV.   This compares favourably
with the suggested experimental value of E_R ~ 2140 to 2170 MeV and Γ/2
of 50 to 100 MeV.   Although the agreement with experiments seems to be
good in the single p-p phase shifts, the situation is not as rosy in
the triplet states.   This might be due to the absence of ρ-exchange
in our N-N scattering, and after all it is the vector exchange that
gives rise to the spin-orbit interaction.   The role of ω- and ρ-
exchange has been investigated by Kloet and Silbar[13] in a similar model
and they find a marked improvement in the agreement with experiment.
    Turning to the final state interaction, I present in Fig. 19 the

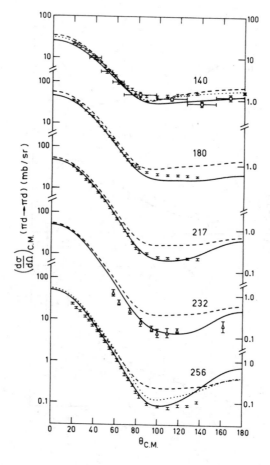

Fig. 19 Differential cross section for πd elastic scattering.
The solid curve is the result for the full calculation
while the dash curves are the result of dropping the
coupling to the N-N channel.

differential cross section for π-d elastic scattering over the energy region of the (3,3) resonance. Here the solid curve is the result of the full calculation, while the dashed curve is the result of the three-body calculation of π-d scattering and does not include the coupling to the N-N channel. I should point out that in general the coupling to the N-N channel improves the agreement with experiment. This is particularly the case at higher energies. Although the present model does give good agreement with the differential cross section, there is a major discrepancy in the polarization cross section indicating that we need more refinement in the present model.

Having established the success of the model in describing π-d and N-N elastic scattering, I would like to turn to pion production. As I have already stated, one expects the pion production to be sensitive to the short range behavior of the deuteron wave function and the πNN vertex. Fortunately one can constrain the short range behavior of the deuteron wave function by examining electron deuteron scattering particularly the tensor polarization data. However, there is no direct experimental constraint on the off-mass shell behavior of the πNN form factor. In the absence of any such constraint I have chosen to adjust the choice of this form factor for the model to fit the pp → πd total cross section at one energy. In Fig. 20, I present

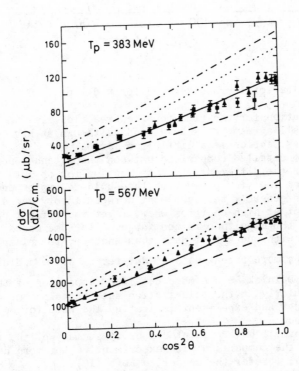

Fig. 20 The sensitivity of the differential cross section for pp → πd to the choice of πNN form factor.

the differential cross section for pp → πd at two different energies
for several choices of the πNN form factor.   Fortunately, the same
form factor gives equally good agreement at low energies as at higher
energies.   With this choice of form factor I present Fig. 21 the

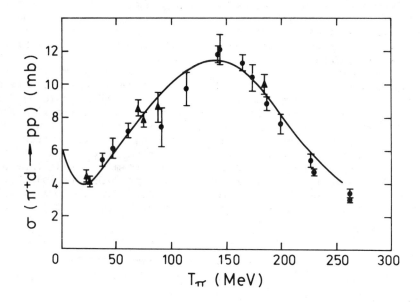

Fig. 21   The total cross section for $\pi^+d \to pp$.

total cross section for the time reversed reaction i.e. πd → pp.   Here
again the overall agreement is very good, although at high energies the
calculated cross section is a little too large.
      For a more detailed comparison between theory and experiment I
have in Fig. 22 the differential cross section for pp → πd for a number
of energies from the pion production threshold to proton energies of
~ 800 MeV.   In general, the agreement between theory and experiment is
very good except at high energies where I get the correct angular distribution, but the magnitude is too large.   In some cases, the agreement is better with one set of data than another, for example at 425.
I agree with the more recent data of Hurster et al.[14] and disagree with
the results of Dolnick[15].   A more detailed comparison of the model with
theory, particularly in the polarization data reveals some discrepancy
between the model and experiment indicating the need for more refine-
ments in the model.
      In the above, I have presented a brief summary of the results for
this model and the comparison with experiment.   For more details of
the calculation, I refer the reader to Ref. (10).
      Although there are several features of the above model that can be
refined, and while I might resolve some of the discrepancy between
theory and experiment, there are two major factors that come out of
the above calculations that can influence the analysis of the (p,π)

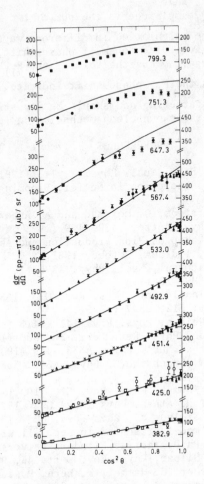

Fig. 22   The differential cross section for pp → π⁺d for
different energies.

data in heavier nuclei:

(i)   The choice of the πNN form factor.   Here the dominant problem
with the πNN vertex in (p,π) reaction has been the non uniqueness
of its non-relativistic reduction.   This can be in principle
overcome by working in a relativistic framework.   The more
serious problem is the sensitivity of pion production to the πNN
form factor and the need to find a means for putting some experimental constraint on at least the range of this form factor.   One
possible theoretical constraint can be the use of the πNN form
factors from bag models of hadrons[16,17].

(ii) In the past, little importance has been given to the inclusion of
the P_{11} interaction in pion scattering of nuclei.   This has been
due to the small phase shifts at low energies in this channel and
the lack of understanding in the dominant mechanism of π-N scattering in this channel and in particular the reason for the change
in the sign of the phase shift.   It is apparent from the present

calculation, that the P_{11} amplitude is the sum of two terms (see Eq. (2)) which enter into the calculation in different ways. Although the full π-N amplitude gives rise to small phase shift, the individual terms in Eq. (2) are large. Thus the role of the P_{11} amplitude is much more important than the P_{11} phase shift indicate. This has been demonstrated for π-d[18] scattering and should be investigated in heavier nuclei, particularly in the two nucleon mechanism in (p,π) reactions.

REFERENCES

1. H.W. Fearing, Prog. Part. Nucl. Phys. 7, 113 (1981); Workshop on Pion Production and Absorption in Nuclei, Bloomington, Indiana, October 1981.
2. S. Dahlgren, P. Grafström, B. Hoistad and A. Åsberg, Nucl. Phys. A204, 53 (1971); Phys. Letters 47B, 439 (1973).
3. J.M. Eisenberg, Workshop on Pion Production and Absorption in Nuclei, Bloomington, Indiana, October 1981.
4. Y. Avishai and T. Mizutani, Nucl. Phys. A326, 352 (1979); A338, 377 (1980); A352, 399 (1981).
5. I.R. Afnan and B. Blankleider, Phys. Rev. C22, 1638 (1980).
6. M.G. Fuda, Phys. Rev. C12, 2097 (1975).
7. D.S. Koltun and A. Reitan, Phys. Rev. 141, 1413 (1966); A.W. Thomas and I.R. Afnan, Phys. Rev. Lett. 26, 906 (1971).
8. J. Chai and D.O. Riska, Nucl. Phys. A338, 349 (1980); O.V. Maxwell, W. Weise and M. Brack, Nucl. Phys. A348, 388 (1980); A348, 429 (1980).
9. I.R. Afnan and A.W. Thomas, Phys. Rev. C10, 109 (1974).
10. B. Blankleider and I.R. Afnan, Phys. Rev. C24, 1572 (1981).
11. I.R. Afnan and A.T. Stelbovics, Phys. Rev. C23, 1384 (1981).
12. The results of this model presented in sec. III are from Ref. (10) and B. Blankleider, Ph.D. Thesis, Flinders University (unpublished).
13. W.M. Kloet and R.R. Silbar, Phys. Rev. Lett. 45, 970 (1980).
14. W. Hürster, Th. Fischer, G. Hammel, K. Kern, P.R. Kettle, M. Kleinschmidt, L. Lehmann, E. Rössle, H. Schmitt and L. Schmitt, Phys. Lett. 91B, 214 (1980).
15. C.L. Dolnick, Nucl. Phys. B22, 461 (1970).
16. S. Theberge, A.W. Thomas and G.A. Miller, Phys. Rev. D22, 2838 (1980).
17. C.E. DeTar, Phys. Rev. D24, 752 (1981); D24, 762 (1981).
18. I.R. Afnan and B. Blankleider, Phys. Lett. 93B, 367 (1980).
19. J. Bystricky, C. Lechanoine and F. Lehar, Saclay Report DPhPE79-01, 1979.

Electron impact ionization at low and intermediate energy. The asymmetric (e,2e)-process

H. Ehrhardt

Department of Physics, University of Kaiserslautern, D-6750 Kaiserslautern West Germany

Summary

The goal of the present paper is the qualitative and quantitative understanding of the electron impact ionization process for impact energies E_0 reaching from threshold (IP) to 500 eV. At low energies IP $\leq E_0 \leq$ 2-3 IP all three scattering amplitudes (direct, momentum exchange, particle exchange) essentially contribute to the scattering and produce rapidly varying triple differential cross sections with respect to scattering angle θ_a and the angle of ejection θ_b of an electron. For intermediate ($3\cdot$IP $\leq E_0 \leq 10\cdot$IP) and for high impact energy ($E_0 \geq 10\cdot$IP) the (e,2e) angular correlation spectrum contains only two peaks, namely the socalled binary and the recoil peak. The angular position and the intensity distribution of the binary peak contains information about the dynamics of the binary collision, the interaction between impinging electron and an atomic electron, and the momentum distribution of the atomic electron, which is ejected by the collision. The recoil peak, which is for intermediate impact energies more intense than the binary peak, is only scarcely understood and seems to be quite sensitive to correlation effects in the scattering system. For the first time a second order scattering approximation has been applied being able to describe quite well intensity, angular position and the shape of the recoil peak for the ionization of helium at E_0 = 500 eV.

In the past a large number of measurements of total electron impact ionization cross sections have been made for many target atoms and molecules.
Those data are of great importance for applications, for example non-equi-
librium gas discharges, terrestrial and extraterrestrial plasma, radiation
chemistry and biology etc., but they reveal only little or none information
on the physics or dynamics of the ionization process itself. Some interesting information has been obtained from single and double differential [1]
cross sections,

$$d\sigma/dE \qquad \text{and} \qquad d^2\sigma/dE\ d\Omega$$

with

$$E_0 = IP + E_a + E_b = \text{const}$$

where E_a and E_b are the energies of the two outgoing electrons after ionization. θ_a and θ_b are the scattering respectively ejection angles with respect to the direction of the incoming electron. For any target system it
is known from single differential cross section determinations, that one
of the outgoing electrons is fast, whereas the other is slow.

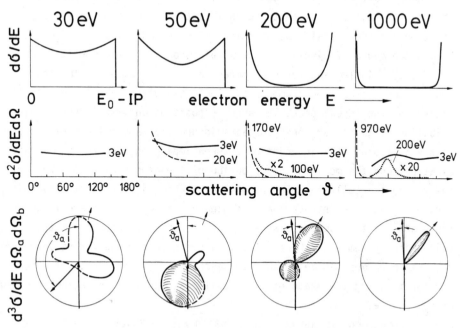

Fig. 1: Schematic diagram of single, double and triple differential cross
sections for electron impact ionization of atoms or molecules. The energy
values at the top of the figure represent the impact energy E_0.

The energy distributions of these two groups are equal, since $E_a + E_b$ = const for a given E_o. The half width of the distribution is only several electron Volt wide and rather independent of the value of E_o, if $E_o \geq 5 \cdot IP$. Furthermore, double differential cross section measurements show, that the slow electrons have a nearly uniform angular distribution, whereas the fast electrons are practically all scattered into a small angular cone in forward direction. Therefore, if (e,2e) reactions are studied with the goal to investigate the electron impact ionization process, extreme asymmetric configurations with respect to angles and energies have to be chosen. Typical values are

> θ_a in the range from 0° to 15°
> θ_b in the range from 30° to 150°
>
>> in both halves of the scattering plane
>
> E_b from 1 eV to 50 eV.

Outside these regions, the contributions to the total cross section are less than approximately 5 to 10% depending somewhat on the target. If the impact energy E_o is lowered below approximately three times ionization energy (binding energy of the atomic electron), then also larger values of θ_a have to be investigated. Close to threshold all θ_a and θ_b values will contribute to the cross section, i.e. $0 \leq \theta_a$, $\theta_b \leq 180°$, if certain angular correlations are allowed for (see below).

In this paper only helium as a target will be discussed, since it is as representative as any other target for the ionization process, but the results will not be masked by low lying electronic excitation states of the remaining ion or by vibrational or rotational effects affecting energy or angular momentum considerations. The process under investigation therefore is

$$e^-(\vec{k}_o) + He(1s^2) \quad He^+(1s) + e^-(\vec{k}_a) + e^-(\vec{k}_b)$$

$\vec{k}_o$, $\vec{k}_a$, $\vec{k}_b$ being the momenta of the electrons involved. The choice of helium should also be an alleviation for theory, since the helium atom as well as the helium ion have only low polarizability, so that in the initial and final scattering channels simple atomic and ionic wavefunctions will be satisfactory.

Another goal, a sideline incentive, of triple differential cross section measurements was to present experimental material for the theoretical treatment of the three particle problem, if all three particles in the final channel are free and in addition have long range forces. Of course, the ionization of hydrogen atoms by structureless particles as electrons, positrons or protons would be a more direct approach, but the expected intensities will be much lower and therefore the errors larger. Asymmetric (e,2e) measurements on H have recently be made by J. Williams [2] and symmetric (e,2e) on H by E. Weigold et al. [3] with the aim to measure the momentum distribution of the bound atomic electron.

Apparatus

Figure 2 shows a schematic diagram of the apparatus used for the measurements of triple differential cross sections.

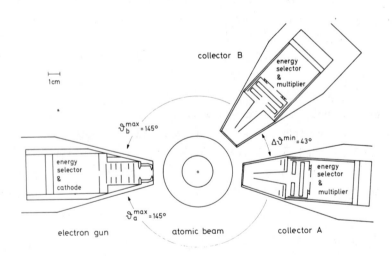

Fig. 2: Diagram of the apparatus used for the measurement of triple differential cross sections.

An electron beam of about $2 \cdot 10^{-7}$ Amps and 25 eV to 500 eV energy is formed in the electron gun and crossed with an atomic beam. The gas pressure inside the beam is about 10^{-2} Torr, outside 10^{-5} Torr, whereas the background pressure (with atomic beam switched off) is approximately 10^{-7} Torr. The collectors A and B are tuned to a certain pair of values E_a and E_b with $E_o - IP = E_a + E_b$. Although in principle only one collector needs to have an energy selecting device, we used two such systems, mainly to reduce background of slow electrons scattered from the walls or diffusing out of the gun and to assure, that only such scattering reactions are recorded, in which the electronic ground state of the helium ion is formed. The count rates in the collectors vary around 10^3 to 10^5, the coincidence rates from 50 - 250 counts per minute. The accidental count rates are varying from 10 - 100 counts per minute depending mainly on the scattering parameters θ_a and E_b. For very small angles θ_a the primary electron beam enters the collector A, but the system in front of the electron optics assures a beam collection efficiency of $10^6:1$, so that only very few stray electrons can enter the electron optical system and reach the multiplier via the energy selector. The time resolution of the coincidence unit is 6 nano seconds.

The measuring procedure is such, that for fixed values E_a, E_b and θ_a collector B is continuously moved in a preset angular range for θ_b and true and false coincidence counts are collected for about one day. Then a new set of fixed parameters E_a, E_b and θ_a is chosen etc. All our measurements have been such, that all momenta $\vec{k}_o$, $\vec{k}_a$, $\vec{k}_b$ are in one plane. Some out-of-plane measurements (4) have been made by Beaty et al.

The triple differential ionization cross section

The first coincidence experiment with the simultaneous determination of all possible (except spin) collision parameters E_o, E_a, E_b, θ_a and θ_b has been made in 1969 by Ehrhardt et al. (5). The results have been surprising (see Fig. 3) and could only be understood qualitatively by comparison with a simple first Born approximation using plane waves for the incoming electron and the two outgoing electrons and simple hydrogenic wave functions for the neutral helium and the helium ion. Figure 3 shows the experimentally obtained data (dots) and the calculated triple differential cross section (full line) in a polar diagram.

188

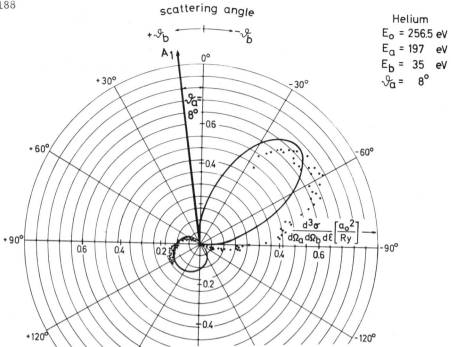

scattering angle

Helium
$E_o = 256.5$ eV
$E_a = 197$ eV
$E_b = 35$ eV
$\vartheta_a = 8°$

Fig. 3: Example of a triple differential cross section for scattering parameters marked in the figure. The dots represent the measurements, the full line represents a theoretical result. A_1 indicates the position of the detector for the measurement of the fast electron with energy E_a.

In this diagram the fast electron ($E_a = 197$ eV) is detected by the collector A under a scattering angle $\theta_a = 8°$ with respect to the direction of the incoming electron ($E_o = 256.5$ eV). The slow electrons ($E_b = 35$ eV) are detected in the angular range of θ_b between $-30°$ and $-90°$ and in the backward direction between $\theta_b = +60°$ and $+125°$. The intensity at each angular position is given by the distance of the experimental point and the origin of the polar diagram. The larger, right hand side lobe is called the binary peak, the somewhat smaller peak in the left half plane is called the recoil peak. The above mentioned simple calculation showed that the binary peak has its origin mainly in the binary electron-electron interaction and the ion being a spectator to the motion of the two outgoing electrons. Furthermore, the shape and especially the width of the binary peak reflects the momentum distribution of the atomic electron before the collision and before its ejection as a result of the collision.

The recoil peak can not be explained by a binary collision, since both outgoing electrons are in the same half plane. For reasons of momentum conservation a third particle, namely the ion has to take up momentum with direction into the right half plane. This has indeed be measured by

McConkey [6]. The collision dynamics therefore must be more complex for electrons which are ejected into the recoil peak than for those, which are ejected into the binary peak. But it should be mentioned already here that some interference occurs between the binary amplitude and the recoil amplitude and that the true nature of one peak can only be observed, if the other peak is vanishing to a high degree. This situation can occur for certain collision parameters.

The binary encounter peak

The binary peak exhibits properties, which can be explained qualitatively and even partly quantitatively by a classical momentum diagram [7] for binary encounters. Figure 4 shows such a diagram.

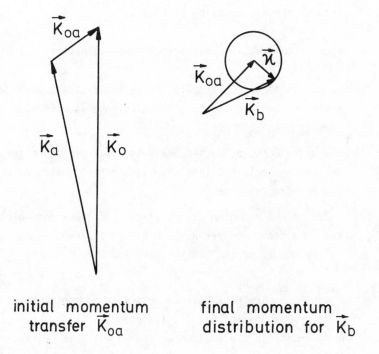

initial momentum
transfer $\vec{K}_{oa}$

final momentum
distribution for $\vec{K}_b$

Fig. 4: Momentum diagram for the binary encounter peak

Since $\vec{k}_o$ and $\vec{k}_a$ are fixed for each experiment, the momentum transfer $\vec{k}_{oa} = \vec{k}_o - \vec{k}_a$ to the atomic electron is preset. In order to make a transition from the ground state of the atom to the ionization continuum, the ejected electron must gain the energy E_b + IP, the absolute value of $\vec{k}_b$ is $|\vec{k}_b| = 2(E_b + IP)^{1/2}$. This value is known from the experiment, since E_b and IP are known. On the other hand, the momentum diagram is only complete if the momentum distribution $\vec{\chi}$ of the atomic electron before its ejection is taken into account. In this way the distribution of $\vec{k}_b$ around $\vec{k}_{oa}$ can be obtained. This we have done for instance for the case of He($1s^2$) and the argon 3p-wave function. In general, the widths and shapes obtained for the binary peak by this procedure fit very well with the experimental results. Some deviations are obtained concerning the angular position of the peak. The experimental peak is always shifted slightly to larger angles with respect to the direction of the momentum transfer vector $\vec{k}_{oa}$. Usually this deviation is larger for small angles θ_a and small E_b and E_a. These observations would be qualitatively in agreement with post collisional correlation between the two outgoing electrons.

Also other experimental results are in accordance with classical principles:

i) because of the inelasticity of the ionization process the angle between the two outgoing electrons is generally smaller than 90°

ii) the angle between the two outgoing electrons increases with increasing angle θ_a. This also is a direct consequence of the momentum diagram (8)

iii) nodes in the momentum distribution $\vec{\chi}$ of the atomic electron due to quantum numbers n, l, m are transformed into minima in the shapes of the binary peaks (9)

The intensity of the binary peak depends very much on the scattering angle θ_a, but relatively much less on E_b. If, for example, θ_a is increased from 4° to 8°, the intensity in the maximum of the binary peak decreases by more than a factor of two for the same value of E_b. It might very well be that low-angle-θ_a-scattering represents distant collisions with small E_b (small momentum transfer $|\vec{k}_{oa}| = |\vec{\chi} + \vec{k}_b|$, $|\vec{k}_b| = 2(E_b + IP)^{1/2}$) with large cross sections, whereas large θ_a indicates close collisions with large momentum transfer and small cross sections. The tendency is clearly in this direction but has to be investigated more closely.

If E_b is increased by a factor of ten (for example from 3 eV to 30 eV), the intensity in the binary peak seems to drop only by about 30%. This number is preliminary, since even relative cross sections for different E_b are difficult to measure and therefore have only the character of order-of-magnitude data.

The formation of the binary peak with respect to the impact energy is well established by now. The following qualitative rules can be presented:

i) For high and intermediate E_o, i.e. $3 - 5 \cdot IP \leqq E_o \leqq \infty$ the binary peak is the dominating feature of the triple differential cross section. This holds for all values θ_a.

ii) If E_b is small, i.e. $1.5 \text{ eV} \leqq E_b \leqq 10 \text{ eV}$, then the contribution of the binary peak with respect to the increasing recoil peak intensity reduces, but always remains slightly dominant.

iii) For high $E_b \geqq 20 \text{ eV}$ the binary peak represents a very high portion of the cross section. The recoil peak practically vanishes.

iv) At low impact energy ($E_o \leqq 3 \cdot IP$) the binary peak gradually vanishes and with it the model of a binary encounter.

The theoretical description [10] of the binary peak goes along the above mentioned lines. At very high energies of both outgoing electrons (and for equal angles $\theta_a = \theta_b \cong 45°$) the impulse approximation gives a very good description and consequently the momentum distribution $\vec{\chi}$ of the target electron can be obtained from such (e,2e) experiments. Also for asymmetric configurations, but high E_o and moderate E_b, any reasonable theory seems to be quite good. Theory fails more and more as the influence of the recoil peak increases, probably because of the increasing interference effects and increasing $e^- - e^-$-correlation of the two outgoing electrons.

The recoil peak

The recoil peak is a scattering feature, which has its origin in the active participation in the collision of all three free particles. The ion acts not anymore as a spectator, but it takes up momentum $\vec{k}_{ion}$ of the order of $|\vec{k}_{oa}|$ to about $2|\vec{k}_{oa}|$ and moves in a direction, which is not too far from the direction of $\vec{k}_{oa}$, mostly into the angular range $-40°$ to $+30°$ around $\vec{k}_{oa}$ (see Fig. 5).

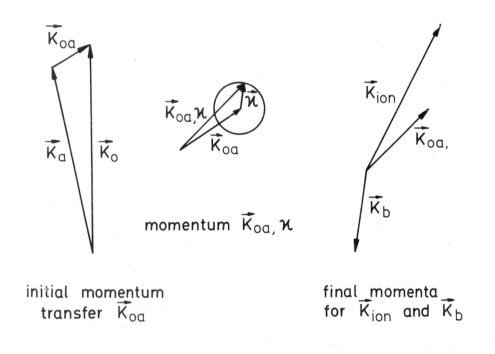

initial momentum
transfer $\vec{K}_{oa}$

momentum $\vec{K}_{oa, \varkappa}$

final momenta
for $\vec{K}_{ion}$ and $\vec{K}_b$

Fig. 5: Momentum diagram for the explanation of the recoil peak

Figure 5 describes only i) in summary the momentum conservation in the recoil collision, ii) which momenta are involved, and iii) which directions and absolute values of the momenta result from a typical recoil collision. From such a vector diagram no conclusions can be drawn with respect to the details or dynamics of a recoil collision. For example, it is interesting to note, that the momentum distribution $\vec{\chi}$ of the target electron before its ejection is not reflected into the intensity distribution of the recoil peak as should be expected from Figure 5. This has been tested extensively for p-electron respectively π-electron ejection from neon and argon respectively molecular nitrogen for many values of the scattering parameters θ_a and E_b. In this cases a node in the target momentum distribution is present, but no indication of such a node is visible in the recoil peak. Also, for the cases of s-electrons (helium) and σ-electrons (molecular hydrogen) the angular width of the recoil peak is much broader than the momentum distribution $\overrightarrow{\chi}$ of the target electron would allow to be. Already these results indicate a more complicated reaction mechanism than suggested from the momentum conservation considerations of Figure 5. We assume that only a multiple scattering mechanism or strong correlations between the three free particles in the exit channel can explain the appearance of the recoil peak. The following experimental results strengthen this assumption:

      a) for high or intermediate impact energy E_0 and all energies of the ejected electron E_b the recoil peak shifts to larger angles θ_b, if θ_a is enlarged, so as if the slow electron shifts away from the fast scattered electron (see Fig. 6)

      b) the recoil peak is large for low E_b (see Fig. 7)

      c) for low E_b there always exists a recoil peak no matter how large the impact energy E_0 may be. Figure 8 shows for $E_0 = 500$ eV and for helium, that 85% of all ionizing collisions contain a not negligible recoil contribution. If E_0 is lowered, than the intensity of the recoil peak increases very much, it exceeds the binary intensity for approximately $E_0 = 80$ eV. The binary intensity finally vanishes practically totally below $E_0 \cong 2 \cdot IP$ and only recoil scattering is present in this energy region

If the impact energy E_0 is around $1.5 \cdot IP$ or lower the patterns of the triple differential cross section can not anymore be described in terms of binary and recoil peaks. This model then is no longer valid. Instead, the particle and momentum exchange amplitudes become important and interfere strongly with the direct scattering amplitude, producing rapidly varying

194

(with varying θ_a) structures in the triple differential cross section (11).

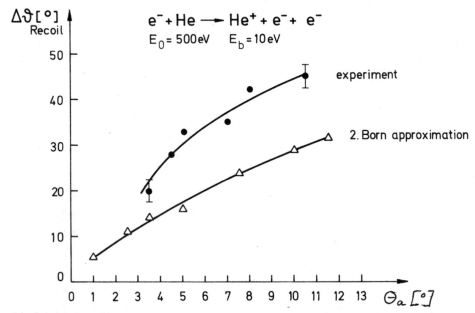

Fig. 6: Deviation of the symmetry axis of the recoil peak from the direction of the momentum transfer vector $\vec{k}_{0a}$ as a function of the angle θ_a.

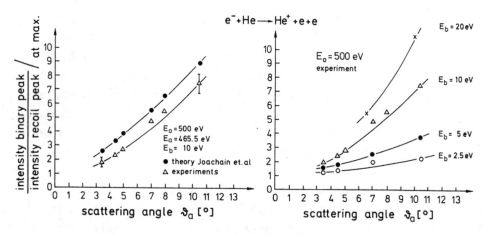

Fig. 7: Ratio of the intensity of the binary peak and intensity of the recoil peak as a function of the scattering angle θ_a for different values of E_b, the energy of the slow ejected electron.

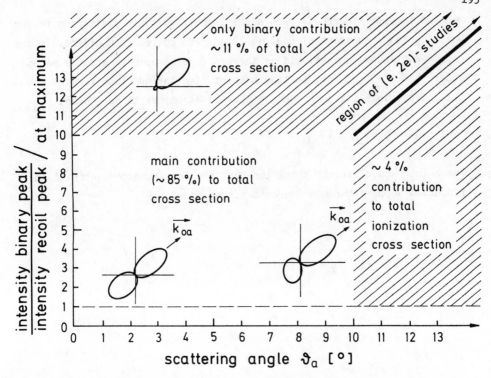

Fig. 8: Same parameter plot as in Figure 7 showing different areas of importance of the electron impact ionisation. The hatched area at the right side of the diagram ($\theta_a > 10°$) contains collisions with large momentum transfer and is difficult to calculate, but its contribution to the total cross section is small. The upper part contains practically only binary collisions unperturbed by recoil scattering. The large majority of ionizing collisions contains strong recoil scattering contributions.

Possibly because of the complicated reaction mechanism which leads to the recoil peak in an ionization process, single collision approximations were not capable of describing this pattern satisfactory, although quite elaborate approaches have been applied. These theories included variational methods, Coulomb waves for the impinging and outgoing electrons, distorted waves, the application of many body Greens functions, etc.

Only very recently, Byron, Joachain, and Piraux [12] used a second order Born approximation for the description of the electron impact ionization of atomic hydrogen and helium [13]. For the first time the calculated triple differential cross sections and their dependencies with E_b and θ_a agreed very well with the experimental results. Although some 20% to 30% deviations to the experimental data still exist, the approach seems to be the right way to describe the more complicated reaction mechanism which leads to the recoil peak. With regard to all observations which have been made up to now, the recoil peak seems to be a multiple scattering feature which contains substantial contributions from correlation effects in the final (continuum) channel. It is therefore reasonable to apply a second order theory. Of course, for the present time the question remains how the intermediate state in such an approximation has to be chosen.

The author would like to thank the Deutsche Forschungsgemeinschaft for financial support through Sonderforschungsbereich 91.

References

1. Ehrhardt, H., Hesselbacher, K.-H., Jung, K., Schulz, M. Tekaat, T. and Willmann, K., Z. Phys. 244, 254 (1971).
    Opal, C.B., Beaty, E.C. and W. K. Peterson, Atomic Data, 4, 209 (1972)
    DuBois, R.D. and Rudd, M.E., Phys. Rev. A 17, 843 (1978)
    Shyn, T.W., Sharp, W.E., Kim, Y.-K., Phys. Rev. A 24, 79 (1981)

2. Williams, J.F., (to be published).

3. Weigold, E., Noble, C.J., Hood, T.S. and Fuss, I., Journ. Phys. B 12, 291 (1979).

4. Beaty, E.C., Hesselbacher, K.-H., Hong, S.P. and Moore, J.H., Phys. Rev. A 17, 1592 (1978).
    Hong, S.P. and Beaty, E.C., Phys. Rev. A 17, 1829 (1978).

5. Ehrhardt, H., Schulz, M., Tekaat, T. and Willmann, K., Phys. Rev. Lett. 22, 89 (1969).

6. McConkey, J., Newell, W.R. and Grove, A., Journ. Phys. B., L 55 (1970)

7. Ehrhardt, H., Hesselbacher, K.H., Jung, K., Schubert, E. and Willmann, K., J. Phys. B 5, 2107 (1972).

8. Ehrhardt, H., Hesselbacher, K.-H., Jung, K., Schulz, M. and Willmann, K., J. Phys. B 5, 2107 (1972).

9. Ehrhardt, H., Jung, K. and Schubert, E., Coherence and Correlation in Atomic Collisions, ed. by Kleinpoppen, H. and Williams, J.F., Plen. Publ. Corp., 1980.

10. Schulz, M., J. Phys. B 4, 1476 (1971).
    Bransden, B.H., Coleman, J.P., J. Phys. B 5, 537 (1972).
    Phillips, D.H. and McDowell, M.R.C., J. Phys. B 6, L 165 (1973).
    Salin, A., J. Phys. B 6, L 34 (1973).

Jacobs, V.L., Phys. Rev. A 10, 499 (1974).

Robb, W.D., Rountree, S.P. and Burnett, T., Phys. Rev. A 11, 1193 (1975).

Baluja, K.L. and Taylor, H.S., J. Phys. B 9, 829 (1976).

Burnett, T., Rountree, S.P. and Doolen, G., Phys. Rev. A 13, 626 (1976).

Madison, D.H., Calhoun, R.V., Shelton, W.N., Phys. Rev. A 16, 552 (1977).

Bransden, B.H., Smith, J.J. and Winters, K.H., J. Phys. B 11, 3095 (1978) and J. Phys. B 12, 1267 (1979).

Smith, J.J., Winters, K.H. and Bransden, B.H., J. Phys. B 12, 1723 (1979).

Balashov, V.V., Grum-Grzhimailo, A.N., Kabachnik, N.M., Magunov, A.I. and Strakhova, S.I., J. Phys. B 12, L 27 (1979).

Tweed, R.J., J. Phys. B 13, 4467 (1980).

11. Ehrhardt, H., Hesselbacher, K.H., Jung, K. and Willmann, K., J. Phys. B 5, 1559 (1972).

Schubert, E., Schnuck, A., Jung, K. and Geltman, S., J. Phys. B 12, 967 (1979).

Schubert, E., Jung, K. and Ehrhardt, H., J. Phys. B 14, 3267 (1981).

12. Byron, F.W., Joachain, C.J. and Piraux, B., J. Phys. B 13, L 673 (1980).

13. Ehrhardt, H., Fischer, M., Jung, K., Byron Jr., F.W., Joachain, C.J. and Piraux, B., (to be published).

HIGH MOMENTUM TRANSFER REACTION MECHANISM
IN NUCLEON-NUCLEUS SCATTERING

H.V. von Geramb

Theoretische Kernphysik, Universität Hamburg
West Germany

and

School of Physics, University of Melbourne, Australia[*]

The experimental and theoretical studies of high momentum components in the nuclear wave functions have opened many new structures in this century. Guided by the famous Rutherford experiment[1], high energy electron and hadron scattering were instrumental in the discovery that hadrons are built out of quarks and gluons[2]. In recent years *classical nuclear physics* extended its projectile energies together with an improvement of the resolution which is required in complex nuclei studies. The medium energy nuclear physics studies aim towards new phenomena associated with mesonic degrees of freedom, properties of metastable superhigh dense nuclear matter and many body physics with relativistic energies. It shall ultimately join nuclear and particle physics results in a unified description.

Low momentum transfer studies of protons and electrons from nuclei have been extensively employed to establish nuclear shapes, optical model potentials and transition potentials to low lying excited states[3]. The underlying reaction picture often assumes quasi free nucleons within the nucleus and the processes sample, closely, within the kinematically allowed NN interaction. As shown in Fig. 1, these methods are useful for studies of low momentum components in the intrinsic nucleon motion in nuclei. Backward scattering is forbidden in free NN collisions. On nuclei, it occurs only because of the *internal momentum* of the nucleons and *multiple scattering mechanism* of various kinds.

0094-243X/860199-12$3.00 Copyright 1982 American Institute of Physics

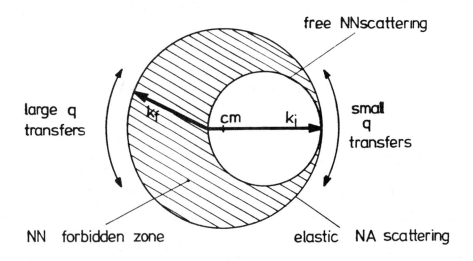

free NNscattering

large q transfers

cm

small q transfers

NN forbidden zone

elastic NA scattering

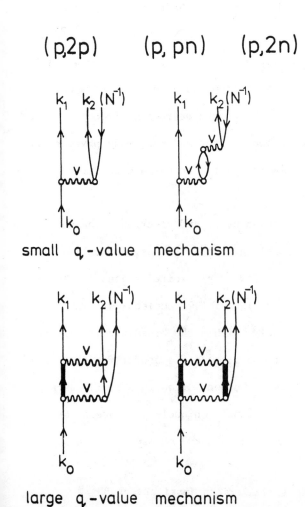

(p,2p)   (p,pn)   (p,2n)

small q-value mechanism

large q-value mechanism

Figure 1: Graphical representation of kinematic regions for small and large momentum transfers. The lower half of the figure shows the dominant dynamics of one and two step processes involving (a) only nucleons, (b) N^* excitations. In low energy scattering with projec energies around 20 - 50 MeV the bubbles represent giant resonance excitations with teir subsequent decay.

Frankel et al.[4] have reported inclusive scattering experiments with 600 and 800 MeV incident proton energies on light and medium heavy targets. They detected high energy ejectiles, protons and light ions up to ^{6}Li, at large angles > 100° with some emphasis on 180° data. A consistent analysis assumes direct one step reactions to prevail rather than statistical processes involving distribution of the incident energy among many particles[4,5]. This explanation rests on the model assumption of a *single scattering* requiring that the struck nucleon is moving backward with high virtual momentum before the collision. The free NN scattering amplitude mediates the transition. See Fig. 2. A rapid fall-off with detected proton energy was observed by Frankel et al.[4] which is interpreted as a direct manifestation of the rapidly falling momentum distribution in the nucleus. With a new phenomenological form of the high momentum distribution, consistent for all nuclei and data, the situation appears explained. The analytic forms for the probability density (of finding a nucleon with momentum k in the target ground state), k > 700 MeV/c are (the subscript s refers to the sine-form and c to the cosine-form)

$$n_s(k) = N_s k \gamma_s / \sinh(\gamma_s k)$$

and

$$n_c(k) = N_c / \cosh^2(\gamma_c k)$$

N is a normalization with $\frac{2}{(2\pi)^3} \int \frac{d^3k \ M \ n(k)}{E(k)} = A$ while γ is a momentum scale. These n(k) are functions of k^2, but for large k fall-off like exp $(-\gamma_s k)$ or exp $(-2\gamma_c k)$ respectively. Amado and Woloshyn[5] fitted the parameters and found γ_s = 2.5 fm, γ_c = 1 fm. Frankel fitted a consistent description

$$n(k) \simeq e^{-k/k_o}$$

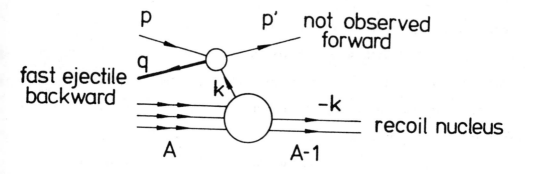

Figure 2

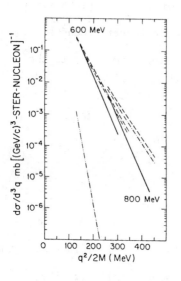

Figure 3: Backward inclusive proton spectrum for 600- and 800-MeV protons on Ta. The solid line is a fit to the data from Frankel et al.. The dashed and dash-dotted lines are calculations based on $n_s(k)$ and $n_c(k)$ respectiv[e]. The -.. line uses the finite temperature Fermi-gas momentum spectrum. This figure is a reproduction of fig.2 in Ref.5.

th k_o^{-1} = 2.46 fm. Fig. 3 shows the Amado Woloshyn result in comparison with

ta from Frankel. Interesting to see is the failure of the finite-temperature

rmi-gas distribution. We see that the exponential momentum distribution

rees qualitatively well with the data, while the Fermi-gas fit is three to

ur orders of magnitude smaller. The study of high energetic inclusive back-

rd scattering is one mean to study high momentum components of nuclear wave

nctions and experimental data are now available up to 400 GeV[2].

At somewhat lower energies, the medium energy field of nuclear

ysics, the situation of high momentum transfers are often much more compli-

ted since the reaction time increases and thermalization becomes more likely.

r present understanding is thereby governed by the explicit inclusion of

ryon resonances on and off the energy shell. Experimentally and theoreti-

lly we approach this field from various directions. A great deal is known

om meson-nucleon and meson-nuclear scattering about the on-shell formation

d decay of baryons. At least a semi-quantitative understanding of the

derlying dynamics can be set forward. Lower energy phenomena deal with the

me isobars but as virtual constituents. It was about 10 years ago that N*

mponents of nuclei were seriously suggested in explainations of high momentum

ta. Complete reviews on these developments are available from Kisslinger[6],

een[7] and Weber-Arenhövel[8]. I'm a novice in this field and have only

cently been faced with it.

Gross sections and analyzing powers of elastic proton scattering

om ^{12}C have been measured recently at Indiana University Cyclotron Facility[9-12].

The data with 495, 570 and 645 MeV/c incident protons momenta cover

angular region from about 5° to 160° and constitute an experimental study

momentum transfers up to 1200 MeV/c. It was natural to look for an optical

del potential which reproduces these data. Various attempts were made but only partial success was achieved since the optical model reproduced only small momentum transfers, < 800 MeV/c and required the inclusion of modified radial form factors[9]. These modifications were not arbitrary, since their inspiration resulted from microscopic calculations[13] and consequences of relativistic formulations[14,15]. Namely, the transformation of a Dirac equation into a Schroedinger equation suggests that the commonly used Woods-Saxon form factors used to be modified. The change alters the momentum transfer behaviour, but not sufficiently. At this point one may hesitate to proclaim the failure of the phenomenological or microscopic optical potential which knows only of nucleons, other improvements may result from better treatments of the NN interaction and/or better treatments of the nonlocality resulting from microscopic many body treatments of the optical potential. After trials in various directions we are now convinced that the answer does not lie within *standard treatments*. With projectile energies around pion threshold it is known that the nucleus should not be considered only as an assembly of nucleons interacting through a two body force. It is certain that baryon resonances as well as mesons exist in nuclei in addition to nucleons.

We approach the analyses by a synthesis of two parallel fields[9]. The first is concerned with a full microscopic treatment of the optical model potential within the nuclear matter approach[13]. It is based on a free NN potential (Hamada-Johnston, Paris) which reproduces practically all NN data below 150 MeV in the two nucleon centre of mass system. This is the subpion threshold region. The second is the explicit incorporation of N* + nucleus doorway states, an approach initiated by Kisslinger[5]. He uses a doorway isobar channel where a *created pion* and a nucleon form an N* baryon, that in particular may be a $\Delta(1236)$. Our synthesis for the purpose of elastic proton scattering consists in treating the nucleus as an assembly of nucleons which have a sizeable coupling to N* intermediate states. This is equivalent to an intermediate pick-up process of a virtual pion into an intermediate baryonic quasi bound state. In a graphical representation this is shown in Fig. 4.

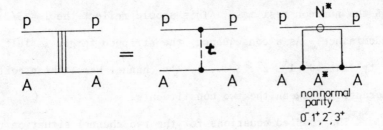

Figure 4

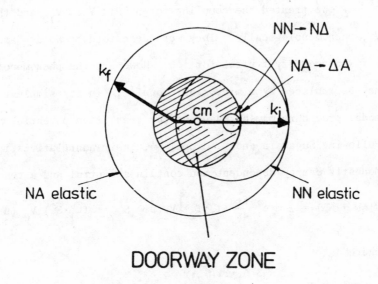

DOORWAY ZONE

Figure 5

We limit the quasi bound particle to spin 3/2, parity (+) and a mass with 1.15 the nucleon rest mass. This should reflect the $I = 3/2^+$ pion nucleo predominance. As a consequence, the stripped target is left in a non no parity state $0^-(1^+, 2^-, 3^+....)$. The channel Hamiltonian contains the ave interaction between the two constituents.

The coupled equations for the two channel situation read

$$Eu_0 - (T_0 + V_{00})u_0 = V_{01} u_1$$

$$(E+Q)u_1 - (T_1 + V_{11})u_1 = V_{10} u_0$$

The value for Q was taken - 140 MeV, equal to the pion rest mass. This i in the spirit of the weakly bound particle model which goes back to the Born Oppenheimer approximation through the original Butler-Peierls stripp model[15].

We treated the coupling potentials V_{01}, V_{10} and the diagonal pote V_{11} phenomenologically. They were described by Woods-Saxon form factors a fixed geometry typical for ^{12}C. However, the phenomenological treatmen may be replaced by a microscopic one[17]. In its simplest form, a folding model procedure may be based on the transition potential $N + N \rightarrow N + \Delta$. Following Sugawara and v. Hippel[16], their nonrelativistic potentials are modestly energy dependent and contain a central and a tensor term.

$$V(NN \rightarrow N\Delta) = \frac{1}{3} m_\pi c^2 \frac{ff}{4\pi} (\tau_1 \cdot \tau_2)\left[V_2(\mu r) S_{12} + (\vec{\sigma}_1 \cdot \vec{S}_2) V_0 (\mu r)\right]$$

where

$$S_{12} = \frac{3(\vec{\sigma}_1 \cdot \vec{r})(\vec{S}_2 \cdot \vec{r})}{r^2} - (\vec{\sigma}_1 \cdot \vec{S}_2)$$

$$V_0(x) = \frac{1}{x} \exp(-x)$$

$$V_2(x) = (1 + \frac{3}{x} + \frac{3}{x^2})V_0(x)$$

f and f* are NNπ and NΔπ coupling constants μ = 0.7 fm, σ,τ are the Pauli spin, isospin operators acting between nucleons and S,T are the spin, isospin transition operators acting between an NΔ. In the higher energy regime of medium energy physics further amplitudes may become important, like NN → ΔΔ and other pion-nucleon resonances. In the above formulae the coupling constant for the NNπ vertex may be taken $f^2/4\pi$ = 0.08. f* is less well known; but values ranging $f*^2/4\pi$ = 0.23 - 0.36 are often used[7]. Unfortunately, the transition potential quoted above has an unpleasant $1/r^3$ singularity (in the tensor term) for small r. Regularization is necessary and several prescriptions are proposed; viz. a short range cut-off $[1 - \exp(-\Lambda r)]$[7]. Green and Niskanen[18] replace the pion propagator

$$q^2/(q^2 + \mu^2) \text{ by } [q^2/(q^2 + \mu^2)][(\Lambda^2 - \mu^2)/(\Lambda^2 + \mu^2)]$$

which has the effect to replace $V_2(\mu r)$ by $V_2(\mu r) - (\Lambda/\mu)^3 V_2(\Lambda r)$ having only a $\frac{1}{r}$ singularity left. At the same time the central potential $V_0(\mu r) \rightarrow V_0(\mu r) - (\Lambda/\mu)^3 V_0(\Lambda r)$ retaining its 1/r singularity. The transition potentials V_{01}, V_{10} derived in a folding model may after all establish only a good quantitative start with the correct selection rules and reasonable strengths. The diagonal potential V_{11} is certainly not as easy to determine microscopically. Some results are available from the Erlangen group[19] which formulates a giant resonance model for *collective* intermediate (Δ-hole) excitation.

We have performed numerical analysis of the 122, 160 and 200 MeV elastic proton data from ^{12}C. The results for the differential cross sections are shown in Figure 6 and Figure 7 shows

the vector analyzing power. In these figures, the dashed lines are obtained by dropping the effects of the mesonic coupling; they represent standard optical model analysis fully microscopically determined. The solid lines were obtained with the unchanged optical model plus the coupling

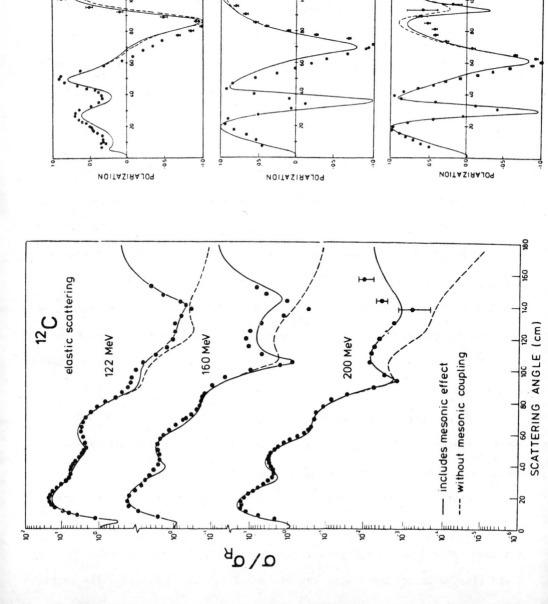

to the N* channel. The latter calculations beyond any doubt, account for the experiment.

Concerning the experimental future we make two remarks. As it became quite apparent from our ^{12}C analysis, the available data are limited to the threshold region where mesonic effects are only marginal. A severe enhancement of intermediate bound state formation, due to the unique πN correlation manifest in the Δ(1236) resonance, will occur with projectile energies ranging 250 - 350 MeV. Since the mesonic effects are essentially a *backangle phenomenon*, data beyond 90° and possibly close to 180° should be taken. The other remark concerns the enrichments of the possible reaction beyond elastic scattering. With minor modifications all what has been said for elastic data applies to inelastic and charge exchange transition. Non normal parity states are most likely privileged.

The author acknowledges efforts put forward by Dr. K.A. Amos and the University of Melbourne to make the overseas travel possible.

REFERENCES

*) Visitor in the period January 15-March 15, 1982.
1. E. Rutherford, Phil. Mag. 21 (1911) 669.
2. L.L. Frankfurt and M.I. Strickman, Phys. Rep. 76 (1981) 215.
3. H. Ueberall, Electron Scattering from complex nuclei, Acad. Press. N.Y. (1971); R.C. Barrett, D.F. Jackson, Nuclear Sizes and Structure, Oxford Clarendon Press (1977); A. Chaumeaux, V. Layly, R. Schaeffer, Phys. Lett. 728 (1977) 33.
4. S. Frankel et al., Phys. Rev. C13 (1976) 737; Phys. Rev. Lett. 36 (1976) 642; Phys. Rev. Lett. 38 (1977) 1338; Phys. Rev. C18 (1978) 1375.
5. R.D. Amado and R.M. Woloshyn, Phys. Rev. Lett. 36 (1976) 1435.
6. L.S. Kisslinger, Ann. Phys. (N.Y.) 97 (1976) 374; Mesons in Nuclei, ed. M. Rho and D. Wilkinson North Holland Publ. (1976); Meson-Nuclear Physics 1979, ed. E.V. Hungerford III. American Institute of Physics (1979).
7. A.M. Green, Pep. Prog. Phys. 39 (1976) 1109.
8. H.J. Weber and H. Arenhövel, Phys. Rep. 36C (1978) 277.
9 H.O. Meyer, P. Schwardt, G.L. Moake and P.P. Singh, Phys. Rev. C23 (1981) 616. H.V. v. Geramb, M. Coz, H.O. Meyer, J.R. Hall, W.W. Jacobs, P. Schwandt, Hamburg preprint (1982).
10. H.O. Meyer, J.R. Hall, W.W. Jacobs, P. Schwandt and P.P. Singh, Phys. Rev. C24 (1981) 1734.
11. J.R. Comfort, G.L. Moake, C.C. Foster, P. Schwandt, C.D. Goodman,

210

J. Rapaport, W.G. Love, Phys. Rev. C24 (1981) 1834.

12. J.R. Comfort, Phys. Rev. C24 (1981) 1844.

13. H.V. von Geramb, F.A. Brieva and J.R. Rook, Microscopic Optical Potentials, Lecture Notes in Physics, Vol. 89 (Springer Verlag N.Y. 1979).

14. M. Jamion, C. Mahaux and P. Rochus, Phys. Rev. Lett. 43 (1979) 1097.

15. L.G. Arnold, B.C. Clark, R.L. Mercer and P. Schwandt, Phys. Rev. C23 (1981) 1979.

16. S.T. Butler, Proc. Roy. Soc. A208 (1951) 559.

17. H. Sugawara and F. von Hippel, Phys. Rev. 172 (1968) 1764.

18. A.M. Green, J.A. Niskanen, Nucl. Phys. A249 (1975) 493, Nucl. Phys. A271 (1976) 503.

19. M. Dillig, H.M. Hoffman and K. Klingenbeck, Meson-Nuclear Physics 1979, ed. E.V. Hungerford III, AIP-conference proc. No. 54 American Inst. Physics 1979.

ADEQUACY OF THE PLANE WAVE APPROXIMATION
IN THE ANALYSIS OF COMPTON PROFILES

Lawrence B. Mendelsohn

Department of Physics, Brooklyn College of CUNY, Brooklyn, N.Y. 11210

ABSTRACT

The errors incurred in Compton scattering calculations by using the "Impulse Approximation" are evaluated. Distorted waves for the ejected electron are calculated in both the Hydrogenic and Hartree-Slater wavefunction approximations, the former analytically and the latter via extensive numerical calculations of the partial wave expansion. Corrections to "Impulse" include the appearance of a Compton defect, Compton profile asymmetry and secondary maxima. A characteristic correction parameter which is (binding energy)/ (momentum transfer)^2 is obtained. Recent X-ray and HEEIS experiments demonstrate the existence of these non-impulse effects. Estimates are also made of non-impulse effects (corrections to PWIA) for (e, 2e) coincidence experiments.

INTRODUCTION

In my talk, I will discuss the errors incurred in calculating the Compton scattering cross-section for free atoms within the so-called "Impulse Approximation." The major constituent of this approximation consists of representing the ejected electron by a plane wave.[1] I will discuss the effects of taking a more accurate distorted wave representation for the ejected electron within two distinct wave function approximations – the Hydrogenic[2,3,4] and Hartree Slater.[5,6,7] These effects include the appearance of a Compton defect (the shift of the profile maximum away from q = 0.0), a Compton profile asymmetry ($J(+q) \neq J(-q)$), the appearance of secondary maxima on the high energy photon side of the profile which are related to the nodal structure of the bound state electron wave function, and a correction to impulse at the Compton profile center (q = 0.0) which is proportional to a characteristic parameter – (binding energy)/(momentum transfer)^2. The coefficients C(n,ℓ) of these correction terms depend strongly on the bound state ionized. The C(n,ℓ) are observed to increase rapidly with increasing principal quantum number n and angular momentum quantum number ℓ. Thus the corrections for outer orbitals are larger than one might assume apriori. These non impulse effects have now been observed in a number of incident X-ray[8,9] and incident high energy electron[10,11] Compton scattering experiments.

Since the Compton profile represents an integration of the (e,2e) coincidence cross-section over all solid angles for the "slow ejected electron", it seems reasonable that one can use the aforementioned Compton corrections to make estimates of the magnitude of the error incurred in using the PWIA to calculate (e,2e) coincidence cross-sections. One should keep in mind however that in obtaining our Compton corrections to "Impulse", we have only treated the "slow ejected

electron" with an improved distorted wave calculation. Clearly a definitive calculation of the correction to PWIA for (e,2e) scattering
would have to treat all three electron continuum wave functions in
the same accurate distorted wave approximation. Since these results
are not available, we cite our Compton corrections which suggest that
a PWIA (e,2e) coincidence cross-section calculation for a 400ev incident electron ionizing a Ne 2s or Ar 3p electron will be in error by
about 20%. The (e,2e) coincidence cross-section experiments on these
bound state atomic levels have already been performed[12,13] and have
been found to be in agreement with PWIA calculations (the experiments
are normalized to the PWIA result). Thus if the Compton correction
estimates are applicable to (e,2e) scattering, it appears that a large
absolute error in the PWIA cross-section has only a small effect on
cross-section shape.

I. THE IMPULSE CALCULATION

Within the impulse approximation, the Compton profile J(q) for
spherically symmetric systems is given by

$$J(q) = 2\pi \int_{|q|}^{\infty} |\chi(p)|^2 p\, dp \qquad (1)$$

where q can be thought of as the displacement in wavelength in momentum
units of the point on the profile from the unperturbed profile center
and $\chi(\vec{p})$ is the momentum space wave function. It is clear that the
impulse J(q) is a symmetric monotonic decreasing function of q with
increasing $|q|$ and therefore can never exhibit any asymmetry or Comp-
ton defect. The maximum of the impulse J(q) always occurs at q = 0.

Since outer electrons contribute most prominently to J(q) at
small values of q, Compton scattering represents a powerful technique
to probe these electrons in atoms, molecules and solids. The outer
electrons are of course the ones which determine the chemical properties of substances. Accurate Compton scattering measurements can
serve as a test of the accuracy of atomic, molecular or solid state
wave functions. That is, correlation effects, exchange effects, relativistic effects, and the accuracy of APW or OPW solid state wave
functions can be tested by comparing theory and experiment. However
as many recent experimental investigations have shown, there are profile asymmetries and Compton defects present when very accurate Compton data is obtained, indicating a violation of the impulse approximation.

The present investigator has published a series of papers showing
that a correct theoretical description of Compton scattering can be
obtained within the Born approximation if binding effects and a more
accurate wave function for the ionized electron, than plane wave, are
included in the theory. We will discuss the theory in the next sections. These improved calculations inherently lead to corrections to
impulse J(0) values and profile normalization integrals, asymmetries
in the profile, Compton defects, and a low lying secondary profile

maximum on the $-q$ side of the profile for 2s or 3p electrons.

From the foregoing, it follows that before one can use Compton scattering to test for correlation, exchange, relativistic or solid state effects, it is essential to obtain an estimate of the corrections to impulse so as to ascertain their importance in analyzing the experimental results and to make the appropriate corrections when they are called for.

II. BORN APPROXIMATION COMPTON PROFILE CALCULATIONS

The non-relativistic first Born approximation expression for the Compton scattering cross-section can be written[1]

$$\frac{d^2\sigma}{d\Omega dE} = \left(\frac{d\sigma}{d\Omega}\right)_T \frac{E_2}{E_1} \sum_f |\langle \psi_f | \sum_{j=1}^N e^{i\vec{K}\cdot\vec{r}_j} | \psi_i \rangle|^2 \delta(e_f - e_i - (E_1 - E_2))$$

$$= (d\sigma/d\Omega)_T \frac{1}{K} \frac{E_2}{E_1} J(q,E) \tag{2}$$

where $(d\sigma/d\Omega)_T$ is the Thomson cross-section, E_1 and E_2 are the incident and scattered X-ray energies respectively, $E = E_1 - E_2$, e_i and ψ_i are the energy and wave function for the initial bound state, e_f and ψ_f are the energy and wave function corresponding to the final state of the electron in the continuum and K corresponds to the momentum transfer. The delta function guarantees conservation of energy. J, the generalized Compton profile, is no longer a function of the single variable q as was the case in the impulse approximation. Choosing free electron values for ψ_f and e_f and taking e_i as the energy associated with a free but moving electron, one can derive the impulse result.[1]

III. THE EXACT HYDROGENIC METHOD

Taking ψ_i and ψ_f as hydrogenic bound state and continuum wave functions respectively, and working in parabolic coordinates along the lines suggested by F. Bloch,[14] we have obtained <u>analytic results</u> for the cross-section and Compton profile for K-shell and L-shell electrons. We refer to this as the Exact Hydrogenic (EH) method. We note that K-shell results had been obtained earlier by P. Eisenberger and P.M. Platzman[15] who did their calculation in spherical coordinates. We have obtained the leading correction to the impulse J(0) value for the 1s and 2s cases analytically. For the 1s case,

$$\frac{J_{IH}(0)}{J_{EH}(0)} \sim 1 + .145 \frac{|e_{1s}|}{K^2} . \tag{3}$$

For the 2s case,

$$\frac{J_{IH}(0)}{J_{EH}(0)} \sim 1 + .704 \frac{|e_{2s}|}{K^2} . \tag{4}$$

In the above, IH stands for the impulse hydrogenic result. Thus (binding energy)/(momentum transfer)^2 is the relevant parameter in determining corrections to impulse in the profile itself. For s states the correction to the impulse J(0) is negative, the EH J(0) lying below the IH J(0). While no analytic evaluation was done, it was observed in all of our EH p state calculations that just the reverse occurred. That is, the EH J(0) value lies above the IH J(0) for p states. In Fig. 1, we **show** calculated EH and IH 2s electron scattering cross-sections for an incident photon of 20kev scattered through an angle of 90° by a hydrogenic ion with a nuclear charge $Z = 5$. This figure demonstrates the non-impulse effects obtained when a distorted wave calculation is done. Note that the solid line curve (EH) has its maximum occurring at nega-tive q (Compton defect), that it is asymmetric with J(-q) $>$ J(+q) for $|q|$ values out to 1.5 or so, and that IH J(0) $>$ EH J(0) for the s state. In addition we observe that the 2s state exhibits a low lying secondary maximum on the negative q side of the profile. It occurs at the value of q for which the IH profile exhibits a plateau. The position of the secondary maximum in momentum space is simply related to the position of the node in the 2s spatial wave function. Therefore we expect to find such secondary maxima in 3p, 4d,..., Compton profiles. Such secondary maxima will be washed out when looking at the total profile. However the use of high intensity sources in coincidence experiments may make the ob-servation of these low lying secondary maxima feasible. We note that because of the choice of effective charge used in the EH ap-proximation, EH profile results become less accurate as we move away from the region near the profile center. This becomes quite clear when we compare our EH Compton profile calculations to the experiments of Eisenberger[16] on MoK$_\alpha$ and AgK$_\alpha$ X-rays scattering from neon. The EH method shows improved agreement, when compared with the best impulse calculations, with experiment in the region about the profile center. However, the EH calculation becomes in-creasing less accurate as we move out onto the profile wings. Therefore we have gone to a more realistic atomic model in order to get accurate profile results over the entire profile.

IV. EXACT HARTREE-SLATER CORRECTIONS TO IMPULSE

Since the EH approach to the correction of impulse profiles utilizes a screened hydrogenic model for both the ground and con-tinuum states, both electron exchange and correlation are ignored. For our next approximation we have used the atomic central field potential model due to Herman and Skillman.[17] It employs the full Coulomb potential with the Slater local approximation to the non-local Hartree-Fock exchange and the so called Latter r^{-1} tail. In the EHS approximation we take ψ_i as a Hartree-Slater bound state atomic orbital and ψ_f as a Hartree-Slater continuum wave

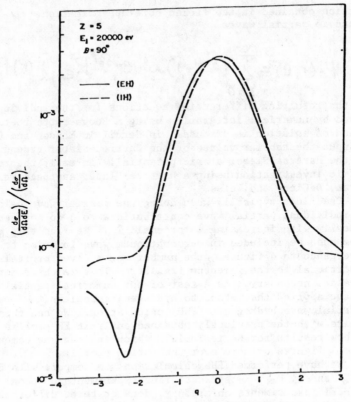

FIG. 1. COMPTON SCATTERING FROM A 2S ELECTRON

	X=-2.6589		X=-5.3643		X=-12.441	
q	EH	PWH	EH	PWH	EH	PWH
-2.5	.0000	.0000	.0000	.0000	.0246	.0245
-2.0			.0654	.0653	.0618	.0617
-1.4	.0000	.0000	.2049	.2046	.1948	.1945
-1.0	.0000	.0000	.4192	.4189	.4028	.4022
-0.7	.6727	.6718	.6588	.6580	.6412	.6405
-0.5	.8314	.8300	.8317	.8309	.8179	.8170
-0.4	.9007	.8995	.9100	.9090	.8999	.8987
-0.3	.9570	.9559	.9759	.9747	.9705	.9691
-0.2	.9961	.9946	1.024	1.023	1.024	1.022
-0.1	1.015	1.013	1.051	1.050	1.057	1.056
0.0	1.011	1.009	1.054	1.053	1.065	1.064
+0.1	.9857	.9841	1.033	1.032	1.049	1.048
+0.2	.9408	.9395	.9894	.9890	1.009	1.007
+0.3	.8798	.8789	.9275	.9271	.9487	.9473
+0.4	.8076	.8065	.8523	.8518	.8739	.8726
+0.5	.7288	.7276	.7689	.7683	.7898	.7886
+0.7	.5686	.5674	.5972	.5969	.6142	.6135
+1.0	.3630	.3626	.3756	.3754	.3846	.3840
+1.4	.1863	.1861	.1870	.1868	.1885	.1882
+2.0	.0682	.0682	.0647	.0646	.0630	.0629
+2.5	.0315	.0315	.0284	.0284	.0267	.0267

Table I. Comparison of AH and PWH Profiles for Helium (Z*(0)=1.59).

function obtained in the frozen core approximation. ψ_f may be expanded in partial waves

$$\psi_f(r,\theta) = \sum_{\ell'=0}^{\infty} (2\ell'+1) i^{\ell'} e^{i\delta_{\ell'}} R_{\epsilon_f,\ell'}(r) P_{\ell'}(\cos\theta) \quad (5)$$

and the resulting differential equations for the individual $R_{ef,\ell'}(r)$ solved by numerical integration using a Runga-Kutta routine. The method of solution is discussed in detail in Manson and Cooper.[8] We input the tabular values of the Hartree-Slater ground state orbital and the Hartree-Slater atomic potential. An early theoretical Compton profile investigation using a Hartree-Slater approach was that of Currat, DeCicco and Weiss.[19]

One finds typically when doing the partial wave calculations that the individual partial wave contributions to J do not decrease monotonically with increasing ℓ for small ℓ. We find that many partial waves must be included in our continuum wave function to obtain profile convergance to 4 figures, the number for a given orgital being roughly proportional to the momentum transfer. In some cases more than 200 waves are necessary. As a test of our numerical partial wave procedures we have applied the method to hydrogenic problems. We refer to this as a partial wave hydrogenic (PWH) calculation. We can then compare our results with the previously obtained analytic EH profile calculations. We show results for He in Table I which indicate agreement of PWH with EH to 3 figures or more over the entire profile.

We have performed EHS calculation[7] to compare with Eisenberger's Mo K_α and Ag K_α X-ray scattering experiments on neon[16] and with the HEEIS experiments on argon[20], done at three different scattering angles (values of momentum transfer). When compared with impulse and EH calculations, EHS results were found to give the closest agreement with the experimental neon profiles. We have also calculated an impulse Hartree-Slater J(0) value of 2.490. Comparing this to the EHS J(0) for the Mo scattering case, we observe that the correction to the impulse J(0) within the Hartree-Slater model is 3%. Within the hydrogenic model the correction was 2.3%, indicating that the magnitude of the correction to impulse is roughly model independent. For the HEEIS experiments on argon, a significant improvement of the EHS profile over impulse Hartree-Fock was observed, the improvement often amounting to a factor of 3 or 4. Thus at a value of q where the IHF (Impulse Hartree-Fock) profile is off by 8% with experiment, the EHS calculated value may only be off by 2%.

We have performed extensive calculations for scattering from krypton as a function of incident photon energy. Table II gives the Compton profile J(0) values for the individual orbitals of krypton at a fixed scattering angle of 170°. Directly under the incident photon energy E, we list the value of the momentum transfer squared χ^2. To the right of each state we give the Hartree-Slater binding energy (B.E.) in atomic units. Included in the 60kev results and to the

Table II. Comparison of Exact Hartree-Slater J(0) with Impulse Hartree-Slater J(0) for the Orbitals and Shells of Krypton; EHS Calculations done for Variable Incident Photon Energy E at θ=170°.

state	B.E.	E(kev) → 17.37	40	60	80	100	120	140	160	IHS J(0)
		χ² → 180	397	841	1414	2099	2879	3745	4688	
				EHS J(0) Values						
2s	135.5		.0724	.0724(-12)	.0803	.0821	.0833	.0842	.0848	.0877
2p	123.3		.0530	.0515(+17)	.0499	.0487	.0478	.0471	.0466	.0441
L-Shell			.463	.464 (+6)	.460	.456	.453	.449	.449	.440
3s	19.73	.184	.206	.213 (-6)	.217	.219	.220	.221	.222	.226
3p	15.23	.154	.150	.146 (+8)	.143	.141	.140	.139	.139	.132
3d	7.098	.106	.0902	.0896(-9)	.0908	.0920	.0931	.0940	.0946	.0984
M-Shell		2.345	2.218	2.197 (-2)	2.202	2.208	2.213	2.219	2.222	2.250
4s	1.946	.622	.650	.659 (-2)						.672
4p	.952	.528	.501	.494 (+2)						.484
N-Shell		4.409	4.308	4.278 (+.7)						4.250
L+M+N			6.99	6.94						6.94

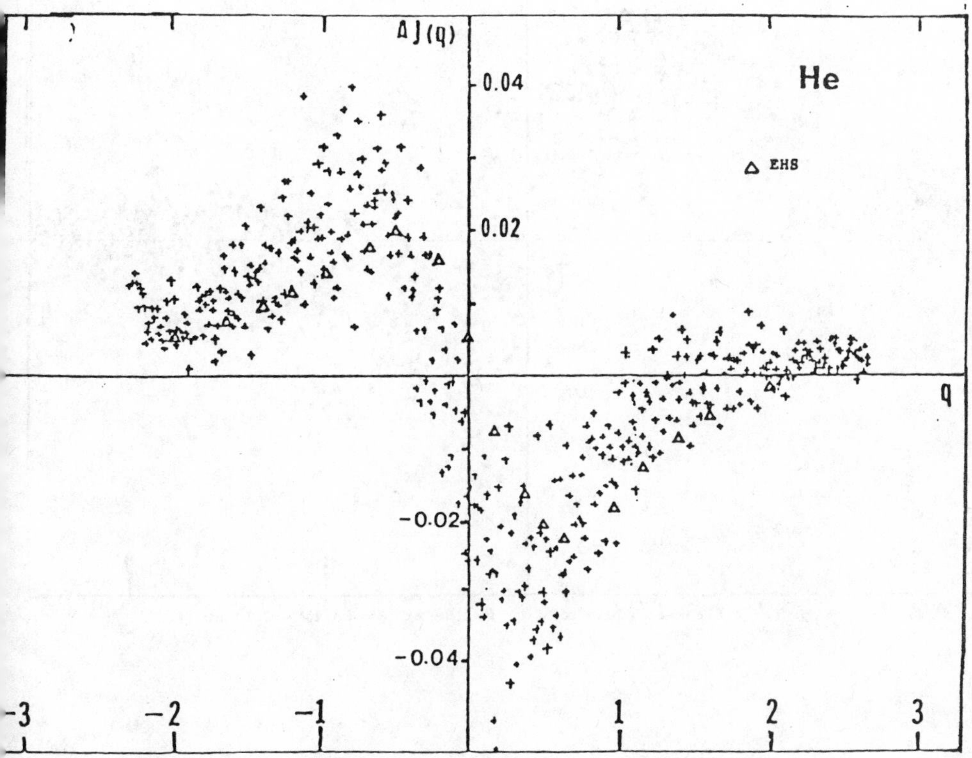

Fig. 2. Difference plots of J(q) for He at θ = 6° (χ = 5.36 a.u.)

Table III. Impulse Correction Coefficients

state	$C(n,\ell)$
2s	+0.8
2p	−1.3
3s	+2.0
3p	−3.7

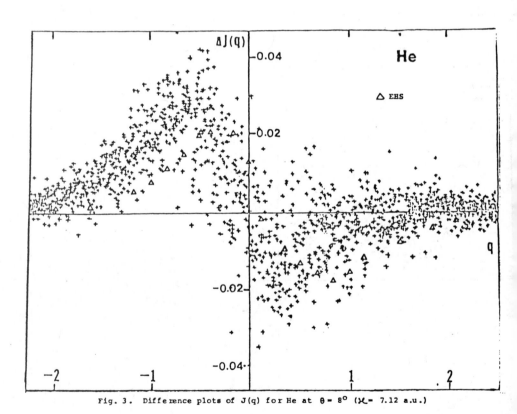

Fig. 3. Difference plots of J(q) for He at $\theta = 8°$ ($\varkappa = 7.12$ a.u.)

right of the J(0) values we give the percentage differences with
the IHS values. The corrections to IHS for the individual orbitals
of the K-shell, L-shell and M-shell are of order 15%, 7%, 2% respectively at this energy. However when one considers the individual
shell contributions, the compensating effect of the IHS J(0) s state
profile values lying too high and the IHS p state profile values
J(0) lying too low are such, that the corrections for the K-shell,
L-shell and M-shell are considerably reduced and are of order 5%,
2%, and .7% respectively. For the closed shell atom itself, the
EHS result for the L+M+N shells appears to have converged to the
impulse result of 6.94. At the lower Mo K_α energy, when only the
M and N shells can be ionized, the IHS J(0) value falls 4% below
the EHS J(0) result for the M+N shells. By analyzing the convergance
of our EHS at high momentum transfers, we obtain constants C(n,ℓ)
appropriate to the Hartree-Slater approximation, which enter the
impulse correction equation with the expected form

$$\frac{J_{IHS}(0)}{J_{EHS}(0)} \sim 1 + C(n,\ell)(B.E./\chi^2). \tag{6}$$

We have obtained approximate values of C by fitting the data in
Table II with Eq. 6. These are given in Table III. Note that the
outer orbitals have significantly larger correction coefficients
than the inner ones. Also our EHS approximate C(2,0) = +.8 is
consistent with our earlier 2s EH result of +.704.

V. COMPARISON OF EHS WITH HEEIS COMPTON ASYMMETRIES AND DEFECTS

H. Wellenstein and coworkers have obtained the experimental
Compton defect for helium at various momentum transfer by determining the center of gravity of the top 80% of the profile for
momentum transfer greater than 3.0 a.u. For smaller values of the
momentum transfer, the defect was obtained utilizing a much smaller
profile region about the center. Analogous defect calculations
have been performed within the EHS approximation and have been
compared with the experimental results. This is shown in Table IV.
Except at the lowest value of momentum transfer, the EHS defects
are negative in sign in agreement with the experimental defects.
The EHS defects also mirror the small experimental defects in
magnitude, giving q in excellent agreement with the experimental
values.

In figures 2 and 3 we show the HEEIS experimental Compton profiles for momentum transfers of 5.36 a.u. and 7.12 a.u. respectively.
Actually these are Compton profile difference plots where the symmetric impulse Compton profile calculated from two electron atomic
wave functions including correlation by Benesch[20] has been subtracted
from the experimental results to amplify the small differences. It
should be pointed out that figures 2 and 3 show the raw data

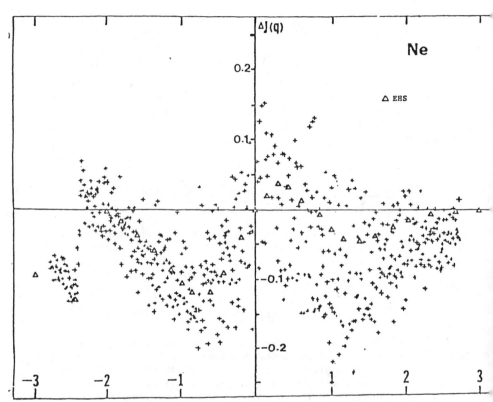

Fig. 4. Difference plots of J(q) for Ne at $\theta = 12°$ ($\mathcal{K} = 10.69$ a.u.)

normalized by the Bethe sum rule without any correction or fitting procedures. Such a procedure makes the original data available to other researchers. The graphs also give the EHS profiles, again with the impulse profile subtracted. The experiments and EHS theory are in reasonable agreement in showing the profile asymmetry with J(-q) larger than J(+q) for all values of momentum transfer. The EHS J(0) values increase with increasing momentum transfer to an impulse Hartree-Slater value of 1.088 at large momentum transfers. As the impulse Hartree-Fock (the reference curve subtracted) calculation gives J(0) = 1.069, the difference value of nearly .02 at q = 0.0 at the larger momentum transfer is readily understandable. Since He is a low Z atom where the use of the approximate Hartree-Slater exchange potential can introduce an error, we find the defect and asymmetry agreement to be quite encouraging. As a futher test of HEEIS and EHS, we have studied neon for 35 keV electrons scattering with a momentum transfer of 10.7 a.u. In Figure 4 we plot the neon results where impulse Hartree-Fock neon profiles have been subtracted from the experiments and the EHS calculations. Both theory and experiment in neon are in excellent agreement and show the asymmetry reversed from the helium case. That is J(+q) is greater than J(-q). We predicted just such a positive defect and asymmetry for rare gases such as neon and argon in our paper[5] presented at the first conference on Momentum Space Wavefunctions held at the University of India ra May 31 - June 4, 1976.

TABLE IV. Comparison of Theoretical (EHS) and Experimental Compton Defects, Δq.

mom transf (κ)	Δq(Exp)[a]	Δq(EHS)[b]	EHS -q→+q range
	Defects		
2.25	+.007	0.00	± .5
2.65	−.018	−0.017	±1.0
3.56	−.028	−0.032	±1.4
4.47	−.027	−0.027	±1.4
5.36	−.028	−0.025	±1.4
7.12	−.019	−0.018	±1.4
8.95	−.020	−0.014	±1.4
10.6	−.019	−0.012	±1.4
12.4	−.016	−0.010	±1.4

a. A. D. Barlas, W. Rueckner, and H. F. Wellenstein, in MOMENTUM WAVE FUNCTIONS, AIP Conf. Proc #36, p. 241 (1977). Results reported give center of gravity of the top 80% of the profile (except for κ<3).

b. EHS results are center of gravity results for the top 80% of the theoretical profile, (except for κ<3). The q range used in the EHS calculations are given in the 4th column.

REFERENCES

1. L.B. Mendelsohn and V.H. Smith, Chapter 5 in COMPTON SCATTERING, edited by B. Williams, McGraw Hill, 1977, p. 102-138.
2. L.B. Mendelsohn and F. Biggs, in Inner Shell Ionization Phenomena and Future Applications, ed. R.W. Fink, S.T. Manson, J.M. Palms and R.V. Rao (U.S. Atomic Energy Commission Conference 720404) Vol. 3, p. 1142-74 (1973).
3. L.B. Mendelsohn, B.J. Bloch, and V.H. Smith, Phys. Rev. Letters 31, 266 (1973).
4. B.J. Bloch and L.B. Mendelsohn, Phys. Rev. A9, 129-155 (1974).
5. H. Grossman and L.B. Mendelsohn, in MOMENTUM WAVE FUNCTIONS, 1976. (AIP Conf. Proc. 36, edited by D.W. Devins), 249-254 (1977).
6. T.C. Wong, L.B. Mendelsohn, H. Grossman, H.F. Wellenstein, Phys. Rev. A26, to be published.
7. H. Grossman and L.B. Mendelsohn, to be published.
8. R.S. Holt, J.L. Dubard, M.J. Cooper, T. Paakkari and S. Manninen, Phil Mag B39, #6, 541 (1979).
9. J.R. Schneider, P. Pattison, H.A. Graf, Phil Mag B38, #2, 144 (1978).
10. W.H. Rueckner, A.D. Barlas and H.F. Wellenstein, Phys. Rev. 18, #3, 895 (1978).
11. A.D. Barlas, W.H. Rueckner and H.F. Wellenstein, J. Phys. B11, #19, 3381 (1978).
12. I.E. McCarthy and E. Weigold, Phys. Reports 27C, 275 (1976).
13. A. Giardini Guidoni, G. Missoni, R. Camillone and G. Stefani, in ELECTRON AND PHOTON INTERACTIONS WITH ATOMS (Plenum Publishing, N.Y. 1976).
14. F. Bloch, Phys. Rev. 45, 674 (1934).
15. P. Eisenberger and P.M. Platzman, Phys. Rev. A2, #2, 415 (1970).
16. P. Eisenberger, Phys. Rev. A5, #2, 628 (1972).
17. F. Herman & S. Skillman, Atomic Structure Calculations (Prentice Hall, Englewood, N.J. 1963).
18. S.T. Manson and J.W. Cooper, Phys. Rev. 165, #1, 126 (1968).
19. R. Currat, P.D. DeCicco and R. Weiss, Phys. Rev. B4, #12, 4256 (1971).
20. T.C. Wong, J.S. Lee, H.F. Wellenstein and R.A. Bonham, Phys. Rev. A12, 1846 (1975).
21. R. Benesch, J. Phys. B: Atom Molec. Phys., 9, 258 (1976).

RESONANCES IN HIGH ENERGY ELECTRON SPECTROSCOPY

R.A. Bonham[†]
School of Physical Sciences, The Flinders University of South Australia,
Bedford Park, South Australia 5042.

ABSTRACT

It has been reported that dipole forbidden transitions can be observed by use of high energy 20-50 keV electron spectroscopy by observation of the angular momentum transfer dependence of the energy loss spectrum. A number of new states in the molecules N_2, O_2, CO and NO were observed at energy losses between 20 and 35eV. These were assigned to bound state excitations originating from inner valence shell orbitals. The experimental energy resolution was ~ 30eV but most energy loss peaks exhibited widths significantly larger. Recently Dillon has observed one of these transitions in O_2 at much lower incident electron energy and an energy resolution of 170 meV. He found a very broad (~ 5eV) band with no observable fine structure. These measurements and the recent theoretical work of Langhoff suggest that these forbidden excitations from inner valence levels may terminate in resonance states of σ_u symmetry. The experimental evidence supporting this suggestion and a complete reassignment of previously observed spectral features is given.

INTRODUCTION

Electron impact spectroscopy as a tool for elucidating the electronic structure of atoms and molecules dates from the Franck-Hertz experiment[1]. Numerous studies have been carried out since that time with the aim of determining the energy level structure of the system being studied[2,3]. In most cases studies have dealt with the targets bound states. This situation is gradually changing however as recent studies have begun to focus on the subject of resonances[4,5]. It is interesting to note that scattering resonances were observed in the early 30's although they were not understood at the time[6]. Now nearly all the analogs to resonance types known in nuclear physics have been documented in the atomic and molecular world[4,5,7].

This work will attempt to focus on a new class of resonances (although known in nuclear physics) which exist in the case of molecules. Recent work has focused on the observation of shape resonances which can be largely described as one electron states imbedded in the continuum[8,9]. Evidence for the existence of such states has been obtained from dipole (e,2e)[10], (e,e+ion)[11], (e,e')[12] and (γ,e)[13] studies on a variety of molecules. All of these studies permit the observation of the excitation of a molecule to a resonance state according to dipole selection rules.

[†] Permanent address: Department of Chemistry, Indiana University, Bloomington, Indiana 47405, U.S.A.
The author wishes to acknowledge support by the National Science Foundation, Grant No. CHE-8000253.

A number of years ago J.S. Lee and co-workers[14,15] reported the existence of a number of new spectroscopic states in the simple molecules N_2, O_2, CO, NO, CO_2 and N_2O. These experiments were of the (e,e') type but differed from the usual energy loss studies in two major respects:

(1) The studies were carried out at incident electron energies of 25keV so that structural information could be extracted within the framework of the first Born approximation;

(2) The spectral studies were carried out as a function of scattering angle to very large momentum transfers (k ≅ 4 a.u.).

It was this second aspect which made these studies unique. The results for a typical case, O_2, are shown in Fig. 1. It will be noted

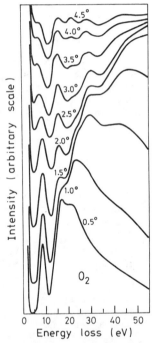

Figure 1. The energy loss spectrum for O_2 obtained with 25 keV electrons over a range of scattering angles for the energy loss region 0 → 50eV.

that a number of new transitions not observed at small momentum transfer become apparent as the momentum transfer is increased. The observed intensity in the framework of the Born approximation can be shown to be proportional to the photo absorption intensity in the limit of zero momentum transfer[16]. Hence the lower diagram at an angle of 0.5° is effectively the same as a photo absorption spectrum with the corresponding selection rules. Presumably what is happening is that dipole allowed excitations, which are known to have intensities which decrease rapidly with angle, according to the first Born approximation[17], disappear with increasing momentum transfer while the intensities of dipole forbidden but quadrupole allowed states increase with increasing momentum transfer (or at least they don't die off as fast). The spectra were obtained with extremely poor energy resolution (2.5eV) necessitated by the lack of scattered intensity

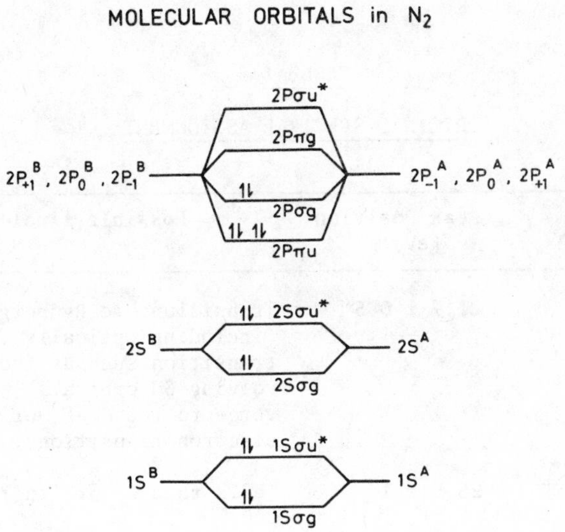

MOLECULAR ORBITALS in N_2

Fig. 2. The standard molecular orbital diagram for molecular nitrogen showing all the molecular orbitals that can be constructed from linear combinations of occupied atomic orbitals.

at high momentum transfers.   In spite of this the breadth of many of the
spectral features exceeds the energy resolution.   It is also apparent
that not all of the features in the spectrum have the same angular dependence.

In the original work by Lee et al. the then known one electron energy
diagram shown in Fig. 2 was employed to propose a tentative assignment of
the observed states.   Note that the molecular orbitals possess cylindrical
symmetry with the σ orbitals being singly degenerate and the π orbitals
doubly degenerate.   These energy levels are obtained by combining all the
occupied atomic orbitals in the atoms B through Ne.   For the molecules N_2

through O_2 the two upper levels will have vacancies in the ground state.

Excitations from filled to unfilled orbitals constitute what are usually
referred to as valence excitations.   Obviously such a simple one electron
picture will have its limitations but it will be interesting to see how far
it will go in providing a simple explanation of the spectrum.   Assignments
were made by considering only dipole forbidden excitations from the occupied
levels to the two lowest unoccupied levels.   Of course in a more exact
picture a complete set of atomic basis set orbitals for each atom in the
molecule must be employed.   This leads to sets of unoccupied molecular
orbitals such as $np\sigma_u$ where $n = 2,3,...$ The transition to the first member

of such a set is called a valence excitation and the remaining are called
excitations to Rydberg states because they generally display an atomic like
series of converging level spacings in the spectrum.   Obviously the picture
is still not complete unless the extended single particle model is improved

by configuration mixing. The results of the first attempt to assign these spectra are given in Table I. An assignment of this type can be easily

TABLE I

ORIGINAL SPECTRAL ASSIGNMENT

	Peak Position (eV)	Possible final states
O_2	21.7 ± 0.5	Transitions to Rydberg states (including optically forbidden transition such as those involving 3d orbitals) which converge to the $C\ ^4\Sigma^-$ of O_2. Two electron transitions.
	28.5 ± 1	($\sigma_g 2s$ to $\pi_g 2p$ orbital) $^3\Pi_g$.
NO	21.3 ± 0.5	Transitions to Rydberg states (including optically forbidden transitions) which converges to the $^3\Pi$ or $^1\Pi$ of NO^+; two electron transitions.
	33 ± 1	(1σ to 2π orbital) $^2\Delta$, $^2\Sigma^-$, $^3\Sigma^+$.
CO	21.7 ± 0.5	Two electron transitions.
	31 ± 1	(1σ to 2π) $^1\Pi$, (1σ to 4σ) $^1\Sigma^+$.
CO_2	21.4 ± 0.5	Two electron transitions.
	34 ± 1	($1\sigma_g$ to $3\sigma_g$) $^1\Sigma_g^+$, ($1\sigma_g$ to $2\pi_g$) $^1\Pi_g$.
NO_2	23 ± 0.5	Two electron transitions.
	33.5 ± 1	(1σ to 3π) $^1\Pi$, (1σ to 5σ) $^1\Sigma^+$, Rydberg transitions originating from the 1σ to 2σ orbitals.

made because of the existence of tables of experimentally determined orbital binding energies[18]. These locate the binding energy of occupied orbitals relative to the single ionization threshold. The location of the unoccupied orbitals must be inferred from bound-bound excitations observed in high resolution photo absorption studies[19]. The location of these orbitals is not always unambiguous. As can be readily observed from the table, a number of the lines could not be assigned to valence excitations from the then known energy level scheme. All the levels assigned to two electron excitations correspond to cases where no single electron excitation seemed possible. Recently photo absorption studies combined with theoretical calculations have shown that the second unoccupied orbital, the $2p\sigma_u$ or $2p\sigma^*$ orbital, appears to lie above the single ionization limit several volts in the continuum[20,21,22]. This means that a new interpretation of Lee's work should be undertaken.

REASSIGNMENT OF THE HIGH MOMENTUM TRANSFER ENERGY LOSS SPECTRA

In Table II the orbital binding energies given by Siegbahn et al.[18] for the molecules N_2, CO, NO and O_2 are listed. These cover all occupied and partially occupied orbitals such as the $2p\pi^*$ and $2p\pi_g$ orbitals in NO and O_2. The first unoccupied orbital in N_2 and CO is located by use of the vertical excitation energy from the high resolution electron impact spectrum[20,21]. The location of the $k\sigma_u$ (k stands for continuum) resonance state is given in the top row of Table II. The positive sign indicates that it lies above the first ionization limit in each case. The set of numbers at the bottom of the top row represent the location of the $k\sigma_u$ resonance in each case as inferred from the calculation and observation of dipole allowed resonance states[22,23,24]. The top numbers were obtained by an empirical fit of the present data. Because NO and O_2 have unpaired electrons in the ground state the binding energy spectrum for some orbitals will exhibit multiplet structure. The splittings involved are less than the energy resolution of the present data so the binding energies in Table II for these cases are given as spin weighted averages of the original numbers given by Siegbahn et al.[18].

By use of Table II it is a simple matter to predict the energy loss spectrum for all possible combinations of valence states as shown in Table III. Comparison with the experimental data of Lee et al. indicates the existence of a state involving only a single electron excitation for every observed spectral feature with the possible exception of a weak band in CO at 18eV. If one assumes that only quadrupole or monopole excitations are allowed at these momentum transfers then about half the bands have possible assignments.

There are however a number of pieces of evidence indicating that dipole allowed bands can contribute to high momentum transfer spectra. Recently Dillon and Spence[25] observed a new Rydberg series in oxygen of the type $2s\sigma_u \rightarrow np\sigma_u$ (n = 3,4,...) beginning at 21.7eV. This is of course a dipole forbidden transition but several previously known dipole allowed series such as the $2p\sigma_g \rightarrow np\sigma_u$ (n = 5,6,...) were clearly observable in their small angle spectrum. They remark in their paper that the feature observed by Lee at 21.7eV is probably unrelated to their $2s\sigma_u \rightarrow 3p\sigma_u$ excitation line because the momentum transfer in the Lee experiment was significantly larger than in theirs. There was no evidence for example of any feature at 28.5eV At our urging Dillon instituted a search for the 28.5eV resonance at larger

TABLE II

BINDING ENERGIES IN SOME FIRST ROW DIATOMIC MOLECULES

Molecule Valence Orbitals	Molecular Binding Energies			
	N_2 (eV)	CO (eV)	NO (eV)	O_2 (eV)
$k\sigma*$	$+13.7^a$ $+11.0^b$	$+9.5^a$ $+10.0^b$	$+8.6^a$ $+5.0^b$	$+1.9^a$ $+2.0^b$
$2p\pi*$	-6.1^c	-5.7^c	-10.0^d	-13.1^d
$2p\sigma^d$	-15.5	-14.5	-17.2^e	-19.6^e
$2p\pi^d$	-16.8	-17.2	-14.0	-17.0
$2s\sigma*^d$	-18.6	-20.1	-22.1^e	-23.8^e
$2s\sigma^d$	-37.3	-38.3	-41.4^e	-40.3^e
Bond^c Length (nm)	0.1094	0.1128	0.1151	0.1207

a) Values inferred from this study. J.S. Lee, T.C. Wong and R.A. Bonham, J. Chem. Phys. 63, 1643 (1975), J.S. Lee, J. Chem. Phys. 67, 3998 (1977).

b) Values suggested by Langhoff and co-workers (P.W. Langhoff, private communications and references 22-24).

c) G. Herzberg "Spectra of Diatomic Molecules" 2nd ed. (Van Nostrand, New York, 1950).

d) K. Siegbahn, C. Nordling, G. Johansson, J. Hedman, P.F. Heden, K. Hamrin, U. Gelius, T. Bergmark, L.O. Werme, R. Manne and Y. Baer, ESCA applied to free molecules (North-Holland, Amsterdam, 1969).

e) These values are spin weighted averages.

Table III Valence Transitions in Diatomic Molecules

Molecule	Transition Energy (eV)	Angular Dependence	Proposed Assignment				
			Energy	2pπ* ← X	Energy	kpσ* ← X	Transition Type
N₂(a)	9.8	A	9.4	$2p\sigma_g$			Q
	13.7	A	10.7	$2p\pi_u$			D
			12.5	$2s\sigma_u$			D
	31.4	A	31.2	$2s\sigma_g$	29.2	$2p\sigma_g$	D
					30.5	$2p\pi_u$	Q
					32.3	$2s\sigma_u$	Q
	Not Observed				51.0	$2s\sigma_g$	D
CO(b)	10	A	8.8	$2p\sigma$			D
	14	A	11.5	$2p\pi$			Q*
	18 (weak)	A	14.4	$2s\sigma^*$			D*
	21.7	B			24	$2p\sigma$	D*
	~26 (shoulder)	(B?)			26.7	$2p\pi$	D
	31.1	A	32.6	$2s\sigma$			Q*
					29.6	$2s\sigma$	Q*
	50	A			51.6	$2s\sigma$	D*
	Not Observed						
NO(b)	8	A	4.0	$2p\pi$			D*
			7.2	$2p\sigma$			D*
	14	A	12.1	$2s\sigma^*$			Q*
	21.7	B			18.6	$2p\pi^*$	Q*
	25.5	B			22.6	$2p\pi$	D*
					25.8	$2p\sigma$	D*
	31.1	A	31.4	$2s\sigma^*$	30.8	$2s\sigma^*$	Q*Q*
	48	A			50.0	$2s\sigma$	D*

Table III (contd.)

Molecule	Transition Energy (eV)	Angular Dependence	Proposed Assignment				Transition Type
			Energy	$2p\pi^* \leftarrow X$	Energy	$kp\sigma^* \leftarrow X$	
O_2(b)	Not Observed						
	7.3	A	3.9	$2p\pi^*_u$	15.0	$2p\pi_g$	D*
	15.5	A	6.5	$2p\sigma_g$	18.9	$2p\pi_u$	Q*
	21.7	B	13.1	$2s\sigma_u$	21.5	$2p\sigma_g$	D*
	28.5				25.2	$2s\sigma_u$	Q*
	(28.95, Dillon (c))	A	27.2	$2s\sigma_g$			D*
	44						Q*
	(43, Dillon (c))	A			42.2	$2s\sigma_g$	D*

(a) J.S. Lee, T.C. Wong and R.A. Bonham, J. Chem. Phys. 63, 1643 (1975).

(b) J.S. Lee, J. Chem. Phys. 67, 3998 (1977).

(c) M. Dillon, Private communication.

angles.   A spectra taken at 200eV incident electron energy at an angle of
18° (K = 1.2 a.u.) revealed the existence of a broad spectral band centered
at 28.9eV[26].   The spectral resolution was about 170meV but the FWHM of the
band was nearly 4eV.   In addition the presence of a very broad band centered at 43eV can be inferred from the spectrum.   What was surprising how-
ever was the fact that the $2s\sigma_u \rightarrow np\sigma_u$ Rydberg series was still prominently
observed.   Hence it seems that Lee was in fact the first one to observe
this spectral feature[15] although in a highly unresolved state.   The most
surprising aspect of the 18° spectrum however was the observation of the
dipole and quadrupole allowed series $2p\sigma_g \rightarrow np\sigma_u$ and $2p\pi_u \rightarrow np\sigma_u$ with very
prominent bands at about 15.8 and 17eV.   The persistence of the dipole ex-
citations to high momentum transfer may be explained by the known behaviour
of the Born series.   In the case of inelastic collisions the scattering
amplitude for a fixed scattering angle tends toward the 2nd Born approximation in the high energy limit[27,28,29].   All dipole and quadrupole transition amplitudes must vanish very strongly with increasing K for sufficiently high K values[17].   On the other hand a multiple scattering process within
the same target involving an inelastic transfer of energy, which always takes
place mainly in the forward direction, followed by an elastic scattering
from the remainder of its surroundings will exhibit a less severe fall off
in K and will display a Rutherford like intensity fall off of the form
$1/k_i K^2$ independently of the transition type.   One can make the statement
that at fixed angle the second Born result is approached in the high energy
limit but at fixed momentum transfer the first Born result is approached in
the high energy limit.   The consequence of this result is that the intensity of all spin allowed excitations will, for sufficiently large momentum
transfer, exhibit the same angular dependence, namely they will all be
proportional to $1/k^2 K^4$.   The problem is that in the absence of detailed
calculations we have no idea at what K value this channel coupling effect
begins to dominate.   On the other hand it has been observed in lower
energy experiments at K values comparable to those encountered here[30].
Hence the continued observation of Rydberg series at high momentum transfer
may simply be a manifestation of channel coupling effects.
     An additional parameter in the Lee experiments is the angular depend-
ence of each spectral feature.   The qualitative behaviour of each line
with angle was arbitrarily classified by comparing it with the angular dependence of the first line in the spectrum.   Lines possessing first line
like angular behaviour were assigned to an A class.   Most lines seemed to
fall into this class with the exception of a few which seemed to increase
rapidly with increasing K relative to A class lines.   These lines were
assigned to a B class category.   In Table IV the transition types from the
simple assignments given in Table III are presented.   In the case of N_2
and O_2, Q and D indicate quadrupole and dipole allowed transitions and for
CO and NO Q* and D* indicate transitions which should be largely quadrupole
or largely dipole dominated.
     The angular dependence should be a very important tool in assigning
the spectra.   The difficulty is that few reliable calculations have been
carried out and these only for bound state excitations[31].

Table IV Possible Classification of Transition Type According to Angular
Behaviour

A		B	
$2s\sigma^* \rightarrow 2p\sigma^*$	(Q*)	$2p\sigma \rightarrow 2p\sigma^*$	(D*)
$2s\sigma \rightarrow 2p\sigma^*$	(D*)	$2p\pi \rightarrow 2p\sigma^*$	(Q*)
$2p\pi^* \rightarrow 2p\sigma^*$	(D*)		

In Table V the known resonance states from the theoretical work of
Langhoff and collaborators[22-24] for dipole spectra are given. In general
all of these resonances match up well with features observed by Lee.
Because of the angular dependence arguments given above and the recent work
of Dillon several possible explanations exist for Lee's spectral features.
These are discussed below on a case by case basis.

(a) Nitrogen

The large angle spectrum of N_2 is the simplest of the three cases.
The first line at 9.34 has been studied previously and the absolute generalized oscillator strength has been determined out to rather high momentum
transfer values[32]. The results of this study are consistent with large
channel coupling effects from the elastic channel at the higher momentum
transfer values encountered in this work.

The 13.7eV line is probably a composite of the dipole allowed $2p\pi_u$
$\rightarrow 2p\pi_g$ at 14.4eV[22] and possibly a $2s\sigma_u \rightarrow 2p\pi_g$ (12.5eV) contribution. In
addition the leading and most intense member of the $2p\sigma_g \rightarrow n\sigma_u$ (13.18eV),
$2p\sigma_g \rightarrow n\pi_u$ (13.20eV), $2p\pi_u \rightarrow n\sigma_g$ (13.34eV), $2p\pi_u \rightarrow n\pi_g$ (14.29eV), $2p\pi_u \rightarrow n\delta_g$
(15.2eV), $2s\sigma_u \rightarrow n\sigma_g$ (15.57eV), $2s\sigma_u \rightarrow n\pi_g$ (13.96eV), $2s\sigma_u \rightarrow nd_\sigma$ (17.1eV)
and $ns\sigma_u \rightarrow ns_\sigma$ (17.3eV) Rydberg series contribute in this region of the
spectrum. There may be other optically forbidden Rydberg series in the
same range.

The band at 31.4eV is also a composite of a $2s\sigma_g \rightarrow 2p\pi_g$ excitation[33]
and the resonance state $2p\sigma_g \rightarrow k\sigma_u$ (28eV)[22] with the possibility that the
optically forbidden resonance transitions $2p\pi_u \rightarrow k\sigma_u$ (30.5eV) and $2s\sigma_u \rightarrow k\sigma_u$
(32.3eV) could also be major contributors.

The predicted optically allowed transition $2s\sigma_g \rightarrow k\sigma_u$ (52eV) is not
observed but this may be because it is beyond 50eV where the experiment
stopped.

(b) Carbon monoxide

The CO spectrum shows greater complexity than N_2. The band at 10eV
is most probably due to the $2p\sigma \rightarrow 2p\pi^*$ (8.8eV) and $2p\pi \rightarrow 2p\pi^*$ (11.5eV)
transitions. In addition the Rydberg series $2p\sigma \rightarrow ns\sigma$ (10.78eV), $2p\sigma \rightarrow np\sigma$
(11.4eV) and $2p\sigma \rightarrow np\pi$ (11.52eV) are all possible contributors.

The band at 14eV may be due to the $2s\sigma^* \rightarrow 2p\pi^*$ (14.4eV) transition with
the Rydberg components $2p\pi \rightarrow n\pi$ (14.07eV), $2p\pi \rightarrow n\sigma$ (13.5eV) and $2p\pi \rightarrow n\delta$
(15.23eV) all possible contributors.

Table V - Known Resonance States in Diatomic Molecules

Molecule	Large Angle Energy	Resonance in the photoionization spectrum	
		Energy	Assignment
N_2(a)	9.8		
	13.7	14.4	$2p\pi_u \rightarrow 2p\pi_g$ (d)
	31.4	28	$2p\sigma_g \rightarrow k\sigma_u$
		52	$2s\sigma_g \rightarrow k\sigma_u$
CO(b)	10		
	14	15	$2p\pi \rightarrow k\sigma^*$
	18		
	21.7	24	$2p\sigma \rightarrow k\sigma^*$
	26		
	31.1	32	$2s\sigma^* \rightarrow k\sigma^*$
	50	52	$2s\sigma \rightarrow k\sigma^*$
NO	8		
	14		
	21.7		
	25.5		
	31.1		
	48		
O_2(c)	7.3	6	$2p\pi_u \rightarrow 2p\pi_g$ $(^3\Sigma_u^-)$
	15.5		
	21.7	21	$2p\sigma_g \rightarrow k\sigma_u$ $(^3\Sigma_u^-)$
		20	$2p\pi_u \rightarrow k\delta_g 3d$ $(^3\Pi_u)$
	28.5	26	$2p\pi_g \rightarrow k\pi_u$ $(^3\Sigma_u^-)$
		26	$2p\pi_g \rightarrow k\delta_u 3d$ $(^3\Pi_u)$
	44	42.2	$2s\sigma_g \rightarrow k\sigma_u$

(a) T.N. Rescigno, C.F. Bender, B.V. McKoy and P.W. Langhoff, J. Chem. Phys. 68, 970 (1978).
(b) N. Padial, G. Csanak, B.V. McKoy and P.W. Langhoff, J. Chem. Phys. 69, 2992 (1978).
(c) A. Gerwer, C. Asaro, B.V. McKoy and P.W. Langhoff, J. Chem. Phys. 72, 713 (1980).
(d) G.R.J. Williams and P.W. Langhoff, Chem. Phys. Letters 78, 21 (1981).

The mystery band at 18eV has no obvious assignment although it could be the remnant of the Rydberg series $2s\sigma \rightarrow ns\sigma$ (18.2eV), $2s\sigma \rightarrow np\sigma$ (17.08 eV) and $2s\sigma \rightarrow n\pi$ (17.3eV).

The 21.7eV band may be the resonance $2p\sigma \rightarrow k\sigma^*$ although Padial et al.[23] find this resonance at 24eV. This would leave the resonance $2p\pi \rightarrow k\sigma^*$ (26.7eV) as the most likely candidate for the possible shoulder at $\sim$ 26eV.

The band at 31.1eV is most likely the $2s\sigma^* \rightarrow k\sigma^*$ (32eV) resonance with a possible contribution from the $2s\sigma \rightarrow 2p\pi^*$ excitation.

The broad band at 50eV is most likely due to the $2s\sigma \rightarrow k\sigma^*$ (52eV) resonance[23].

(c) Nitric Oxide

Detailed calculations are not yet available here but all bands appear to have at least one plausible explanation. The $2p\sigma \rightarrow 2p\pi^*$ (7.2eV) matches the observed feature at 8eV while the $2s\sigma^* \rightarrow 2p\pi^*$ (12.1eV) is not far off the band observed at 14eV. The resonance excitations $2p\pi^* \rightarrow k\sigma^*$ (18.6eV) and $2p\pi \rightarrow k\sigma^*$ (22.6eV) match up fairly well with the peak observed at 21.7 eV. The resonance transition $2p\sigma \rightarrow 2p\sigma^*$ (25.8eV) matches the band at 25.5 eV while the resonance $2s\sigma^* \rightarrow 2p\sigma^*$ (30.8eV) and the bound state excitation $2s\sigma \rightarrow 2p\pi^*$ (31.4eV) match up very well with the band at 31.1eV. The remaining band at 48eV corresponds well with the $2s\sigma \rightarrow 2p\sigma^*$ (50eV) resonance. Rydberg excitations should be concentrated between $5 \rightarrow 10$eV and $14 \rightarrow 20$eV[34].

(d) Oxygen

Oxygen is a slightly simpler system than NO. The 7.3eV band has known contributions[24] at low energy of $2p\sigma_g \rightarrow 2p\pi_g$ (6.5eV) and $2p\pi_u \rightarrow 2p\pi_g$ (6eV). The band at 15.5 has the known small angle Rydberg contributions $2p\pi_u \rightarrow ns\sigma_g$ (12.35eV), $2p\pi_u \rightarrow nd\pi_g$ (14.48eV), $2p\pi_u \rightarrow nd\delta_g$ (14.58eV), $2p\sigma_g \rightarrow np\sigma_u$ (15.28eV) and $2p\sigma_g \rightarrow np\pi_u$ (15.55eV) and one would also expect contributions from the transitions $2p\pi_g \rightarrow k\sigma_u$ (15eV) and $2s\sigma_u \rightarrow 2p\pi_g$ (13.1eV).

It should be pointed out that two Rydberg series starting near 10eV are not observed namely the $2p\pi_g \rightarrow np\pi_u$ (10.5eV) and $2p\pi_g \rightarrow np\sigma_u$ (9.97eV) series.

The band at 21.7eV has $2s\sigma_u \rightarrow np\sigma_u$ (21.7eV), $2p\sigma_g \rightarrow k\sigma_u$ (21eV), $2p\pi_u \rightarrow k\delta_g 3d$ (20eV), along with the $2p\pi_u \rightarrow k\sigma_u$ (18.9eV) as prime candidates. The 28.9eV band has the known resonance possibilities $2p\pi_g \rightarrow k\pi_u$ (26eV), $2p\pi_g \rightarrow k\delta_u 3d$ (26eV) and the suspected $2s\sigma_g \rightarrow 2p\pi_g$ (27.2eV) and $2s\sigma_u \rightarrow k\sigma_u$ (25.2eV) contributors. The broad band at 44eV on the other hand has only the $2s\sigma_g \rightarrow k\sigma_u$ (42eV) as a possibility.

SUMMARY AND DISCUSSION

The recent findings of Langhoff and co-workers[22-24] of resonance states in the photoionization continuum has prompted the reinterpretation of the earlier work by Lee[14,15] in terms of the possibility of excitations to resonance states of $k\sigma_u$ symmetry. The recognised importance of coupling to the elastic channel in all inelastic collisions[27,28] complicates the

reassignment since it cannot be assumed that features at high momentum transfer are due only to dipole forbidden processes.   Recent unpublished work
by Dillon suggests that dipole processes still yield relatively large contributions to the spectrum even at large momentum transfers (K ~ 1 a.u.).
Hence strong lines in Rydberg series, excitations to resonance states and
excitations to bound valence states of every symmetry are all possible candidates for contributions to high momentum transfer spectra.   It is of
course true that higher resolution studies, such as those of Dillon and
Spence[25], can distinguish between Rydberg and resonant excitation.

In the case of the 28.95eV excitation band in O_2 which from Dillon's[26]
work is clearly a resonance, it would seem that the only reasonable assign-
ment would be $2s\sigma_u \rightarrow k\sigma_u$.   From the chemist's point of view it is tempting
to relate this $k\sigma_u$ resonance state to the $2p\sigma_u$ orbital in the molecular
orbital diagram given in Fig. II.   Such a simple-minded view is prompted by
the fact that Dillon and Spence found no evidence of valence state ($2p\sigma_u$)
perturbation on their newly discovered $2s\sigma_u \rightarrow np\sigma_u$ ($n \geqslant 3$) Rydberg series.
This suggests that the valence state must be far removed from the spectral
region occupied by the series.   It seems unlikely that the valence state
lies below the series since the spectrum up to 21.7eV is now well studied
and accounted for[25,34].   Hence the idea that the $2p\sigma_u$ orbital is a bound
state embedded in the continuum at least seems to be a useful construct for
the correlation of spectral data.   The danger in the use of such a simple
model lies in the fact that the continuum can and does exhibit resonances
of other molecular symmetry types such as the $2p\pi_g \rightarrow k\pi_u$ transition at 26eV
in O_2[24].   However in the case of dominant resonances of the same symmetry
as a missing orbital from the n=2 shell it would seem that it would be safe
to utilise the simple picture.   Further, other experiments appear to
correlate with such a picture.   For example the total cross section for
electron scattering by N_2 at low energy shown in Fig. III is dominated by a
resonance at 2.4eV and a broad resonance at 22eV which Dill and Dehmer[36]
have assigned π_g and σ_u symmetry types respectively.   Note that the $2p\pi_g$ -
$k\sigma_u$ spacing is 19.8eV while the π_g - σ_u resonance separation is almost
identical.   This is in keeping with the simple view of fixed one electron
energy levels with one being the $2p\pi_g$ and the other the $2s\sigma_u$.   A similar
situation exists in the valence[22] and K shell regions[37] of the photo absorption cross section for N_2.

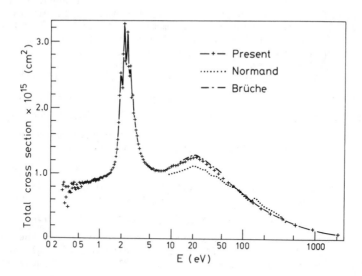

Fig. 3. The absolute total cross section for N_2 determined by Kennerly (Ref. 35). Note the prominent π_g resonance at 2.4eV and the broad σ_u resonance at ~ 22eV.

REFERENCES

1. J. Franck and G. Hertz, Verhandl. deut. physik. Ges. 16, 457; 512 (1914).
2. R.J. Celotta and R.H. Huebner, Electron Spectrosc.: Theory, Tech. Appl. 3, 41 (1979) eds. C.R. Brundle and A.D. Baker (Academic Press, London, 1979).
3. R.A. Bonham, in: Electron Spectroscopy: Theory Tech. Appl. 3, 127 (1979) eds. C.R. Brundle and A.D. Baker (Academic Press, London, 1979).
4. G.J. Schulz, Rev. Mod. Phys. 45, 378 (1973).
5. G.J. Schulz, Rev. Mod. Phys. 45, 423 (1973).
6. E. Brüche, Ann. Phys. (Leipzig) 82, 912 (1927); C.E. Normand, Phys. Rev. 35, 1217 (1930).
7. K. Rohr and F. Linder, J. Phys. B9, 2521 (1976).
8. P.W. Langhoff, in: Electron-molecule and photon-molecule collisions, eds. T.N. Rescigno, B.V. McKoy and B. Schneider (Plenum Press, New York, 1979) pp.183-224.
9. P.W. Langhoff, in: Theory and application of moment methods in many-fermion systems, eds. B.J. Dalton, S.M. Grimes, J.P. Vary and S.A. Williams (Plenum Press, New York, 1980) pp.191-212.
10. A. Hamnet, W. Stoll and C.E. Brion, J. Electron Spec. Rel. Phen.

8, 367 (1976).

11. Th. M. El-Sherbini and M.J. Van der Wiel, Physica 59, 433 (1972).

12. R.E. Kennerly, Phys. Rev. A21, 1876 (1980).

13. D.W. Turner, C. Baker, A.D. Baker and C.R. Brundle, Molecular Photoelectron Spectroscopy (Wiley, New York, 1970).

14. J.S. Lee, T.C. Wong and R.A. Bonham, J. Chem. Phys. 63, 1643 (1975).

15. J.S. Lee, J. Chem. Phys. 67, 3998 (1977).

16. E.N. Lassettre, A. Skerbele and M.A. Dillon, J. Chem. Phys. 50, 1829 (1969).

17. A.R.P. Rau and U. Fano, Phys. Rev. 162, 68 (1967).

18. K. Siegbahn, C. Nordling, G. Johansson, J. Hedman, P.F. Heden, K. Hamrin, U. Gelius, T. Bergmark, L.D. Werme, R. Manne and Y. Baer, ESCA applied to free molecules (North Holland, Amsterdam, 1969).

19. G. Herzberg, Spectra of Diatomic Molecules, 2nd ed. (Van Nostrand, New York, 1950).

20. E.N. Lassettre and M.E. Krasnow, J. Chem. Phys. 40, 1248 (1964); E.N. Lassettre, A. Skerbele, M.A. Dillon and K.J. Ross, J. Chem. Phys. 48, 5066 (1968); A. Skerbele and E.N. Lassettre, J. Chem. Phys. 53, 3806 (1970).

21. E.N. Lassettre and S.M. Silvermann, J. Chem. Phys. 40, 560 (1968); ibid 54, 1597 (1971).

22. T.N. Rescigno, C.F. Bender, B.V. McKoy and P.W. Langhoff, J. Chem. Phys. 68, 970 (1978).

23. N. Padial, G. Csanak, B.V. McKoy and P.W. Langhoff, J. Chem. Phys. 69, 2992 (1978).

24. A. Gerwer, C. Asaro, B.V. McKoy and P.W. Langhoff, J. Chem. Phys. 72, 713 (1980).

25. M.A. Dillon and D. Spence, J. Chem. Phys. 74, 6070 (1981).

26. M.A. Dillon, Private communication.

27. S. Geltman and M.B. Hidalgo, J. Phys. B4, 1299 (1971). M.B. Hidalgo and S. Geltman, J. Phys. B5, 617 (1972).

28. W. Huo, J. Chem. Phys. 57, 4800 (1972).

29. R.A. Bonham, J. Elec. Spec. and Rel. Phen. 3, 85 (1974).

30. W. Eitel and J. Kessler, Phys. Rev. Lett. 24, 1472 (1970).

31. C.W. McCurdy, Jr., and B.V. McKoy, J. Chem. Phys. 61, 2820 (1974).

32. T.C. Wong, J.S. Lee, H.F. Wellenstein and R.A. Bonham, J. Chem. Phys. 63, 1538 (1975).

33. H.L. Hsu and R.M. Pitzer, Private Communication. These authors calculated the transition energy and line width for the $2s\sigma_g \rightarrow 2p\pi_g$ transition to be 31.4eV and 2eV respectively. Interaction with the underlying continuum was neglected.

34. F.R. Gilmore, J. Quant. Spectrosc. Radiat. Transfer 5, 369 (1965).

35. R.E. Kennerly, Phys. Rev. A21, 1876 (1980).

36. D. Dill and J.L. Dehmer, Phys. Rev. A16, 1423 (1977). J.L. Dehmer, J. Siegel, J. Welch and D. Dill, Phys. Rev. A21, 101 (1980). D. Dill, J. Welch, J.L. Dehmer and J. Siegel, Phys. Rev. Lett. 43, 1236 (1979).

37. G.R. Wight, C.E. Brion and M.J. Van der Wiel, J. Electron Spectrosc. 1, 457 (1972/3).

MOMENTUM-SPACE APPROACH TO ELECTRON-ATOM SCATTERING

A.T. Stelbovics

Institute for Atomic Studies, The Flinders University of South Australia,
Bedford Park, S.A. 5042, Australia.

ABSTRACT

The properties of ab-initio momentum-space optical potentials tailored for intermediate-energy electron-atom scattering are discussed. Their use is illustrated in elastic electron-hydrogen scattering.

INTRODUCTION

Traditionally it seems, most of the calculations of electron-atom scattering are carried out in coordinate space where one solves the Schrödinger equation in some approximation and then extracts the scattering amplitude from the scattering wave function. However, as is well known, we can also work in an integral-equation framework. If one defines a T-matrix operator whose value on the energy shell is the scattering amplitude then it is possible to write a Lippmann-Schwinger equation for the T-matrix. The matrix elements of the kernel and inhomogeneous term contain the momentum space potentials. If we further project the T-matrix equation onto a small subset of channels in whose scattering we are most interested in, the channel potentials are folded into a more complicated effective potential which is energy dependent and complex. This "optical" potential is the object we wish to study in some detail.

Historically our interest in the momentum representation of this potential came about through the consideration of the effects of continuum contributions to it. Rather than trying to treat the effect of continuum states by replacing them with a set of discrete pseudo levels we approached the problem directly[1] by using continuum coulomb functions and evaluating their overlaps and performing the integrals in momentum space. This lead us to look at other properties of the optical potential and to establish interesting correspondences with its coordinate-space counterpart. In this workshop talk I will loosely outline the concept of the optical potential and indicate the features of it important in atomic scattering. Then I will discuss its momentum-space properties and illustrate its application in recent calculations.

CONSTRUCTION OF THE OPTICAL POTENTIAL

In order to have a framework upon which to construct equations for T-matrices we will use the close-coupling equations which have proven to be successful in atomic-physics applications[2,3]. The idea behind the equations is to expand the total electron-atom scattering wavefunction in terms of the complete set of target states. To avoid inessential complexity we will write the equations down for electron-atomic hydrogen scattering. Generalisation to arbitrary atoms where target states are described by independent particle wave functions is straightforward[4] but tedious[5].

The Hamiltonian for the electron-atom system is

$$H = K_1 + v_{e_1p} + v_{ee} + H_T. \quad (1)$$

H_T is the target Hamiltonian. In the case of Hydrogen we have the advantage that the target wave functions are known exactly. K_1 is the kinetic energy operator for the incident electron. The target states satisfy the equation

$$(\varepsilon_i - H_T)|\phi_i\rangle = 0 \quad (2)$$

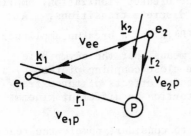

Figure 1. Kinematic notation.

where the wave function $|\phi_i\rangle$ for a general target is properly antisymmetric in electron coordinates. We can also assume to a good approximation that the target is static and centred at the origin of the coordinate system. The total wave function must be antisymmetric as well so if we factor out the spin coordinates the two-electron wave function can either be in a state of total spin s=0 or 1, i.e. singlet or triplet which we denote by + or - respectively. Thus expanding in terms of target states we have

$$\langle 1,2|\psi_n^{\pm}\rangle = \Sigma_i(\langle 1|\phi_i\rangle\langle 2|F_{in}^{\pm}\rangle \pm \langle 2|\phi_i\rangle\langle 1|F_{in}^{\pm}\rangle). \quad (3)$$

This wave function is a solution of the Schrödinger equation with the Hamiltonian of equ. (1). Now if we fold the Schrödinger equation on the left with target states we are left with an effective equation in the coordinates of a single particle, namely

$$\Sigma_i\{(E-\varepsilon_j-K_1)\delta_{ij} - V_{ji}^{\pm}(E)\}|F_{in}^{\pm}\rangle = 0 . \quad (4)$$

The potentials are given by

$$V_{ji}^{\pm} = \langle\phi_{2j}|v_{e_1p}+v_{ee}|\phi_{2i}\rangle \pm \langle\phi_{2j}|\varepsilon_i+\varepsilon_j-E+v_{ee}|\phi_{1i}\rangle$$
$$= V_{ji}^D \pm V_{ji}^E . \quad (5)$$

The subscripts on the target states refer to the electron from which they are constructed. As Sloan and Moore[6] showed the T-matrix equations corresponding to the system of close-coupling equations (4) are

$$T_{ji}^{\pm}(E) = V_{ji}^{\pm}(E) + \Sigma_\ell V_{j\ell}^{\pm}(E)G_{o\ell}(E)T_{\ell i}^{\pm}(E) . \quad (6)$$

The Free Green's function is defined by

$$G_{o\ell}(E) = (E-\varepsilon_\ell-K_1)^{-1} . \quad (7)$$

In momentum space equs. (6) form a set of coupled linear integral equations for the matrix elements $\langle k'|T_{ji}^{\pm}(E)|k\rangle$. The scattering amplitudes for the transitions are formed from these by taking the momenta on the energy shell, i.e. requiring that

$$E = \tfrac{1}{2}k^2 + \varepsilon_i = \tfrac{1}{2}k'^2 + \varepsilon_j . \quad (8)$$

The derivation of the set of equs. (6) is a little cavalier because
the channel indices range over the continuum as well. Physically
speaking this is due to the fact that the breakup (ionization) amplitudes are coupled to the amplitudes for discrete transitions. A
problem arises in interpreting $<\underline{k}'|T^{\pm}_{\underline{k}''i}(E)|\underline{k}>$ as a breakup amplitude.
The long range of the coulomb potential means that the asymptotic
breakup state cannot be described in the close-coupling picture by a
plane wave $<\underline{k}'|$. This is because screening effects of three free-
coulomb particles lead to differing distortion in different kinematic
regimes[7].

However in the application we wish to consider, namely the reaction associated with a few low-lying channels we can project them out
explicitly and fold the continuum-channel effects into optical potentials which we model presently. To treat a set of channels $\phi_o,...,\phi_j$
explicitly we define an operator P such that

$$Pi = i \qquad i \in o,...,j$$
$$= 0 \qquad \text{otherwise} . \qquad (9)$$

The operator P then is a projection and we also define a Q operator by
Q = 1-P. Using standard identities satisfied by these operators[8] one
gets from equ. (6) the equation

$$PT^{\pm}P = Z^{\pm} + Z^{\pm}G_oPT^{\pm}P \qquad (10)$$

where by $PT^{\pm}P$ we mean in an obvious notation the set of $T^{\pm}_{ji}$ such that
$i,j \in P$. The $Z^{\pm}$ are effective potentials over the P-space and are
defined by

$$Z^{\pm} = PV^{\pm}P + PV^{\pm}QG^{\pm}_QQV^{\pm}P , \qquad (11)$$

where $G^{\pm}_Q$ satisfies the equation

$$G^{\pm}_Q = G_oQ + G_oQV^{\pm}QG^{\pm}_Q \qquad (12)$$

which may be used as a basis for generating approximations. All the
complexities of transitions to the continuum are now contained in the
optical potential $Z^{\pm}$ and are embedded in the operator $G^{\pm}_Q$. The key to
providing good optical potentials is to derive sensible approximations
to $G^{\pm}_Q$. To do this we need to establish the features we require the
optical potential to possess. One of the most important features is
a proper description of the large range behaviour. The incident
electron distorts the charge cloud of the target and this gives rise
to a potential tail of the form $-\alpha/r^4$. α is the dipole polarisability
and takes the value 9/2 for hydrogen[9].
One can derive this result from the representation (11). We
choose P to project on the ground state. The first term is the static
exchange potential. The second term can be reduced under some approximation[10] to the $-\alpha/r^4$ form. (The optical potential is non-local so
approximation is always necessary). To illustrate the effect of the

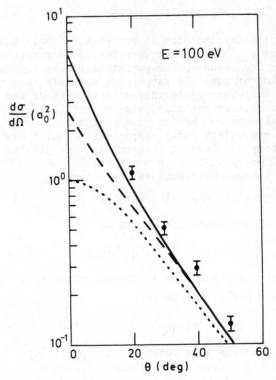

E = 100 eV

$\frac{d\sigma}{d\Omega}$ (a_0^2)

θ (deg)

Figure 2. e-H elastic differential
cross-section.

potential tail we show in fig. 2 the elastic cross-section for the Born approximation (.....), a model which has $\alpha = 9/2$ (———) and one in which the polarisation comes only from the 1s-2p dipole and contributes 66% of the total polarisability (---). If we include all the bound states in the Q-space we account for 81.4% of total polarisability[11]. The remaining strength comes from the continuum. This means that in order to obtain good answers one is forced to include a large number of bound states and to approximate the continuum effects. Two different methods have evolved for this purpose. One developed by P. Burke and co-workers[12,13,14] is to approximate the Q-space by a pseudo-state basis which is tailored to give the correct asymptotic form for the polarization potential. This approach which involves solving the full close-coupling equations becomes very time consuming with increasing energy because one must solve the equations for a large number of partial waves. For example, at 50eV the best calculation of this type[14] is still significantly low in the forward direction.

The second method is based upon solving the elastic-channel equation (10) with the second-order optical potential

$$Z^{\pm} = V^{\pm}_{1s1s} + \Sigma_{q\neq 1s} V^{\pm}_{1sq} G_{oq} V^{\pm}_{q1s} .$$ (13)

This may be regarded as deriving from the Born approximation to equ. (12) and is reasonable at intermediate energies. If one makes a closure approximation (i.e. replaces $G_o(E-\varepsilon_q)$ by $(\varepsilon_o-\varepsilon_c)^{-1}$) the sum in the polarisation term can be performed. The closure energy ε_c is not a free parameter since it is adjusted to yield the correct polarizability. Calculations of this type have proven useful at intermediate energies[15,16] because they do approximate cross-section well in the forward direction. However the shape at backward angles is not reproduced as well.

An improvement to the second-order model may be expected if the closure approximation is not implemented. Recent calculations by our group without resorting to closure have met with success[17]. Before turning to results let us mention the properties of the momentum space potential that we have deduced.

THE EQUIVALENT LOCAL SECOND-ORDER POTENTIAL

Although exchange is explicitly included in our work we shall omit it in the sequel for simplicity. The reader is referred to references 1 and 17 for its discussion. The second-order-potential matrix elements $Z(\underline{q}',\underline{q})$ needed for the solution of equ. (6) range over all (off- and on-shell) values of q' and q. Therefore since the exchange part of the static potential and the polarisation potential are non-local the calculation of Z for each partial wave can become time consuming. For this purpose we introduce an equivalent local approximation, which has proved to give very satisfactory results[1] at intermediate energies, to the second-order potential. We split Z up into its static, and second order discrete and continuum contributions respectively:

$$Z(\underline{q}',\underline{q}) = Z_S(\underline{q}',\underline{q}) + Z_D(\underline{q}',\underline{q}) + Z_C(\underline{q}',\underline{q}) \quad . \tag{14}$$

Then we introduce the following approximation for $Z_{D,C}$:

$$Z_{D,C}(\underline{q}',\underline{q}) \sim Z_{D,C}^{local}(\underline{q}'-\underline{q}) = Z_{D,C}(\underline{q}'-\underline{q}+\underline{k},\underline{k}) \tag{15}$$

where $\underline{k}$ is the on-shell momentum (8) and lies in the direction of $\underline{q}$. $Z_{D,C}^{local}(\underline{q}'-\underline{q})$ is a function of P and u only where

$$\underline{P} = \underline{q}'-\underline{q} , \quad u = \frac{\underline{P}.\underline{k}}{Pk} \tag{16}$$

and we can therefore further simplify this potential by spherically averaging over u, to give the final form of the potential that we generally use in calculations:

$$Z(\underline{q}',\underline{q}) = Z_S(\underline{q}',\underline{q}) + Z_D^{local}(P) + Z_C^{local}(P) \tag{17}$$

with

$$Z_{D,C}^{local}(P) = \tfrac{1}{2}\int_{-1}^{1} du \, Z_{D,C}(\underline{P}+\underline{k},\underline{k}) \quad . \tag{18}$$

The effects of the approximations are illustrated in table I. We have calculated the elastic differential cross-sections using a second-order potential made up from the restricted q-space of 2s and 2p states with and without the equivalent local approximation.

The explicit evaluation of $Z_{D,C}$ in equ. (14) involves computing the following integrals:

$$Z_D(\underline{q}',\underline{q}) = \int d^3\underline{q}'' \Sigma_{m\in Q_D} F_m^*(\underline{q}'',\underline{q}') \frac{1}{E+io-\varepsilon_m-\tfrac{1}{2}q''^2} F_m(\underline{q}'',\underline{q}) \quad , \tag{19a}$$

and

$$Z_C(\underline{q}',\underline{q}) = \int d^3\underline{q}'' \int d^3\underline{p}'' F^*(\underline{q}'',\underline{p}'',\underline{q}') \frac{1}{E+io-\tfrac{1}{2}(q''^2+p''^2)} F(\underline{q}'',\underline{p}'',\underline{q}) \tag{19b}$$

Table I Elastic differential χ-sections for e-H scattering at 100eV.
Second-order potentials using q = 2s,2p with and without
equivalent local approximation.

θ(deg)	1s-2s-2p	1s-2s-2p(local)
5	2.05	1.92
10	1.27	1.24
20	0.723	0.72
40	0.212	0.21
60	0.715-1	0.67-1
80	0.301-1	0.28-1
100	0.163-1	0.15-1
120	0.103-1	0.95-2
140	0.782-2	0.69-2
160	0.653-2	0.57-2
180	0.643-2	0.54-2

where $F_m(\underline{q}'',\underline{q}')$, $F(\underline{q}'',\underline{p}'',\underline{q}')$ are the Born matrix elements for inelastic scattering and ionization respectively. The analytic formulae for their calculations in the e-H problem are given in ref. 17. We notice that the evaluation of the continuum term Z_C in the spherically averaged equivalent local form (18) requires the computation of a 7-dimensional integral. Fortunately the overlaps F_m, $F_{\underline{p}''}$ are relatively smooth functions in momentum space and upon proper treatment of the Green's function singulari-

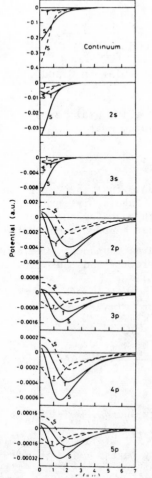

Figure 3. Contributions from various channels to the coordinate-space optical potential. Solid line: imaginary part. Dashed line: real part. Singlet and triplet potentials for discrete excitations (indicated by S and T) must be divided by .25 and .75 respectively.

ties[1,18] the integrals can be computed to better than 1% error.

The imaginary parts of the equivalent local potentials $V_{D,C}(P)$ in the forward direction P=0 are, apart from some kinematic factors, the sum of the Born total cross sections for inelastic scattering and ionization. This can be seen by taking the imaginary (δfn) part of the Green's function in equs. (19) and using equ. (15).

The contributions to the total optical potential at 100eV from the different q-space levels are shown in figure 3. All excitations from discrete states which contribute to the 1% level are present.

The real parts of the optical potentials also have an interesting property in the forward direction. As we showed earlier the dipole-transition matrix elements in the second-order potential were most important in their effect on the elastic cross-section at forward angles and we related this to their long range coordinate space behaviour and the polarizability. In momentum space we can interpret this behaviour in terms of the integral representations (19) for the F's arising from dipole transitions[18]. For them the F_m are of the form

$$F_m(\underline{q}'',\underline{q}) \sim \frac{Y_{1M}(\underline{Q})}{Q} \, G_m(Q^2) \quad , \tag{20}$$

$$\underline{Q} = \underline{q} - \underline{q}'' \quad ,$$

where $G_m(o)$ is finite. (M refers to the magnetic sublevel of the dipole). If we sum over M and Taylor expand the $\underline{q}''$ integral about $\underline{q}'' = \underline{q}$ and $\underline{q}'$ then it can be shown that the m'th dipole contribution to the equivalent local form (18) is

$$Z_{D_m}(P) = G_m(o)G_m(P^2)\tfrac{1}{2}\int_{-1}^{1} du \int d^3q'' \frac{\underline{Q}.\underline{Q}'}{Q^2 Q'^2 (E+io-\varepsilon_m-\tfrac{1}{2}q''^2)}$$

$$+ \text{ integrals which are non singular in } Q \text{ and } Q'. \tag{21}$$

(For non-dipole transitions there are no singular-integral terms). We can define the polarisability associated with level m as

$$\alpha_m = - 8\pi \frac{d}{dP}\bigg|_o Z_m(P) \quad . \tag{22}$$

After a proper analysis[18] it is possible to deduce that the total contribution to the polarizability comes from the singular integral in equ. (21). The same argument can be carried through for the continuum overlaps $F_{p''}$. It is interesting to see what happens when one fourier transforms the integral (21) into coordinate space. One can in fact carry out the evaluation of (21) analytically and thus find the fourier transform. The result is

$$V_m^{pol}(r) = - \frac{1}{r^4}\left\{ \frac{3G_m^2(o)}{\varepsilon_m-\varepsilon_o} + \frac{k_o-W^{\frac{1}{2}}}{k_o} \frac{3G_m(o)G_m[(k_o-W^{\frac{1}{2}})^2]\cos(k_o-W^{\frac{1}{2}})r}{(k_o-W^{\frac{1}{2}})^2} \right.$$

$$\left. + \frac{k_o+W^{\frac{1}{2}}}{k_o} \frac{3G_m(o)G_m[(k_o+W^{\frac{1}{2}})^2]\cos(k_o+W^{\frac{1}{2}})r}{(k_o+W^{\frac{1}{2}})^2} \right\} + 0(r^{-6}) \quad . \tag{23}$$

Here $\tfrac{1}{2}k_o^2$ is the incident electron kinetic energy and $W = k_o^2 + 2(\varepsilon_o-\varepsilon_m)$. The coefficient of the r^{-4} term contains the static polarizability α_m (the first term) and in addition two oscillatory terms. The third

term goes to zero with increasing energy while the second term does not
fall off but oscillates. The reason for the oscillatory terms appear-
ing is due to the fact that no adiabatic approximation has been made to
the Green's function of equ. (21) as is usually the case in the derivation of the $-\alpha/r^4$ potential. Essentially the same oscillatory
behaviour has been deduced in a completely coordinate space approach by
W. Huo[19].

STATUS OF THEORETICAL CALCULATIONS

In the previous two years several detailed calculations have
appeared which have looked at intermediate energy scattering in the
e-H system. In addition to those of refs. 14, 15, 17 there are the
calculations of Kingston and Walters[20] and Scott and Bransden[21].
All the calculations are slightly different in that they do make approximations as to which terms of the multiple-scattering series dominate.
Each calculation shown in tables II, III and IV is the best of its
type.

Table II Differential elastic, total elastic and total reaction cross-
sections for e-H scattering at 50eV compared with experiment (Williams[22],
Van Wingerden et al.[23] (asterisked)) for the five different calculations reported in the text. (The POM numbers have been read from a
published curve). For each calculation the column gives the ratio of
theory to experiment. In the experimental column the numbers in
parentheses are the total errors in the last significant digits.

θ(deg)	Expt.(a_0^2)	CCE	COM	DWSBA	EOM	POM
10	5.04(51)	0.76	1.24	1.01	0.83	0.99
15	3.18(37)	-	1.28	-	0.87	1.04
20	2.17(23)	0.78	1.22	1.04	0.91	1.01
30	1.12(12)	0.77	1.02	0.96	0.95	0.98
40	0.551(59)	0.89	0.97	1.01	1.06	0.98
50	0.308(27)	0.98	0.90	1.00	1.10	0.97
60	0.205(19)	0.96	0.78	0.89	0.98	0.88
70	0.146(14)	0.92	0.68	-	0.88	0.82
80	0.0993(121)	0.97	0.69	-	0.92	0.81
90	0.0716(82)	0.99	0.70	0.80	0.92	0.81
100	0.0558(66)	0.99	0.69	-	0.89	0.81
110	0.0421(43)	1.06	-	0.83	0.91	0.86
120	0.0349(33)	1.06	0.73	0.84	0.91	0.86
130	0.0288(30)	1.10	-	-	0.94	0.90
140	0.0273(26)	1.04	0.72	0.82	0.87	0.84
σ_E	3.83*	-	4.16	3.84	4.30	3.83
σ_R	6.57*	-	9.3	-	6.11	8.04

The column labelled CCE in table II refers to the 6-state close-
coupling-equation solution of Fon et al.[14]. It includes coupling to
the physical 1s,2s,2p states and in addition three pseudo states to
take account of the other excited states and the continuum. The

multiple-scattering-series terms one generates by solving equ. (10) for the T-matrix contain contributions from rescattering in the Q-space (e.g. terms of the type $V_{1s,2p}G_0V_{2p,2p}G_0V_{2p,1s}$) which are not taken

Table III Differential elastic, total elastic and total reaction cross-sections for e-H scattering at 100eV.   Details given for Table II.

θ(deg)	Expt.(a_0^2)	COM	DWSBA	EOM	POM
20	1.10(10)	0.82	0.77	0.81	0.73
25	0.692(71)	0.83	0.82	0.89	0.77
30	0.509(49)	0.74	0.75	0.86	0.73
40	0.288(27)	0.61	0.65	0.72	0.62
50	0.132(12)	0.70	0.75	0.79	0.76
60	0.0722(71)	0.74	0.78	0.85	0.75
70	0.0491(46)	0.68	-	0.80	0.69
80	0.0295(30)	0.76	-	0.87	0.78
90	0.0209(20)	0.77	0.78	0.87	0.81
100	0.0155(15)	0.77	-	0.85	0.84
110	0.0115(12)	-	0.81	0.89	0.87
120	0.0092(9)	0.82	0.82	0.90	0.87
130	0.0078(7)	-	-	0.87	0.87
140	0.0065(7)	0.85	0.86	0.90	0.92
σ_E	1.75	1.54	1.45	1.50	1.45
σ_R	5.53*	6.14	-	5.28	5.87

account of in the second-order theories.   The columns labelled COM contain the results of the most recently developed form (1981) of the eikonal-Born-series approach of Byron and Joachain[15] which uses closure to sum over intermediate states in the second-order potential of equ. (13).   In this approach the full T-matrix equation is not solved. Rather the first few dominant terms obtained by iteration of equ. (10) are retained.   One expects this approximation to improve with increasing energy.   The DWSBA column shows the calculation of Kingston and Walters[20] which is a distorted wave approach using the second-order potential.   In this approach one treats the rescattering effects in the elastic channel to all orders.   However terms of the form $V_{1s,1s}(G_0 V_{1s,2p}G_0V_{2p,1s})^n$ where n > 1 are not contained in this type of calcula-tion.   Again their neglect should be justified in the first approximation at intermediate energies.

Our explicit second-order optical model results are presented in the column labelled EOM.   In this model we have evaluated the optical potential (13) explicitly (and including all exchange terms) in the equivalent local approximation for the second-order terms.   The contributions from the continuum, 2s,3s,2p,3p,4p and 5p states shown in fig. 3 were included.   With this explicit second-order potential the full T-matrix equation (10) was solved.   Thus the additional approx-imations made in the DWSBA and COM approaches are avoided.   The final columns, POM, represent the calculation of Scott and Bransden which are most similar in spirit to ours.   They include the 2s and 2p contribu-tion in the second-order potential exactly as in the EOM model.   The

Table IV Differential elastic, total elastic and total reaction cross-sections for e-H scattering at 200eV. Details given for Table II.

θ(deg)	Expt.(a_o^2)	COM	DWSBA	EOM
15	1.22(12)	0.53	-	0.95
20	0.419(40)	0.92	0.92	1.04
30	0.172(17)	0.85	0.87	1.01
40	0.0706(68)	0.88	0.90	1.04
50	0.0314(32)	0.96	0.97	1.06
60	0.0187(19)	0.88	0.88	0.98
70	0.0125(14)	0.78	-	0.89
80	0.00859(92)	0.74	-	0.83
90	0.00584(61)	0.75	0.74	0.83
100	0.00412(42)	0.78	-	0.85
110	0.00323(31)	-	0.76	0.84
120	0.00272(35)	0.72	0.72	0.82
130	0.00199(25)	-	-	0.93
140	0.00178(26)	0.80	0.79	0.88
σ_E	0.669*	0.631	0.641	0.672
σ_R	3.77*	3.75	-	3.27

effects of the higher-order states and continuum are mimicked by choosing a large number of variationally chosen pseudo-states. The resultant potential is then used to solve equ. (10) just as for our model. The experimental results shown are the absolute results of Williams[22]. The absolute experiment of Van Wingerden et al[23] yields similar results but slightly above those of Williams.

In all the tables we have presented the ratio of theory to experiment rather than the theory numbers. In this way the differences between experiment and theory are more easily highlighted. A few comments are in order. At 50eV the CCE calculation does very well for angles $\gtrsim 50°$ but is too low in the forward direction. On the other hand while all the second-order theories do much better in the forward direction they fall below experiment at backward angles. This of course reflects the different emphasis on reaction mechanisms of the two approaches. In the forward direction the cross-section is dominated by dipole excited channels and we need to incorporate a large number of them. This is the forte of the second-order models. Clearly the CCE model requires more pseudo channels to reproduce the forward-angle behaviour. At backward angles the CCE does better. It includes rescattering effects in the Q-space which are neglected in the other models. At 50eV these seem to be significant at the 10% level. The COM model appears to give the poorest representation of the reaction mechanism. It is too large in the forward direction and has the wrong shape, falling significantly below experiment and the other models at backward angles. DWSBA and POM are very similar. Both represent forward scattering excellently but seem to have slightly the wrong shape at backward energies. EOM is a little below experimental error at 10° but thereafter gives a good description of experiment and lies within the error bounds. Again it is consistently on the low side of the experimental results.

At 100eV there is much more disagreement with experiment for all models. This is surprising since all the approximations in the theories are tailored so that they become better at higher energies. Our EOM model does better at forward angles than the other theories and moreover the shape is preserved over the whole angular range. Both DWSBA and POM have different shape profiles compared to their 50eV results. The COM results are much closer to DWSBA and POM than they were at 50eV.

At 200eV there are only three models to compare. COM and DWSBA give very similar answers and if anything their shape falls off at backward angles. The EOM does much better in absolute terms than either of its competitors. Practically all the numbers are within the error tolerances of the experiment. Again the shape does fall off at larger angles.

What can we conclude from the above comparison of theory and experiment? First and foremost there appears to be an inexplicable difference between theory and experiment at 100eV. This discrepancy cannot be explained on the basis of the approximations used in the theoretical calculations. It is well known that absolute measurements with atomic hydrogen have great difficulties associated with them. Perhaps it is time to do this experiment again! At 50eV and 200eV the agreement with theory and experiment is much better. However fine details in shape for the theories can still account for 10% differences. It appears that the calculations with the greatest detail in the reaction mechanism (CCE and EOM) do slightly better overall. We feel that theory is reaching the stage where a complete set of experimental measurements for e-H is needed with much smaller error bars. This will allow us to discriminate between the theories and the approximations they make in the modelling of the reaction mechanism.

ACKNOWLEDGMENTS

I would like to thank the Australian Research Grants Committee for a Research Fellowship. I would also like to thank Prof. R.H. Bonham for informing me of the existence of the work of W.M. Huo (ref. 19).

REFERENCES

1. I.E. McCarthy and A.T. Stelbovics, Phys. Rev. A22, 502 (1980).
2. B.H. Bransden and M.R.C. McDowell, Phys. Rep. 30, 207 (1977).
3. P.G. Burke and W.D. Robb, Advances in Atomic and Molecular Physics 11 (New York: Academic, 1975) p.143.
4. M.J. Seaton, Phil. Trans. Roy. Soc. (London) 245, 469 (1953).
5. K. Smith and L.A. Morgan, Phys. Rev. 165, 110 (1968).
6. I.H. Sloan and E.J. Moore, J. Phys. B. 1, 414 (1968).
7. M.R.H. Rudge, Rev. Mod. Phys. 40, 564 (1968).
8. H. Feshbach, Ann. Phys. NY5, 357 (1958); ibid 19, 287 (1962).
9. R.M. Sternheimer, Phys. Rev. 96, 951 (1954).
10. M.H. Mittleman and K.M. Watson, Phys. Rev. 113, 198 (1959).
11. L. Castillejo, I.C. Percival and M.J. Seaton, Proc. Roy. Soc. A254, 259 (1960).
12. P.G. Burke, Proc. Phys. Soc. 82, 443 (1963).
13. P.G. Burke, D.F. Gallaher and S. Geltman, J. Phys. B 2, 1142 (1969).
14. W.C. Fon, K.A. Berrington, P.G. Burke and A.E. Kingston, J. Phys. B 14, 1041 (1981).
15. F.W. Byron, Jr. and C.J. Joachain, Phys. Rev. A 15, 128 (1977);

ibid J. Phys. B 14, 2429 (1981).

16.   B.H. Bransden and J.P. Coleman, J. Phys. B 5, 537 (1972).

17.   I.E. McCarthy, B.C. Saha and A.T. Stelbovics, Phys. Rev. A 23, 145 (1981); ibid FIAS-R-91 (submitted to J. Phys. Lett.) (1982).

18.   I.E. McCarthy, B.C. Saha and A.T. Stelbovics, J. Phys. B 14, 2871 (1981).

19.   W.M. Huo, J. Chem. Phys. 56, 3468 (1972).

20.   A.E. Kingston and H.R.J. Walters, J. Phys. B 13, 4633 (1980).

21.   T. Scott and B.H. Bransden, J. Phys. B 14, 2277 (1981).

22.   J.F. Williams, J. Phys. B 8, 2191 (1975).

23.   B. Van Wingerden, E. Weigold, F.J. de Heer and K.J. Nygaard, J. Phys. B 10, 1345 (1977).

HYPERFINE STRUCTURE AND ZERO FIELD QUANTUM BEATS

P.J.O. Teubner

School of Physical Sciences, The Flinders University of South Australia,
Bedford Park, S.A. 5042, Australia.

ABSTRACT

The consequences of zero field quantum beat studies in electron impact to the measurement of scattering parameters in electron sodium collision is explored. The beats in sodium, which were first observed in an electron photon coincidence experiment, arise from the decay of coherently excited hyperfine levels in the J=3/2 level of the 3^2P state.

The phenomenon about which I will talk today, zero field quantum beats, has its origin in the hyperfine structure of atomic systems. One could construe from the title that one can hope to gain information about the hyperfine structure of atomic states by studies of the electron impact excitation of beats. I will show that this cannot be done but given the hyperfine structure one can deduce information on the scattering amplitudes involved in the collision. These experimentally determined quantities can then be used to provide a sensitive test of the theory of electron-atom collisions. The theory can then be used with more confidence in say momentum density calculations. The phenomenon of beats arises from the overlap between the nuclear and extra-nuclear properties of the atom. Thus, even though we can provide no new information on hyperfine structure, it is of relevance to this workshop.

The modulation of light arising from the decay of coherently excited atomic states has been observed in beam foil spectroscopy where the phenomenon is known as zero field quantum beats. Extensive studies of the beat patterns have yielded information on both the fine and hyperfine structure of excited states in atoms and ions[1]. The process has also yielded information on alignment and orientation in charge transfer collisions of protons with carbon foils[2]. Beats have also been observed in sodium atoms excited by fast laser pulses[3].

We have recently observed zero field quantum beats in a state which has been populated by electron impact. The beats arise from the interference of radiation from coherently excited levels of the hyperfine multiplet. The timing resolution of our experiments cannot compete with that available in beam foil experiments thus, at present, the technique cannot be used to gain information on the hyperfine structure of atomic states. Nevertheless we will show that the amplitude of the beat pattern depends on the scattering amplitudes involved in the collision. Thus analysis of the beat pattern promises to yield information on the collision which can then be used to provide a critical test on the theory of the interaction.

Whilst it is possible to gain this information using other techniques this workshop is not necessarily the place to discuss the relative merit of such other processes.

The introduction of oscillatory terms in the radiative decay of excited atoms was first described by Macek[4]. Subsequently the theory was expanded such that it applied to electron-photon coincidence experiments[5]. This theory treated the time dependence in detail and showed that the measurement of angular correlations would yield information on the magnitude and phases of the scattering amplitudes for the coherent excitation of degenerate magnetic substates of an excited state.

The Macek and Jaecks theory[5] has been applied with great success[6,7,8] to systems in which the natural width of the states is large compared to their splitting. Thus the phenomenon of the electron impact excitation of beats has been ignored for a decade.

In the main most experiments have been carried out on atomic hydrogen[7] and on helium[6,8]. Helium has no hyperfine structure since its nuclear spin is zero and in the case of singlet excitation there is no fine structure. In the case of hydrogen 2^2P excitation the natural width of the level $\gamma \sim 625$ MHZ whilst the fine structure separation is $\sim 1.1 \times 10^5$ MHZ. The possibility of observing fine structure beats in electron photon coincidence experiments is extremely remote.

On the other hand the properties of sodium are such that the natural width of the resonant 3^2P state is comparable to the HFS splitting in the J=3/2 level. The energy levels for the first excited state in sodium are shown in figure 1. The mean energy of the state and the fine structure

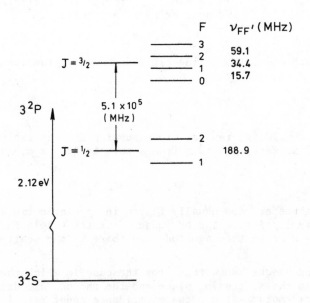

Fig. 1 The energy levels for the first excited state of sodium.

separation is taken from the tables of Moore[9] whilst the HFS splitting has been derived from the a and b coefficients of Deech et al.[10]. Intuitively one can say that the beat period will be of the order of 17 nsecs which should be observable with current technology.

In order to pursue the matter further we must refer explicitly to the theory of Macek and Jaecks[5]. The coincidence count rate depends on three different terms namely scattering amplitudes which describe the excitation process; dipole matrix elements describing the decay and a time dependent factor. The last factor describes the time dependence of the emitted light thus it consists of an exponential decay plus an oscillatory term which describes the beats.

The scattering amplitudes appear as the density matrix elements

$$\langle a_{M_L} a_{M_{L'}} \rangle$$

which are related to differential cross sections σ_{M_L} through the relationship

$$\langle a_{M_L} a_{M_{L'}} \rangle = |a_{M_L}|^2 \qquad M_L = M_{L'}$$

$$= \sigma_{M_L}$$

where σ_{M_L} is the differential cross section for the excitation of the degenerate magnetic substate M_L.

For a P state, which is excited by an electron such that the LS coupling approximation can be used to describe the excitation, there is reflection symmetry in the scattering plane. Thus

$$a_1 = - a_{-1}$$

It is fashionable to define two independent parameters λ and χ. Where

$$\lambda = \frac{\sigma_o}{\sigma_o + 2\sigma_1} = \frac{\sigma_o}{\sigma} \qquad (1)$$

and σ is the total differential cross section for the excitation of the state. The parameter χ arises from a term in Re $\langle a_o a_1 \rangle$ and is given by

$$\chi = \arg a_o - \arg a_1 \qquad (2)$$

These parameters are usually fitted in angular correlation experiments. Their importance can be judged from the simple fact that together they measure the magnitude and phase of the scattering amplitudes.

Macek and Jaecks[5] show that, for the case in which the photon detector is in the scattering plane and the photons are detected without regard to their polarisation, the coincidence count rate is

$$dN_e = v \, n_e n_A \gamma \frac{3}{8\pi} [A_{oo} + A_{11} + (A_{11} - A_{oo}) \cos^2\theta_\gamma + \sqrt{2} \, Re \, A_{o1} \sin 2\theta_\gamma$$

$$+ A_{1-1} \sin^2\theta_\gamma] d\Omega_e d\Omega_\gamma \qquad (3)$$

n_e, n_A are the number densities of incoming electrons and target atoms, v is the velocity of the electrons and $d\Omega_e d\Omega_\gamma$ are the solid angles of th two detectors.   The coefficients $A_{qq'}$ are given by

$$A_{qq'} = \sum_{JJ'FF'M_L M_{L'}} \cup \langle a_{M_L} a_{M_{L'}} \rangle \int_0^{\Delta t} dt \; \exp[-(\gamma+i\omega)t \qquad]_{JFJ'F'} \qquad (4)$$

In this expression the coefficients $\cup$ are given by equation (20) in Macek and Jaecks[5].   They contain all of the angular momentum algebra for the problem.   We have calculated the $\cup$'s for the 3^2P state of ^{23}Na which has the following properties: $I = 3/2$, $F = I+J$, $J = L+S$, $L = 1$ and $L_o = 0$.   We have assumed that LS coupling can be used to describe the excited state.

The beat frequencies

$$\omega_{JFJ'F'} = \frac{E_{JF}-E_{J'F'}}{\hbar}$$

If one considers the predicted time development of the coincidence peak for the special case $\theta_\gamma = 90°$, equation (3) then becomes

$$dN_c = K[A_{oo} + A_{11} + A_{1-1}]e^{-\gamma t}$$

where the mean lifetime of the state

$$\tau = 1/\gamma$$

This has the form

$$dN_c = K'[\alpha + \sum_i \beta_i \; \cos(\omega_{FF'}t'+\phi)]e^{-\gamma t} \qquad (5)$$

Explicitly the time dependence of the data in the coincidence peak is given by

$$I(t) = k \; e^{-\gamma t}[0.6 + 0.09\lambda + (3\lambda-2) \times 10^{-2} \sum_i \beta_i \; \cos(\omega_i t+\phi)]$$

$$(6)$$

In this expression the factors β_i are determined from the coefficients $\cup$ in equation (4) and the frequencies ω_i are known from the HFS splittings of the atom.   The phase factors ϕ_i are introduced by the integration over the timing resolution of the circuits.

Ideally the only unknown in equation (6) is the value of λ at the particular electron scattering angle under consideration.   Therefore a fit of the data to the function shown in equation (6) should yield λ. The situation is however complicated by the phase factors ϕ_i which depend explicitly on the way in which the time integration in equation (4) is done.   At present we are investigating three possible alternatives namely that given by Macek and Jaecks[5], a slightly different form given by Parker et al.[11] and the introduction of a Gaussian apparatus function.   The last of these procedures offers the most realistic alternative for our present experimental conditions.

For the commonly occurring isotopes of the alkalis and for which LS coupling is valid, the angular momentum determines that only the 3-1, 3-2, 2-0 and 2-1 HFS states in the J=3/2 level contribute to the beats.   The periods of these contributions are 16.9, 10.7, 20.0 and 29.1 nsecs respectively.

According to Macek and Jaecks, radiation from the 3^2P state should be modulated if observed in an electron-photon coincidence experiment. The question is, can this modulation be observed?

We note that at θ_γ = 90° the beat pattern does not depend on χ. The reason why we choose θ_γ = 90° is because the non trivial algebra of equation (4) is simplified at this angle.

The apparatus which was used to observe the beats is shown schematically in figure 2.   A beam of sodium atoms was formed in a Joule heated

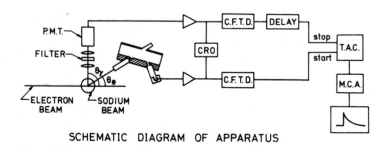

SCHEMATIC DIAGRAM OF APPARATUS

Fig. 2   A schematic diagram of the apparatus

Monel oven and entered the interaction region where it was intersected by a well focussed electron beam of energy 100eV.   A cylindrical mirror electron spectrometer viewed the interaction region and analysed the energy of electrons which had been scattered through a scattering angle of 3°.   Electrons which had excited the 3^2P state and had lost 2.1eV of energy were detected with a channel electron multiplier.   Pulses from detector were processed with standard fast-timing electronics and acted as the start pulse for a time to amplitude converter (TAC).   The energy resolution of the electron channel was 0.9eV which was sufficient to ensure that only those electrons which had excited the 3^2P state were detected.

A photomultiplier tube (PMT) viewed the interaction region and detected photons which were emitted at 90° to the incident beam direction in the scattering plane.   The 589 and 589.6 nm photons which arose from the decay of the 3^2P state passed through a 10 nm bandwidth interference

ilter and were detected by the photomultiplier tube. Pulses from the
tube were amplified, delayed and acted as the stop pulse for the TAC.
The output of the TAC was analysed with a multichannel analyser (MCA).

Pairs of Helmholtz coils and magnetic shielding material reduced
the magnitude of the magnetic field in the interaction region to less
than 2×10^{-6}T. All surfaces which bounded the interaction region were
earthed and coated with colloidal graphite. The local electric field
in the interaction region was therefore that due to space charge in the
electron beam which was estimated to be less than 0.3 Vm^{-1}.

The output of the TAC is shown in figure 3. The line is a fit to

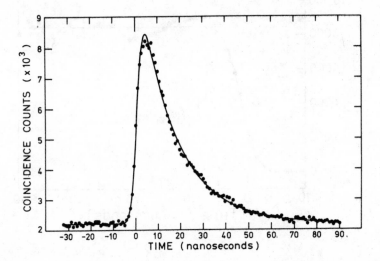

Figure 3. The coincidence peak showing a fit to the data of the
form of equation (7).

data of a function of the form

$$I(t) = \left[b + a \ \text{erf}\left(\frac{t-t_o}{s\sqrt{2}} - \frac{s\gamma}{\sqrt{2}}\right) \right] \exp\left[\frac{s^2\gamma^2}{2} - \gamma(t-t_o)\right] \qquad (7)$$

where b was the height of the random background, erf the normalised
error function, to the prompt response time, s the standard deviation
of the Gaussian, γ the reciprocal lifetime of the state and 2a the
amplitude of the decay function. This procedure has been used by
Read and coworkers to determine the lifetime of atomic states[12]. Over
many runs we have found that the lifetime of the 3^2P state was 16.30 ±
0.15 nsecs which was in excellent agreement with the preferred value
given by Fischer[13].

The beat pattern can be seen as an oscillation of the data about the fit in figure 3. The beats are more apparent if one follows the suggestion of Blum and Kleinpoppen[14] namely the data at each time interval are multiplied by $\exp[\Gamma t]$ where

$$\Gamma t = -\tfrac{1}{2}s^2\gamma^2 + \gamma(t-t_o) \tag{8}$$

The resulting time variation is shown in figure 4 where the error bars

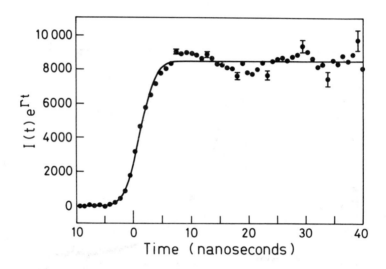

Fig. 4 Deconvolution of the coincidence spectrum where Γt is defined in equation (8).

correspond to plus or minus one standard deviation.

Figure 5 shows a fit of the beat pattern to a function of the form

$$I(t) = C'[A + \sum_i \beta_i \cos(\omega_i t + \phi_i)] \tag{9}$$

as predicted from equation (5). The beat frequencies ω_i which have been used are $\omega_{32} = 371.3$, $\omega_{31} = 587.5$, $\omega_{20} = 314.8$ and $\omega_{12} = 216.1$ radians per second.

The timing resolution for the data in figure 3 was 9.4 nsecs full width at half maximum which influenced significantly the relative amplitudes of the component frequencies. Clearly an improvement in the timing resolution is required and we are working on that problem at the moment. We have ascertained that of the several factors which influence the timing resolution the most critical effect comes from the photomultiplier tube.

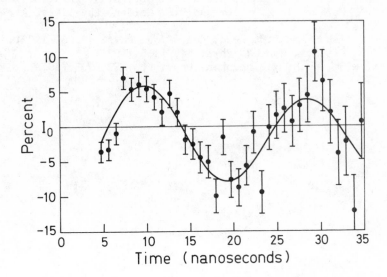

Fig. 5 A fit of the beat pattern to a sinusoidal function of the
form of equation (9).

With our current timing resolution it is not possible to analyse
the beat pattern for any greater detail.   Nevertheless we are pointed
in the right direction and it is envisaged that the obvious improvements
will yield values of the scattering parameter λ which can be used to
compare with theoretical predictions.

ACKNOWLEDGMENTS

The work described in this talk was carried out in collaboration
with J. Furst and J. Riley.   It was supported in part by grants from
the Australian Research Grants Committee.

REFERENCES

1. H.J. Andrä, Phys. Sci. 9, 257 (1974).
2. P. Dobberstein, H.J. Andrä and W. Wittman, Z. Physik, 257, 272 (1972).
3. S. Haroche in Topics in Applied Physics - High Resolution Laser
   Spectroscopy ed. K. Shumoda (Springer-Verlag, Berlin, 1976) 13,
   p.253.
4. J. Macek, Phys. Rev. Letters, 23, 1 (1969).
5. J. Macek and D.H. Jaecks, Phys. Rev. A 4, 2288 (1971).
6. M. Emmyan, K.B. MacAdam, J. Slevin and H. Kleimpoppen, J. Phys. B
   12, 1519 (1974).
7. S.T. Hood, E. Weigold and A.J. Dixon, J. Phys. B. 12, 631 (1979).
8. N.C. Steph and D.E. Golden, Phys. Rev. A 21, 759 (1980);
   M.T. Hollywood, A. Crowe and J.F. Williams, J. Phys. B 12, 819 (1979).
9. C.E. Moore, Atomic Energy Levels, U.S. National Bureau of Standards,

National Standard Reference Data Series - 35 Vol. 1.

10. J.S. Deech, P. Hannaford and G.W. Series, J. Phys. B. 7, 1131 (1974)

11. G.A. Parker, T.M. Miller and D.E. Golden, Phys. Rev. A 25, 588 (1982).

12. R.E. Imhof and F.H. Read, Rep. Prog. Phys. 40, 3 (1977).

13. C.F. Fischer, Can. J. Phys. 54, 1465 (1975).

14. K. Blum and H. Kleinpoppen, Phys. Rep. 52, 203 (1979).

METASTABLE ION LIFETIMES MEASURED BY PHOTOION-PHOTOELECTRON COINCIDENCE

GARY D. WILLETT

Department of Theoretical Chemistry
University of Sydney 2006
Australia

ABSTRACT

The Photoion-Photoelectron coincidence (PIPECO) technique, in which nominally zero-kinetic energy electrons are detected in coincidence with their concomitant parent or fragment ions, can be used to measure the absolute fragmentation rates and kinetic energy release distributions (KERDS) of gas phase state selected ions. This lecture will be used to review the technique and discuss results for the unimolecular decomposition of the following metastable molecular ions: $C_6H_5X^+(X=Cl,Br,I)$, $C_8H_8^+$, $C_6H_6^+$, $C_4H_4X^+(X=O,NH)$. Analysis of their associated photoionization efficiency curves and threshold photoelectron spectra together with a comparison of their measured dissociation rates with those calculated using statistical theories will be used to demonstrate the versatility of the technique in establishing metastable molecular ion dissociation mechanisms.

I INTRODUCTION

The combination of the photoelectron and mass spectrometric experiments over the last ten years has opened the way to a more thorough study of a variety of ion molecule reactions with a precision greater than that offered by the two separate experiments. Despite the fact that photoelectron spectroscopy (PES) gives accurate information relating to the electronic and vibrational states of molecules and their associated ions, it usually only deals with direct transitions and provides little information regarding the subsequent fate of the associated parent ion.[1,2] On the other hand, mass spectrometry (MS) has provided a wealth of information about molecular identity and hence fragmentation channels, appearance energies, ionic and neutral heats of formation and in some cases even phenomenonological ionic dissociation rate constants.[3] Unfortunately though, ions prepared in MS, whether by electron impact, photon impact or the more exotic techniques such as plasma desorption or field ionization cannot be prepared easily in selected internal energy states. Prior to the development of photoion-photoelectron coincidence (PIPECO) then, little direct experimental evidence regarding the nature of ionic reactions has been reported.[4]

Two forms of the PIPECO technique are currently in use: one form utilizes a fixed wavelength photon ionizing source (for example, HeI, 21eV) and the internal energy of the ion is scanned by using an electron analyser to select electrons of different energies[5]. The second method uses a steradiancy electron energy analyser to select zero kinetic energy or threshold electrons and the ion internal energy is scanned by varying the energy of the monochromatic photons produced from a H_2 or He continuum discharge source[4].

The state selection of ions is based upon the conservation of energy and momentum according to the following expression:

$$E_i = h\nu - E_e(1 + \frac{m_e}{m_i}) \qquad (1)$$

Where E_i and E_e represent the internal energy of the state selected ion and the associated electron kinetic energy respectively, $h\nu$ the photon energy and m_e and m_i the respective masses of the electron and ion. Thus for the latter technique, where ions are detected in delayed coincidence with the threshold electrons, only ions with energies equal to that of the ionizing radiation are selected from all generated ions.

The examples discussed later in this paper deal exclusively with the second PIPECO technique as it offers one particular advantage in the measurement of metastable (10^{-4}-10^{-7}sec^{-1}) ion lifetimes. As the incident photon energy is varied, ion states produced by both direct and autoionization (ionization from superexcited neutral molecules) are populated; in other words, ions are produced over the whole range of photon energies used and not just those in energy regions which have favourable Franck Condon factors[6]. It is important to generate ions in these so called "Franck Condon gap regions" as this is where many fragmentation channels onset.

Unimolecular reactions, that is, reactions where the probability of a reactant species remaining undissociated at a time t after the reaction commences, is given by exp(-kt) with k as the unimolecular rate constant, have long formed an important part of chemistry. The development of the PIPECO technique now provides the starting basis for studies of more complicated reactions in the condensed phases where, for instance, interactions such as solvent effects often complicate reaction mechanisms.

Since the inception of MS considerable effort has been directed towards the characterization and reaction dynamics of ionic unimolecular reactions. A complete characterization would include determination of the following:

1) Determination of the forward and reverse activation energy barriers for dissociation and isomerization reactions.
2) A knowledge of the internal energies of the reactants and products.
3) How the excess internal energy of the molecular ion is partitioned between the fragmentation, fluorescence and isomerization reactions.
4) The various rates of isomerization, fluorescence, fragmentation and radiationless transitions between electronic and vibrational states as a function of parent ion internal energy.

A careful comparison of data obtained from PIPECO experiments together with results from statistical theories such as RRKM or QET[7] are bringing us closer to achieving such a complete characterization.

Before experimental unimolecular reactions, in contrast to hypothetical systems, can be treated in terms of any statistical theory we must know some details about the mechanism for the reaction. It is worthwhile then to address the following questions:

1) Are there any simple unimolecular reactions of molecular ionic systems?

2) What are the criteria for recognizing complex reaction mechanisms?

Baer has supplied a partial answer to these questions in a recent PIPECO review where he suggests the following[4]:

"Although the statistical theory can be used to calculate rates which are in agreement with experimental rates, it is not yet possible to predict the rate of a reaction. The reason has less to do with the theory than with the fact that ionic dissociations are often complex sequences of reactions, and before the theory can be applied to the reaction in question, the sequence of steps must be known with a considerable degree of certainty".

We discuss in this paper systems where simple unimolecular decompositions have been confirmed by PIPECO experiments. It is noteworthy though, that in other PIPECO studies reactions have been studied which show apparent non statistical features[4,8,18]. Such results however, need to be treated with caution as they could simply reflect the inadequacy of our understanding of the reaction mechanism to properly account for it with the use of statistical theory.

A summary of the experiments that may be carried out on the PIPECO spectrometer are shown in Figure 1. The non-coincidence experiments such as threshold PES and MS are routinely used to obtain detailed information on molecular ionization energies and fragment ion appearance energies respectively. When these two techniques are combined in the coincidence experiment to examine the unimolecular or bimolecular reactions of state selected ions, the applications are wide ranging and include the determination of the following:

a) Predissociation of di- and triatomics.

b) Breakdown diagrams of polyatomics.

c) Kinetic energy releases in polyatomic dissociations.

d) Lifetimes of metastable ions.

e) Cross-sections of ion-molecule reactions.

These applications are treated in a review by Baer[4] and only point (d) will be discussed in detail in this paper.

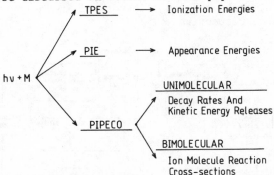

Figure 1. Experiments that may be carried out on the Photoion-Photoelectron Spectrometer.

II EXPERIMENTAL DISCUSSION

The threshold photoion-photoelectron spectrometer, shown as a schematic in Figure 2, has been developed by Baer[4]. UV radiation from a low pressure H_2 resonance light source is dispersed by a 1 metre normal incidence monochromotor with a wavelength of $2A^0$ (18 meV at $1000A^0$). The photons enter a 5.8 cm long ionization/acceleration region 1 cm from the positive voltage end.

In the applied field of about 17V/cm electrons are accelerated towards a steradiancy electron energy analyser with consists of a set of 1 mm long collimated holes with a length to diameter ratio of 40. A schematic of this analyser, as shown in Figure 3, rejects energetic electrons which are ejected with velocity components perpendicular to the electric field because they hit the walls or the apertures. Even though angular discrimination is achieved, there are complications arising from electrons which are ejected parallel to the analyser holes. It is possible, however, to distinguish these hot electrons from the threshold electrons on the basis of their different times-of-flight[6].

By collecting threshold electrons as a function of the incident photon energy it is possible to measure threshold photoelectron spectra such as the one shown in Figure 4a for furan. These spectra are similar to the more usual HeI photoelectron spectra except that the intensities are often perturbed by ionizations from superexcited (autoionizing states) neutral states which give added intensity to the ionization process.

In the electric field of the ionization region, ions are accelerated 4.8 cm in the opposite direction to that of the electrons until they enter a set of collimated holes ($\ell/d=10$) which serve to pressurize the photoionization region. Mass analysis is achieved either by time-of-flight with a 9 cm long drift region and 60V drawout voltage giving unit mass resolution at m/e=30. When better resolution is required a quadrupole mass filter is used. Photoionization efficiency curves such as those shown in Figure 4b are recorded when ions are collected as a function of photon energy. Standard pulse counting equipment is used to detect both the ions and electrons. Typically the electron and ion flight times in the spectrometer are ~0.1 and ~10μ sec respectively.

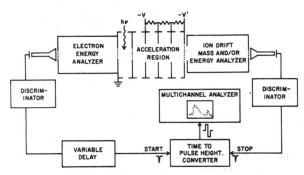

Figure 2.   Schematic of the Photoion-Photoelectron Spectrometer.

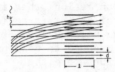

Figure 3.   Threshold electron energy
            analyser (see reference 4).

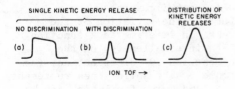

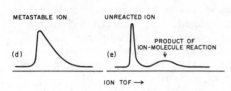

Figure 5.   Schematic ion time-of-
     flight distributions obtained in
     the PIPECO experiment. (see ref.4)

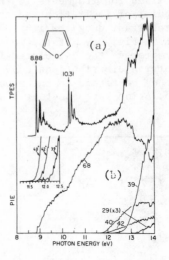

Figure 4. Furan: (a) Threshold
     Photoelectron spectrum
     (b) Photoionization efficiency
     curves (see reference 9).

In the coincidence experiment the internal energies of the
parent ions are selected by collecting only those ions which are
assoicated with the zero kinetic energy electrons. This is accompli-
shed by utilizing the delayed electron and ion signals as start and
stop inputs, respectively, to a time to pulse height converter,
whose output is fed into a multichannel analyser. When total ionization rates are below 1,000 cps (this corresponds to a chamber
pressure of $\sim 10^{-6}$ torr), the detection of the ion and electron pairs
is essentially unambiguous. The resulting ion time-of-flight (TOF)
distributions then for the parent molecular ions of known internal
energy contain all the dynamical information of the associated ion-
molecule reactions. Figure 5 illustrates a variety of such TOF
distributions often observed in the PIPECO experiment.

Kinetic energy released in the fragmentation process results in
symmetrically broadened fragment ion TOF distributions such as those
shown in Figure 5 a-c. The precise shape of such distributions
depends upon whether the apparatus discriminates against ions with
initial velocity components perpendicular to the experimental axis
(cf. Fig.5a and b). For polyatomic systems where a distribution of
energies may be released and in favourable cases determined[9], a
Gaussian shaped peak such as that shown in Fig.5c is usually observed.

264

If an ion fragments in the acceleration region, that is between the point of ionization and the beginning of the drift region then the resultant daughter ion TOF distribution will be asymmetric as shown in Fig. 5d. If the reaction can be assumed to be a unimolecular process the calculated TOF distributions of either Gaussian or Lorentzian distributions convoluted with a single exponential exp(-kt) may then be fitted to the experimental data to obtain the unimolecular dissociation rate constant k or ion lifetime.

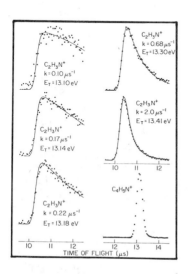

Figure 6. Coincidence time-of-flight distributions for the $C_2H_3N^+$ fragment ions. Taken from reference 11.

Figure 6 illustrates the effect of the variation of the parent ion lifetime as a function of its internal energy on the fragment ion TOF distributions for pyrrole . The solid lines represent the fitted curve from which the dissociation rate constants are determined. It is with this type of process that this paper is primarily concerned.

Finally, Figure 5e demonstrates the effect of ion molecule interactions occurring in the acceleration region. Charge exchange between accelerating ions and neutral molecules result in a bimodal distribution of unreacted and charge transfer ions. The product ion TOF is a function of both the collision dynamics and the position of the reaction in the acceleration region

III THEORETICAL CONSIDERATIONS

The statistical theories of mass spectrometry propose that the fragmentation behaviour of large polyatomic molecules, upon ionization, can be viewed as a two stage process:
1) The "Randomization" of electronic and vibrational energy initially imparted to the molecular ion by the ionization process into all regions of phase space.
2) If some of these conformations are favourably disposed towards unimolecular bond cleavage, then they act as sinks for irreversible dissociation processes.

Historically it is of interest that Rosenstock, Wallenstein, Wahrhaftig and Eyring in the early 1950's incorporated the RRK concepts of the transition state theory of Eyring and applied this new formalism called the Quasi-Equilibrium Theory to mass spectrometry[12]. Their work parallels the work of Marcus and Rice in their formulation of the RRKM theory which has been traditionally applied to neutral systems[13]. This transition state model is cast from the perspective of the fragmenting molecule or ion. There is of course a second perspective, that of the products being formed, which is often referred to as the Phase Space Approach[14].

In this paper we compare results from the transition state model with experiment as it usually predicts better the rates of reactions. The phase space approach on the other hand, works best for estimates of the partitioning of excess internal energy between the fragment products.

A simple statement of the transition state dissociation rate constant, k(E), for a molecular ion of internal energy E, is given by

$$k(E) = \frac{\sigma}{h} \left[\sum_{\varepsilon=0}^{E-Eo} \frac{\rho^{\ddagger}(\varepsilon)}{\rho(E)} \right] \qquad (2)$$

where: $\rho^{\ddagger}(\varepsilon)$ represents the density of vibrational states of the transition state or activated complex at energy ε. The transition state represents some intermediate structure between the products and precursor ion. $\rho(E)$ represents the density of vibrational states of the precursor ion at energy E, σ is the reaction path degeneracy and h is Planck's constant.

The input parameters for such a calculation are:
1) Activation energy for the dissociation reaction,
2) Excitation or excess internal energy of the molecular ion,
3) Vibrational frequencies and moments of inertia of the precursor ion and transition state and
4) Reaction path degeneracy.

The determination of the first two input parameters for the RRKM calculation is relatively straightforward. The activation energy may be estimated from the tables of ionization energies and ionic heats of formation determined by PES and MS [3]. Whereas the state selection obtained in the PIPECO experiment allows a precise measurement of the excess internal energy of the parent molecular ion.

On the other hand, the determination of the vibrational frequencies and moments of inertia of the molecular ion and transition state is considerably more difficult. From PES it is possible to demonstrate that upon ionization of large polyatomic molecules only a small number of frequencies change dramatically from those of the neutral precursor. Hence it is a reasonable approximation that the neutral precursor frequencies serve as a zeroth order approximation of the molecular ion frequencies. To estimate the transition state frequencies one must make certain assumptions, often based on the structure of the products or ab initio molecular orbital calculations, about its geometry. The rates determined from calculations based on the above rather tenuous assumptions have been used to support the implications about the nature of gas phase unimolecular ionic reactions described at the beginning of this section. The validity of these implications relies heavily upon the accuracy of the transition state geometry.

266

IV DISCUSSION OF EXAMPLES

[A] Unimolecular Dissociation of the Halobenzenes

One of the simplest unimolecular dissociations of large polyatomic systems studied by PIPECO is that of the halobenzenes[5].
Cleavage of the R-X bond in this diatomic like molecule according
to the reaction:

$$C_6H_5X^+ \rightarrow C_6H_5^+ + X \quad (X = Cl, Br, I) \tag{3}$$

minimizes the number of assumptions one needs to make about the
transition state geometry. It is reasonable to assume that the
transition state will have a similar geometry to that of the molecular ion, but with an extended R-X bond length. A comparison of the
experimental rates (points) with rates determined using a statistical
calculation (solid line) is made in Figure 7 and demonstrates that the
unimolecular reaction is statistical in nature over the range of rates
studied.

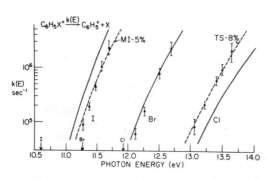

Figure 7. Unimolecular fragmentation
rates (Experimental points and calculated
solid lines) of $C_6H_5X^+$ where X represents
Cl, Br, I. Taken from reference 15.

This comparison supports
the notion that the excited
state molecular ions undergo ratiationless transitions
to high   vibrationally
excited states of the
ground electronic state of
the molecular ion prior to
dissociation. The
prerequisite, of course,
for such statistical
redistribution of electronic
and vibrational energy
prior to dissociation is
that the dissociation rate
from the excited electronic
states of the molecular ion
be considerably slower
than radiationless
transitions to the ground
ionic state.

[B] Isomerization and a Single Unimolecular Reaction

From the variety of PIPECO studies shown in Table 1 it has now
become obvious that the rate of isomerization of molecular ions is
often faster than the rate of direct dissociation reactions. We are
guided to this conclusion by the observation of identical dissociation
rates of metastable ion isomers of widely varying structures. A phase
space representation of this is shown in Figure 8. In 8c the different ions A and B are produced in energy states below both
isomerization and fragmentation channels. Figure 8b has the parent
ion internal energy increased to a point where isomer A can inter-
change to isomer B which subsequently dissociates. Finally in
Figure 8a the internal energy is raised to a point where both fragmentation channels and the isomerization channel are all accessible
and all can compete

Table I   Isomerization-Unimolecular reactions
           observed for state selected metastable ions

1,3 butadiene; 1,2 butadiene cyclobutene; methylene-cyclopropene 2 butyne; 1 butyne	$\rightarrow$ 1,3 butadiene $\rightarrow C_3H_3^+ + CH_3$
1-butene; methyl-cyclopropane cis-butene; isobutene trans-butene	$\rightarrow$ trans-butene $\rightarrow C_3H_5^+ + CH_3$
2,4 hexadiyne 1,5 hexadiyne benzene	$\rightarrow$ benzene $\rightarrow \begin{cases} C_6H_5^+ + H \\ C_4H_4^+ + C_2H_2 \\ C_3H_5^+ + C_3H_3 \end{cases}$
cyclooctatetraene styrene	$\rightarrow$ styrene $\rightarrow C_6H_6^+ + C_2H_2$
pyrrole methacrylonitrile allyl cyanide cyclopropylcyanide	$\rightarrow$ Pyrrole $\rightarrow \begin{cases} C_2H_3N^+ + C_2H_2 \\ C_3H_4^+ + HCN \end{cases}$

(a)

(b)

(c)

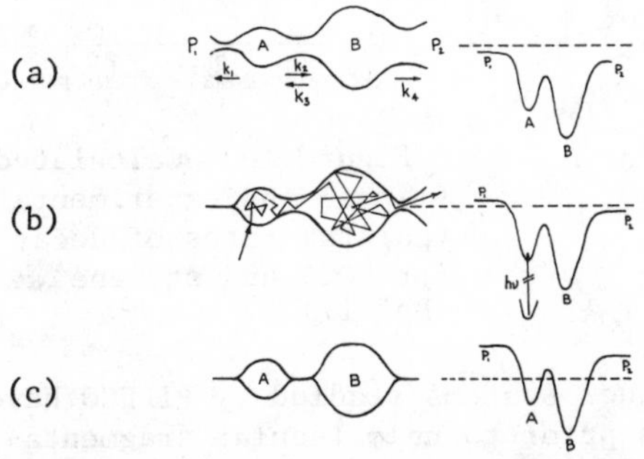

Figure 8. Phase space and potential energy surfaces for an isomeriza- tion - unimolecular reac- tion as a function of parent ion internal energy. Taken from ref. 16.

A simple example of a unimolecular-isomerization reaction studied by PIPECO is observed for isomers of $C_8H_8^+$ [17]. The measurement of identical dissociation rates for the reaction,

$$C_8H_8^+ \rightarrow C_6H_6^+ + C_2H_2 \qquad (4)$$

where $C_8H_8^+$ are the isomers styrene$^+$ (S) and cyclooctatetraene$^+$ (COT$^+$), indicates that an isomerization reaction preceeds the dissociation reaction. A comparison of the dissociation activation energies shown in Fig.9 for styrene$^+$ of 2.51eV and 1.4eV for cyclooctatetraene$^+$ sug- gest that COT$^+$ought to dissociate at a considerably faster rate than styrene. The QET calculation shown as the solid line CC (COT$^+$ mole- cular ion, COT‡ transition state) in Fig.10 confirms this prediction. The comparison also demonstrates clearly that the best agreement between calculated (SS) and experimental rates of dissociation for both isomers is obtained with a styrene molecular ion and, similar to the halobenzene case, a styrene transition state with an extended $C_6H_5-CHCH_2$ bond as inputs for the statistical calculation. Of course this limited study does not preclude the existence of another isomer with an energy lower than that of styrene[+], but because of the excellent agreement between theory and experiment it certainly seems that styrene[+] must certainly be close in energy to it.

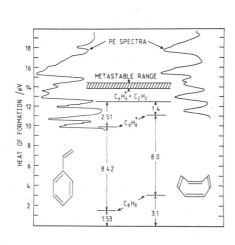

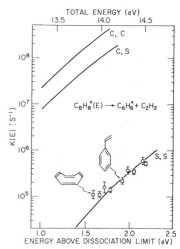

Figure 9. Heats of formation for cyclooctatetraene and styrene. Taken from reference 17.

Figure 10. Calculated (lines) and experimental (points) rates of decay for COT[+] and styrene[+]. (Ref.17)

As shown in Table I, numerous other systems studied by PIPECO have demonstrated isomerization reactions prior to unimolecular fragmentation[11]. One such system, [11] $C_4H_5N^+$, has part of its potential energy surface illustrated in Figure 11. The precise height of the isomerization barriers are unknown but they must be less than the dissociation energy for the identical fragmentation rates of each $C_4H_5N^+$ isomer to be observed.

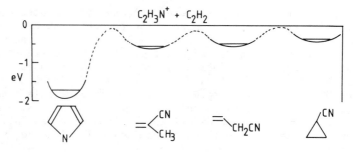

Figure 11. Potential energy diagram for the $C_4H_5N^+$ isomers.

[C] Isomerization and Competing Unimolecular Reactions

A further step in the complication of an ionic unimolecular reaction mechanism is to introduce competing unimolecular reactions along with the isomerization reaction. The dissociation reaction of state selected $C_6H_6^+$ ions represents one such case[18].

The metastable dissociation channels observed in the PIPECO study of $C_6H_6^+$ are:

$$C_6H_6(E) \underset{k_3}{\overset{k_1}{<}} \begin{matrix} C_6H_5^+ + H & (5) \\ k_2 \rightarrow C_4H_4^+ + C_2H_2 & (6) \\ C_3H_3^+ + C_3H_3 & (7) \end{matrix}$$

If all the above reaction channels compete then each reaction channel ought to dissociate with the same total unimolecular rate constant, k_T, given by: $k_T = k_1 + k_2 + k_3$.

The measurement of identical rates for the production of $C_3H_3^+$ and $C_4H_4^+$ from benzene+, 1,5- and 2,4-hexadiyne+ indicate isomerization, most likely to benzene+, occurs prior to fragmentation.

Baer et al.[18] have utilized the experimental fragmentation branching ratios and total rates of decay for $C_6H_6^+$ to determine the relative rates for each fragmentation channel. These rates are compared with the statistically calculated rates for each channel in Figure 12. Thus, although no direct PIPECO measurements of the rate of production of $C_6H_5^+$ were reported, from the above indirect experimental measurements and the RRKM calculations

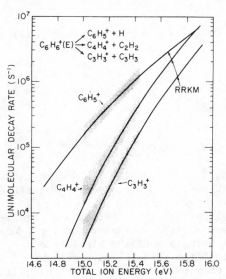

shown in Figure 12, it was demonstrated that all $C_6H_6^+$ metastable isomers dissociate via at least three fragmentation channels with a single decay rate.

The complex nature of the $C_6H_6^+$ reaction mechanism is well illustrated by considering Fig.13 which is a plot of the $C_3H_3^+/C_4H_4^+$ ratio as a function of the $C_6H_6^+$ internal energy. In section I of this figure all three precursor $C_6H_6^+$ ions dissociate with the same $C_3H_3^+/C_4H_4^+$ ratio indicating that they isomerize to a common precursor prior to dissociation. In section II, the ratio remains the same, but the sudden decrease at ~15.6 eV indicates that a new $C_4H_4^+$ channel has now become competitive.

Figure 12. Experimental (shaded) and RRKM calculated (lines) decay rates of $C_6H_6^+$. Taken from reference 18.

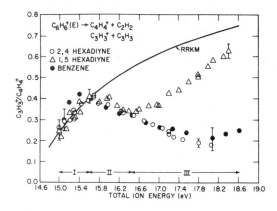

Figure 13. Ratio of the $C_3H_3^+$ and $C_4H_4^+$ intensities plotted as a function of the total internal energy of the state selected $C_6H_6^+$. (Taken from reference 18.)

In section III the 1,5-hexadiyne ratio increases demonstrating that it no longer completely isomerizes to the common precursor ion, whereas the 2,4-hexadiyne$^+$ and benzene$^+$ still appear to decay via a common pathway.

[D] Rate determining isomerization reactions

The PIPECO study on the metastable dissociation of furan$^+$ represents a case where an ion does not completely isomerize to the lowest energy isomer prior to dissociation[19]. The dissociation pathways of three $C_4H_4O^+$ isomers are shown below.

$$I \qquad \begin{cases} C_3H_4^+ + CO \\ C_2H_2O^+ + C_2H_2 \\ C_3H_3^+ + HCO \end{cases} \qquad (9)$$

$$II \quad H_2C=C-C=C=O \qquad \{ \ C_3H_4^+ + CO \qquad (10)$$

$$III \quad HC\equiv C-\overset{O}{\overset{\|}{C}}-CH_3 \qquad \begin{cases} C_3HO^+ + CH_3 \\ C_3H_4^+ + CO \\ C_3H_3^+ + HCO \end{cases} \qquad (11)$$

where $\Delta H_f(I)=197$, $\Delta H_f(II)=194$, $\Delta H_f(III)=247$ kcal/mol. On the basis of earlier discussions one might expect then that isomers I and III will isomerize to isomer II prior to dissociation.

The CO loss channel for vinyl-ketene$^+$ (VK$^+$) occurs via a pathway which has a small or no reverse activation energy barrier[20]. RRKM rates calculated for this channel using a modified vinyl-ketene molecular ion and transition state (line, VK$^+$,VK‡) are shown in Figure 14 to be about two orders of magnitude larger than the PIPECO rates (points). It is only when a modified furan molecular ion and transition state (F$^+$, F‡) are used that there is agreement between the calculated and experimental rates. Further, a calculation based on the assumption that all fragments are formed from an isomerized structure such as vinyl-ketene$^+$ leads to a $C_3H_4^+$ and $C_3H_3^+$ or $C_2H_2O^+$ ratio that is larger than the PIPECO ratio by about three orders of magnitude. In addition, the dissociation fragments observed for another $C_4H_4O^+$ isomer, 3-butyne-2-one$^+$ are quite different from those of furan$^+$.

These last results, coupled with the kinetic energy release measurements made by Holmes et al.[20] on the dissociation of a variety of $C_4H_4O^+$ isomers which indicated that the different isomers had different kinetic energy releases and that furan$^+$ had a two component kinetic energy release for the CO loss channel, suggest that isomerization amongst the $C_4H_4O^+$ isomers is hindered. In other words, the isomerization step is the rate determining step in the isomerization/unimolecular dissociation reaction of furan$^+$.

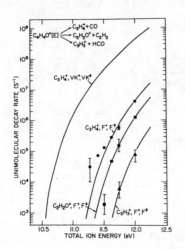

Figure 14

Experimental (points) and RRKM calculated (lines) unimolecular fragmentation rates for state selected furan$^+$ versus total ion energy. (Taken from reference 19)

F=Furan   VK=Vinylketene.
(+)≡Molecular ion, (‡)≡Transition state.

V   CONCLUSION

Despite the apparent statistical nature of the dissociation and isomerization reactions of the metastable ions described in this paper there are other PIPECO studies which indicate that low energy dissociations can often be comprised of a series of competing reaction channels involving direct dissociation, fluorescence, isomerization and radiationless transitions[4, 8, 18]. In certain cases where activation energy barriers are high or molecular ion symmetries may be important, the rates of any one of these reactions may become rate determining or may participate in the overall reaction mechanism to varying degrees. It has been the aim of this paper to systematically increase the complexity of metastable ionic unimolecular reaction mechanisms and demonstrate how the photoion-photoelectron coincidence technique may be used to characterize the reaction dynamics of such processes.

At higher internal energies where many decay channels will become competitive the dissociation of molecular ions may well become a non random process.

VI   ACKNOWLEDGEMENTS

The work described in this paper was carried out in Professor Baer's laboratory at the University of North Carolina at Chapel Hill. Many helpful discussions with Prof. Tom Baer, Dr. Ueli Buechler, Dr. Terry Murray, M/s Debbie Smith, and Mr. Jim Butler are gratefully acknowledged.

   GDW is also indebted to the CSIRO for a research fellowship held in 1978.

VII   REFERENCES

1. J.W. Rabalais, Principles of Ultraviolet Photoelectron Spectroscopy (Wiley Interscien., NY., 1976).
2. T. Baer, J. Electron Spectry. $\underline{15}$, 225 (1979).
3. H.M. Rosenstock, K. Draxl, B.W. Steiner and J.T. Herron, J.Phys.Chem.Ref. Data $\underline{6}$, 1 (1977).
4. T. Baer, Gas Phase Ion Chemistry, (M. Bowers, ED., Acad.Press, NY., 1975) Chapter 5.
5. J. Dannacher, Chem.Phys. $\underline{29}$, 339 (1978).
6. P.T. Murray and T. Baer, Int.J.Mass Spectrom. $\underline{30}$, 165 (1979).
7. W. Forst, Theory of unimolecular Reactions (Acad.Press., NY., 1973).
8. T. Baer, A.S. Werner, and B.P. Tsai, J.Chem.Phys. $\underline{62}$, 2497 (1975), $\underline{63}$, 4384 (1975).
9. D.M. Mintz and T. Baer, Int.J.Mass Spectrom.$\underline{25}$, 39 (1977).
10. L. Squires and T. Baer, J.Chem.Phys.$\underline{65}$, 4001 (1976).
11. G.D. Willett and T. Baer, J.Am.Chem.Soc.$\underline{102}$, 6769 (1980).
12. H.M. Rosenstock, M.B. Wallenstein, A.L. Warhaftig and H.Eyring Proc.Natl.Acad.Scien. $\underline{38}$, 667 (1952).
13. R.A. Marcus and O.K. Rice, J.Phys.Colloid Chem.$\underline{55}$, 894 (1951).
14. W. Chesnavich and M.T. Bowers, J. Am.Chem.Soc. $\underline{99}$, 1705 (1977).
15. T. Baer, B.P. Tsai, D. Smith and P.T. Murray, J.Chem.Phys. $\underline{64}$, 2460 (1976).
16. T. Baer, Proceedings of the 27th Annual Conference of the American Society for Mass Spectrometry, p. 576.
17. D. Smith, T. Baer, G.D. Willett and R.C. Omerod, Int.J.Mass Spectrom. $\underline{30}$, 155 (1979).
18. T. Baer, G.D. Willett, D. Smith and J.S. Phillips, J. Chem. Phys. $\underline{70}$, 4076 (1979).
19. G.D. Willett and T. Baer, J. Am. Chem. Soc. $\underline{102}$, 6769 (1980).
20. J.L. Holmes and J.K. Terlouw, J.Am.Chem.Soc., $\underline{101}$, 4973 (1979)

THE INTERPRETATION OF HIGH ENERGY ASYMMETRIC (e,2e) STUDIES

R.A. Bonham[†]
School of Physical Sciences, The Flinders University of South Australia,
Bedford Park, South Australia 5042.

ABSTRACT

Recently asymmetric (e,2e) experiments carried out with 8keV incident energy electrons have been reported. The high incident energy suggests that the first Born approximation should be a good starting point for investigating the dynamics of the collision process. Calculations were carried out in the framework of the first Born theory neglecting exchange and utilizing plane wave, Coulomb wave and Bates-Damgaard descriptions of the ejected electron wavefunction. The experimental data suggests the existence of small non first Born effects and suggest that the Bates-Damgaard description of the ejected wave is better than the other alternatives. Effects from the orthogonalization of the ejected wave to the bound state orbitals was shown to be negligible at the 0.5% level providing $K > 3\xi_{1s}$ where K is the momentum transferred by the incident electron to the target and ξ_{1s} is the largest orbital exponent in the wave function describing the 1s orbital.

A plane wave Born approximation has been utilized to calculate (e,2e) cross sections for the ejection of electrons from the 1s, 2s and 2p orbitals of Ne using very high incident electron energies (8 keV). This study has been motivated by the recent experimental measurements of the (e,2e) cross sections for the 2s and 2p orbitals in Ne by Benani and Wellenstein[1]. The purpose of the study is to attempt to sort out and identify the major characteristic features of the (e,2e) spectra. The experiments were carried out with 8 keV electrons with the fast electron (~ 7.8 keV) observed in the forward direction (θ_s ~ 5°-15°) in coincidence with a 200eV ejected electron[1]. Measurements were carried out as a function of the angle between the forward direction and the slow electron detector. The data were plotted in the manner first used by Ehrhardt and co-workers[2]. Principle distinguishing features of the spectra were:

(1) the observation of small recoil peaks for both the 2s and 2p orbitals;
(2) the observation of slight tilts of the binary and recoil lobes from the binary encounter direction ($\perp$ to the scattered direction); and
(3) the observation of a deep valley between the two lobes in the case of the 2p orbital (e,2e) diagram.

The (e,2e) cross sections were calculated in the plane wave Born approximation utilizing both a plane wave and a Coulomb wave description of the ejected electron. The orbitals were hydrogen like functions with the 2s function constructed to be orthogonal to the 1s electron for independently chosen orbital exponents. Several variations using the Coulomb representation for the ejected electron were carried out. First the

[†] Permanent address: Department of Chemistry, Indiana University, Bloomington, Indiana 47405, U.S.A.
The author wishes to acknowledge support by the National Science Foundation, Grant No. CHE-8000253.

Coulomb wave function was taken for an ion of charge +1 which is the correct form for the ejected electron wave function at large distances from the ion. Such a choice is not orthogonal to the bound state orbital so the (e,2e) cross section was calculated for the 1s orbital (worst case) with and without Schmidt orthogonalization to check for orthogonalization effects. The Coulomb wave function can also be used with the ion charge set equal to the orbital exponent. This guarantees orthogonalization between the ejected electron and bound state orbitals. The rational for such a choice is that the ejected electron wave function appears in a matrix element multiplied by the usual Born scattering operator and the bound state orbital. These functions will give maximum weight in the integration at a certain radius in the atom. The effective Coulomb potential which the ejected electron sees at this radius should be approximately $-\xi/r$ where ξ is the orbital exponent. This approach first employed by Mendelsohn[3] is in some sense the continuum analog of the Bates-Damgaard[4] method.

The details of the calculations for obtaining the cross sections are given elsewhere[5]. The results for an 8 keV incident electron beam with the fast electron observed at an angle of 11° in coincidence with a 200eV ejected electron as a function of the ejection angle are shown in Figs. 1-3.

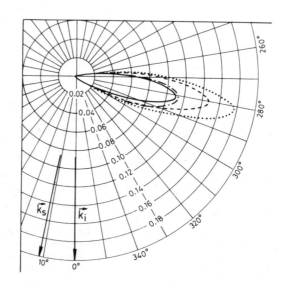

Fig. 1. The (e,2e) cross section for 8 keV electrons ionizing the 1s orbital of Ne with a scattering angle of 11° and an ejected electron energy of 200eV. The lobe symmetry axis is tilted below the binary encounter direction by 14°. The solid curve is a plane wave Born approximation (PWBA) result with the ejected electron wave function taken as a plane wave. The dotted outer curve is a PWBA calculation utilizing a hydrogenic Coulomb wave to describe the ejected electron wave function for asymptotic boundary conditions (ion charge = +1). The inner dash-dot curve comes from a calculation identical to that used in the outer case except that the ion charge was set equal to the orbital exponent and the orbital exponent was selected

to optimize the energy.   This is referred to as the hydrogenic approximation.   All ejected electron wave functions used were orthogonal to the
1s core function.   No recoil peak was observed for ejected wave functions
not orthogonal to the 1s core.

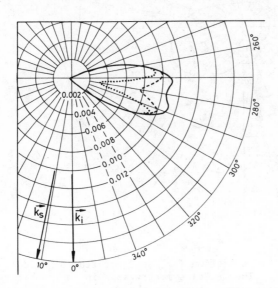

Fig. 2. The (e,2e) results for the 2s state of Ne.   See the legend for
Figure 1 for experimental details.   The outer dotted curve, not explained
in the Fig. 1 legend, is a hydrogenic case with ion charge set equal to the
orbital exponent which was selected to produce an exact fit to the Hartree
Fock impulse compton profile for the 2s orbital at its center.   Note that
in this case the lobe symmetry axis is close to the binary encounter direction of 281° and that no recoil peak is observed even when the ejected wave
is orthogonal to the bound state orbital (hydrogenic case).

    For the 1s case shown in Fig. 1 it was found that no recoil peak could
be observed if the 1s wave function was not orthogonal to the ejected
electron wave function.   A small recoil peak was observed when orthogonality was imposed.   Use of the asymptotic form of the Coulomb wave normalized to the 1s or the Mendelsohn hydrogenic Coulomb wave yield similar
shapes (as does the plane wave choice) but the different choices differ
greatly in the magnitude of the maximum lobe value.
    For the 2s and 2p cases no recoil lobe was observed even when the hy-
drogenic choice guaranteeing orthogonality was employed.   Choosing the
orbital exponent to give the optimum energy or to give a best impulse
Compton profile at the profile maximum produced no qualitative change in
shape but did yield results differing significantly in the magnitude of the
maximum lobe value.
    The depth of the valley between the two lobes proved to be very inter-
esting in the 2p case.   The valley should of course go to zero if the ion
recoil momentum goes to zero.   However for the conditions of the experiments discussed here q varies between two large values.   Hence the observed

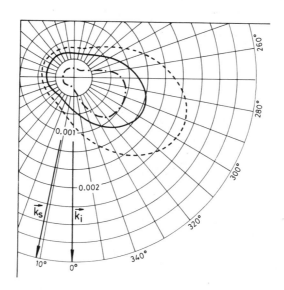

Fig. 3. The (e,2e) cross section for the 2p orbital of Ne calculated in the PWBA approximation. The (e,2e) symmetry axis is in the binary encounter direction. Deepest minima occurs for the hydrogenic Coulomb case and no results exhibited a recoil peak. See the legend for Figs. 1 and 2 for details.

deep valley depends on factors other than the p character of the orbital. The plane wave case yielded only a slight valley while the hydrogenic case yielded the deepest valley. It would thus appear that the valley depth is a direct measure of the effective nuclear charge which the ejected electron sees in its most probable orbital radius at the time ejection takes place. Finally no deviations were observed between the symmetry axis of the lobes and the binary encounter direction except for the 1s case where the lobe was tilted 14° toward the forward direction.

Recently Ehrhardt et al.[6] have pointed out that the recoil peak in the case of He can be explained by the 2nd Born calculations of Byron et al.[7]. The same calculations also seem to account for the observed tilts of the lobe symmetry axis from the binary encounter direction. These observations coupled with lack of a recoil peak or lobe axis tilts in the plane wave Born calculation strongly suggests that experimentally observed axis tilts and recoil structure can be used to determine the extent of 2nd Born contributions to the (e,2e) scattering process.

The magnitude of the observed non 1st Born contributions to the 2s and 2p (e,2e) diagrams in Ne is qualitatively in keeping with the predictions of Ehrhardt et al.[6]. If the experimental high energy (e,2e) diagrams can be placed on an absolute scale they should provide a sensitive test for the detailed shape of the bound state orbital being excited.

REFERENCES

1. L. Benanni and H.F. Wellenstein, Conference on electron and momentum

distributions.   June 7-13, Metz, France, 1981.

2.   H. Ehrhardt, K.H. Hesselbacher, K. Jung and K. Willmann, J. Phys. B$\underline{5}$, 1559 (1972).

3.   B.J. Bloch and L.B. Mendelsohn, Phys. Rev. A$\underline{9}$, 129 (1974).
      L.B. Mendelsohn and B.J. Bloch, Phys. Rev. A$\underline{12}$, 551 (1975).

4.   D.R. Bates and A. Damgaard, Phil. Trans. R. Soc. A242, 101 (1949).

5.   R.A. Bonham, P. Pattison and W. Weyrich, to be published.

6.   H. Ehrhardt, M. Fischer and K. Jung, Z. f. Phys. (in press).

7.   F.W. Byron, Jr., C.J. Joachain and B. Piraux, J. Phys. B$\underline{13}$, 2673 (1980).

MOMENTUM DENSITY MAPS FOR MOLECULES

J.P.D. Cook

School of Physical Sciences, The Flinders University of South Australia,
Bedford Park, S.A. 5042 Australia

C.E. Brion

Department of Chemistry, University of British Columbia,
Vancouver, B.C., Canada

ABSTRACT

Momentum-space and position-space molecular orbital density functions computed from LCAO-MO-SCF wavefunctions are used to rationalize the shapes of some momentum distributions measured by binary (e,2e) spectroscopy. A set of simple rules is presented which enable one to sketch the momentum density function and the momentum distribution from a knowledge of the position-space wavefunction and the properties and effects of the Fourier Transform and the spherical average. Selected molecular orbitals of H_2, N_2 and CO_2 are used to illustrate this work.

1. INTRODUCTION

Until recently there has been little attempt to rationalize the shapes of binary (e,2e) momentum distributions on a chemical basis. The physical basis for binary (e,2e) spectroscopy - the plane wave impulse approximation and the one-electron Hartree-Fock molecular orbital - has been carefully studied by several workers over the last ten years[1,2], but the chemistry has been restricted to simple classifications of momentum distributions as "s-type" or "p-type" (a nomenclature which arose out of measurements on atoms, and is inadequate for results on molecules), and to some remarks on the relative size of the outer lone pair orbitals in some simple hydrides[3]. And this is understandable; it has taken some time to surmount experimental difficulties, to establish the validity of the theory, to make studies of various theoretical wavefunctions including development of the generalised overlap method, to explore the problem of the one-particle model breakdown, and to build up a body of data on the more complex molecules. And so it is that we are only now beginning to look more closely at the information carried in the momentum distribution: studies are being done on the auto-correlation function $B_i(r)$ of the momentum distributions[4], on correlation plots relating $<p>$ to ionization potentials[5], and, in this work and others[6,7], on density maps as a means of understanding the shape of momentum distributions.

Briefly reviewing the theory of binary (e,2e) spectroscopy, the cross-section in the plane wave impulse approximation (PWIA) is:

$$\frac{d^5\sigma}{dEd\Omega_A d\Omega_B} \propto \frac{p_A p_B}{p_0} S_i \ |T(\underline{p}_A,\underline{p}_B;\underline{p}_0,\varepsilon_i)|^2 \ F_i(p) \qquad (1)$$

By a suitable arrangement of the experimental apparatus it is possible to keep the first three factors in equation 1 essentially constant, and so measure $F_i(p)$ directly to within a factor. $F_i(p)$ is the momentum distribution (MD) (also called the form factor or structure factor) and is the spherical average of the 3-dimensional momentum density function $\rho_i(\underline{p})$, of the electron in the ith molecular orbital (MO):

$$F_i(p) = (4\pi)^{-1}\int d\Omega\, \rho_i(\underline{p}) \qquad (2)$$

The momentum density is the square of the momentum-space (P-space) wavefunction $\psi_i(\underline{p})$:

$$\rho_i(\underline{p}) = \psi_i(\underline{p})\psi_i^*(\underline{p}) \qquad (3)$$

The P-space wavefunction $\psi_i(\underline{p})$ is related to its R-space counterpart by the Fourier Transform (FT):

$$\psi_i(\underline{p}) = (2\pi)^{-3/2}\int d\underline{r}\, e^{-i\underline{p}\cdot\underline{r}}\psi_i(\underline{r}) \qquad (4)$$

In words, the amplitude at a point p in P-space is given by the overlap of a plane wave $e^{-i\underline{p}\cdot\underline{r}}$ with the wavefunction $\psi_i(\underline{r})$ over all R-space. This concept of overlap of the FT wave with $\psi_i(\underline{r})$ will be used several times in this discussion.

The representation of the wavefunction ψ_i is equally valid in either R-space or P-space, and neither could be said to take precedence over the other, but for practical purposes (e.g. solution of the Schrodinger equation, conceptualization of electronic structure, etc.) we have come to use the former almost exclusively.

In summary, equations 2-4 form a set of relations which can be pictured schematically as shown here:

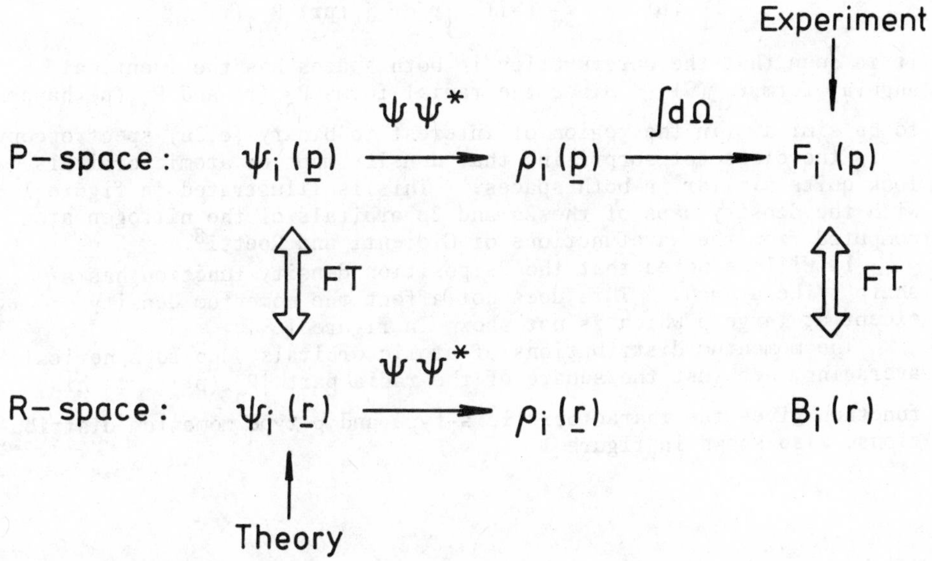

If one would like to understand better the shape of momentum distributions $F_i(p)$, it is helpful to look at the momentum density function $\rho_i(\underline{p})$, which in turn is better understood when one knows something about $\psi_i(\underline{r})$ and the nature of the Fourier Transform. While it is a fairly straightforward matter to write a computer program to generate the momentum distribution from a theoretical wavefunction and to display the momentum and position density functions as contour maps, this can be a time-consuming and expensive proposition. It is the aim of this work to present a set of principles and semi-empirical guidelines which not only rationalize the shape of such functions, but also allow one to quickly sketch them by hand intuitively, without recourse to a computer.

The density maps shown in this work were obtained by computing the function on a plane mesh of 100x100 points. Contours, at values indicated in the figure captions, were then interpolated on the mesh. In the molecular density maps the molecules are aligned with the vertical axis of the figure and the location of the atoms in the position density maps, though not indicated, is obvious by inspection.

2. ATOMS

Let us start out very simply with the density maps of atoms. Atomic wavefunctions can be expressed as:

$$\phi_{n\ell m}(\underline{r}) = N_{n\ell m} R_{n\ell}(r) Y_{\ell m}(\Omega) \tag{5}$$

It can be easily shown that:

$$\phi_{n\ell m}(\underline{p}) = N_{n\ell m} P_{n\ell}(r) Y_{\ell m}(\Omega) \tag{6}$$

where

$$P_{n\ell}(p) = \sqrt{\frac{2}{\pi}} (-i)^\ell \int r^2 dr \, j_\ell(pr) \, R_{n\ell}(r)$$

It is seen that the wavefunction in both spaces has the identical angular form $Y_{\ell m}(\Omega)$. Since the radial forms $R_{n\ell}(r)$ and $P_{n\ell}(p)$ happen to be similar (in the region of interest to binary (e,2e) spectroscopy) it is therefore not surprising that density maps of atomic orbitals look quite similar in both spaces. This is illustrated in figure 1 with the density maps of the 2s and 2p orbitals of the nitrogen atom computed from the wavefunctions of Clementi and Roetti[8].

It will be noted that the 2s position density function has a sharp spike at r=0. This does not affect the momentum density except at large p which is not shown in figure 1.

The momentum distributions of atomic orbitals, due to spherical averaging, are just the square of the radial part $|P_{n\ell}(p)|^2$. This function gives the characteristic s-type and p-type momentum distributions, also shown in figure 1.

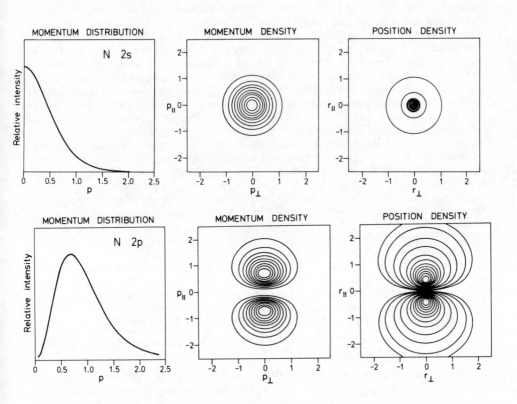

Figure 1. Momentum distributions and momentum and position density maps for the nitrogen atom 2s and 2p orbitals, computed from the wavefunctions of Clementi and Roetti[8]. Momentum density contours are at 10%, 20%, 30% ... 90% of the maximum density. Position density contours are at the same values, and 1%, 2%, and 5% contours have also been added.

3. MOLECULES

In molecules one no longer has a centrosymmetric problem, and the wavefunction cannot be written in the simple form of equations 5 and 6. Instead we use LCAO-MO-SCF wavefunctions, written as:

$$\psi_i(\underline{r}) = \sum_{\text{Atoms A}} \phi_A(\underline{r}-\underline{R}_A) \tag{7}$$

where

$$\phi(\underline{r}) = \sum_{\substack{\text{Basis} \\ \text{functions } j}} c_j \, \chi_{s,p,d\dots}(\underline{r})$$

That is, the ith molecular orbital $\psi_i(\underline{r})$ is a linear combination of atomic wavefunctions $\phi(\underline{r})$ situated on different centres A, and each $\phi(\underline{r})$ is a linear combination of basis functions $\chi(\underline{r})$ having different s,p,d... angular forms. The coefficients c_j are determined by variationally minimizing the total energy of the molecule, until a self-consistent field is reached.

There are two problems with this type of wavefunction in momentum space. The first is that there is no obvious angular form to the wavefunction due to the combination of s,p,d and higher forms on several centres, and certainly whatever angular form there is will not be the same in P-space as it is in R-space. The second problem is what to do with the nuclear coordinates in P-space. In R-space we are used to plotting the nuclei at their equilibrium geometry positions and then arranging the electron density around this nuclear skeleton. In P-space there are no obvious places to plot the nuclear coordinates: if one considers the nuclei to be fixed in R-space, then they have zero momentum and should be plotted at the P-space origin; if one considers that the nuclei move around the centre of mass then one can conceive of some sort of nuclear momentum density function, but this is not directly related to the geometry; one can also consider that when the electrons approach the nuclei they attain very high momenta, which suggests that the nuclei should be plotted at infinity in P-space. None of these concepts tells us directly how the nuclear geometry affects the momentum density, and the only sensible conclusion is to avoid plotting nuclear coordinates in P-space entirely.

Despite these problems all is not lost. There are three principles arising out of the nature of wavefunctions and the properties of the Fourier Transform which can be applied to the understanding of momentum density functions: they are the Symmetry property, the Inverse Scale relation, and the Interference effect.

In the following matter the symbols $r_\|$ and $p_\|$ are used to refer to the components of $\underline{r}$ and $\underline{p}$ in the direction parallel to the bond, $r_\perp$ and $p_\perp$ refer to the perpendicular component, and r and p refer to the magnitude. The subscript "max" attached to these symbols denotes the point where $\rho_i(\underline{r})$, $\rho_i(\underline{p})$, or $F_i(p)$ has reached a local maximum. The subscript "½max" denotes the point where the function has dropped to half its intensity at the origin.

3.1 The Symmetry Property

The Fourier Transform preserves all aspects of the symmetry of the R-space wavefunction over into P-space, and, in addition, introduces the inversion symmetry element i. Inversion symmetry is necessary in $\psi_i(\underline{p})$ to ensure a wavefunction with no net translational motion.

An immediate consequence of the symmetry property is this: all nodal symmetry planes are present in both R-space and P-space. This is a very useful tool if one is trying to sketch the momentum density function, as it indicates immediately where momentum density will not appear. A little thought will convince the reader of this: if $\psi_i(\underline{r})$ has a plane through which the function is symmetric and changes sign, then the overlap of all FT waves travelling in all directions parallel to this plane must be zero due to the cancellation of equal positive and negative parts.

3.2 The Inverse Scale Relation (ISR)

If the wavefunction ψ_i is contracted in a given dimension in R-space, then there will be a proportionate dilation in the corresponding dimension in P-space, and vice versa. This principle can be justified by several arguments. One is the form of the momentum operator in R-space:

$$\underline{P} = -i\underline{\nabla} \tag{8}$$

The momentum is given by the gradient of the wavefunction, so if $\psi_i(\underline{r})$ is dilated in some dimension its slope in that dimension is everywhere reduced, and so $\psi_i(\underline{p})$ is contracted.

This agrees with our intuitive concepts of orbitals. We say that electrons in compact inner shell orbitals move very fast and so have extended momentum distributions, whereas electrons in diffuse outer valence orbitals move relatively slowly and so have momentum distributions confined near the P-space origin.

One must be careful not to use the ISR to relate the density at specific points in R-space to points in P-space, without further consideration. It is usually true to say that density near the nuclei in R-space appears at large p in P-space only because $\psi_i(\underline{r})$ is usually changing rapidly near the nuclei. Rigorously speaking though, the amplitude at a point in one space is given by the overlap of the entire FT wave with the function in the other space, as described in the Introduction.

3.3 The Interference Effect

It has been shown earlier that the nuclear coordinates are meaningless in P-space, but since the FT does not destroy any of the information carried in $\psi_i(\underline{r})$ the geometry must still be contained in $\psi_i(\underline{p})$; the question is: how is it manifested in the momentum density function $\rho_i(\underline{p})$ and the momentum distribution $F_i(p)$?

Consider a wavefunction $\psi(\underline{r})$ made of s-type orbitals on two centres A and B. The momentum wavefunction is not just the sum of the two s-type functions as is the position wavefunction, because the fact that they are on two centres causes them to interfere: it is seen from figure 2 that there are certain frequencies of FT waves which overlap constructively with the wavefunction, and certain frequencies which overlap destructively. For example if the two areas are of the same sign and are separated by $\underline{R}$ then the P-space wavefunction will be reinforced at $\underline{p} = 2\pi/\underline{R}$, and reduced at $\underline{p} = \pi/\underline{R}$, relative to the atomic P-space wavefunction. Similarly one can work out other situations, and the general result is given in table 1. It is seen that the P-space location of these areas of reinforcement and reduction of $\psi_i(\underline{p})$ depend directly on the internuclear separation $\underline{R}$, and in this way the nuclear geometry information is preserved. The interference effect is strong when the system has high symmetry, when R is significant relative to the size of the constituent atomic orbitals, when the n value in table 1 is small, and when the effect is repeated over several areas of charge all separated by R.

The mathematical basis for the above argument is as follows. The Fourier Transform of equation 7 gives:

$$\psi_i(\underline{p}) = \sum_{\text{Atoms A}} e^{-i\underline{p}\cdot\underline{R}_A} \phi_A(\underline{p}) \tag{9}$$

284

The P-space molecular orbital wavefunction is a combination of atomic orbitals $\phi(\underline{p})$ located on the P-space origin, but multiplied by a phase factor $e^{-i\underline{p}\cdot\underline{R}_A}$ which carries the geometry information $\underline{R}_A$. The momentum density function $\rho_i(\underline{p})$ arising from this wavefunction can be separated into one-centre and two-centre parts, $\rho_i^{(1)}(p)$ and $\rho_i^{(2)}(p)$:

$$\rho_i(\underline{p}) = \psi_i(\underline{p})\psi_i^*(\underline{p}) = \rho_i^{(1)}(\underline{p}) + \rho_i^{(2)}(\underline{p}) \qquad (10)$$

where

$$\rho_i^{(1)}(\underline{p}) = \sum_{\text{Atoms } A} |\phi_A(\underline{p})|^2$$

$$\rho_i^{(2)}(\underline{p}) = 2\text{Re} \sum_{\substack{\text{Atom} \\ \text{pairs}}}^{A<B} e^{i\underline{p}\cdot\underline{R}_{AB}}\phi_A^*(\underline{p})\phi_B(\underline{p})$$

$$\underline{R}_{AB} = \underline{R}_A - \underline{R}_B$$

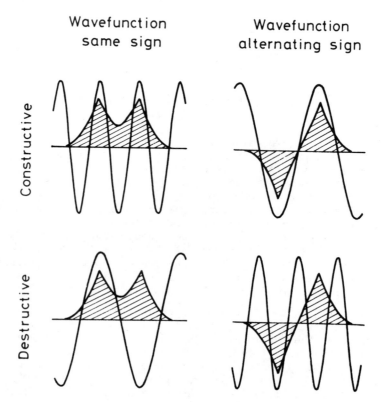

Figure 2. Schematic diagram showing constructive and destructive overlap of the FT wave with a wavefunction which has two extrema separated by R.

Table 1. Interference between areas of charge density separated by $\underline{R}$ resulting in constructive and destructive effects on the momentum density at $\underline{p}$.

	Wavefunction same sign	Wavefunction alternating sign
Constructive	$p = 2n\pi/R$	$p = (2n+1)\pi/R$
Destructive	$p = (2n+1)\pi/R$	$p = 2n\pi/R$
	$n = 0,1,2 \ldots$	

The one-centre or atomic part is the total atomic density summed at the P-space origin. This function is everywhere positive. The two-centre or interference part is the sum of cross-terms at the P-space origin, but modulated by a plane wave of period $2\pi/R_{AB}$. This function may have positive and negative regions and so can reinforce or reduce the one-centre part. Thus the nuclear geometry is present in the momentum density function, but in a rather obscure form.

Consider a simple but concrete example: the σ and σ^* molecular orbitals formed from the combination of identical 2s atomic orbitals on two centres A and B. The wavefunction is:

$$\psi(\underline{r}) = \frac{1}{\sqrt{2}}[\phi_{2s}(\underline{r}-\underline{R}_A) \pm \phi_{2s}(\underline{r}-\underline{R}_B)] \tag{11}$$

Substituting this into equation 10 gives

$$\rho(\underline{p}) = |\phi_{2s}(\underline{p})|^2[1 \pm \cos \underline{p}.\underline{R}_{AB}] \tag{12}$$

The shape of the various functions in the internuclear direction is shown in figure 3.

It is seen that the most noticeable effect of the internuclear separation is to introduce a set of nodal planes into the momentum density function perpendicular to the A-B axis, spaced at $2\pi/R_{AB}$, going out to infinity. In the σ-bonding case the interference effect has contracted the bulk of the density to lower $p_{\|}$ relative to the atomic density, leaving only a small lobe at larger $p_{\|}$. This contraction is seen in all molecular orbitals which are pure σ-bonding in character. Examples will be discussed below. This contraction can also be justified in a general way using the ISR: a σ-bonding orbital in R-space is elongated in the bond direction, and so the P-space wavefunction must show a contraction in this direction.

In the σ^*-antibonding case the $1-\cos\underline{p}.\underline{R}_{AB}$ factor has generated the nodal plane at $p_{\|}=0$, as required by the symmetry property, and has also introduced a maximum in the density which in a crude approximation is given by $p_{\|max} \sim \pi/R_{AB}$. The position of this maximum will in fact be shifted significantly to lower $p_{\|}$ by the slope of $|\phi_{2s}(\underline{p})|^2$ in that region, and so is not a completely accurate reflec-

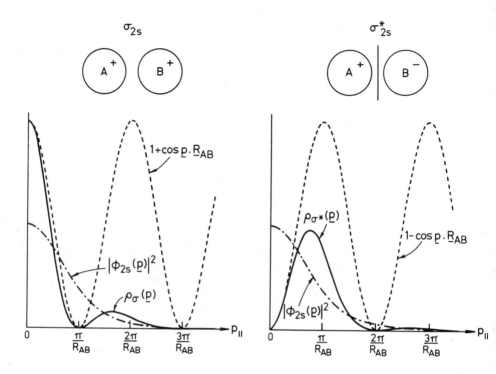

Figure 3. $\rho(\underline{p})$, $|\phi_{2s}(\underline{p})|^2$, and $1 \pm \cos \underline{p}.\underline{R}_{AB}$ evaluated in the $p_{\parallel}$ direction for the σ- and σ^*-bonding cases.

tion of the internuclear separation. Only in core σ orbitals is $|\phi(\underline{p})|^2$ flat enough near the first maximum of $1 - \cos \underline{p}.\underline{R}_{AB}$ so that $p_{\parallel \max} = \pi/R_{AB}$.

The effect of going from a homonuclear to a similar heteronuclear system can be illustrated by taking a slightly asymmetric wavefunction:

$$\psi(\underline{r}) = \frac{1}{\sqrt{2}}[\phi_{2s}(\underline{r}-\underline{R}_A) \pm \phi_{2s'}(\underline{r}-\underline{R}_B)] \qquad (13)$$

which gives the following density function:

$$\rho(\underline{r}) = \tfrac{1}{2}(|\phi_{2s}(\underline{p})|^2 + |\phi_{2s'}(\underline{p})|^2) \pm \phi_{2s}(\underline{p})\phi_{2s'}(\underline{p})\cos\underline{p}.\underline{R}_{AB} \qquad (14)$$

Since $\phi_{2s}\phi_{2s'} < \tfrac{1}{2}(\phi_{2s}^2 + \phi_{2s'}^2)$ the result is that the nodal planes are filled in somewhat. This gives us the useful result that the momentum density functions for an asymmetric molecule are like those of a similar more symmetric molecule, except that the nodal planes are filled in to some extent. Note that this filling-in applies only in the direction where the symmetry is reduced.

It is seen by these simple examples that the molecular geometry may have a significant effect on the momentum density function via the interference effect, and, provided one can interpret the "finger-print" left by this effect, one can go some way toward recovering the geometry information contained in the P-space wavefunction.

4. THE Q-PROJECTION FUNCTION

Along with the principles described in the previous three sections, it is also possible to characterize the wavefunction $\psi_i(\underline{r})$ empirically by the number and arrangement of positive and negative lobes.

Define the Q-projection as the projection of $\psi_i(\underline{r})$ onto a line of direction $\underline{r}'$:

$$Q_i(\underline{r}') = P(\psi_i(\underline{r});\underline{r}') \tag{15}$$

The squared Fourier Transform of $Q_i(\underline{r}')$ gives the momentum density along a line with the same direction as $\underline{r}'$, passing through the P-space origin:

$$Q_i(\underline{p}') = (2\pi)^{-1} \int dr' \, e^{-ip'r'} Q_i(\underline{r}') \tag{16}$$

$$\left| Q_i(\underline{p}') \right|^2 = \rho_i(\underline{p}) \Big|_{\underline{p}=\underline{p}'} \tag{17}$$

If $\left| Q_i(\underline{p}') \right|^2$ is known in two or three directions it is usually a simple matter to sketch in the full density function $\rho_i(\underline{p})$.

In practice it has been found that Q_i usually falls into one of a small number of different classes, depending on the number of positive and negative lobes. These classes are illustrated schematically in figure 4.

To use this figure first choose a direction $\underline{r}'$ in $\psi_i(\underline{r})$ and sketch the projection $Q(r')$. Then determine which class is closest by comparing the sketch with the $Q(r')$ functions in figure 4. The momentum density along p' is then given to a first approximation by the corresponding $\left| Q(p') \right|^2$. This process will give a crude idea of the shape of the momentum density function. Further discussion in section 5 will clarify the use of the Q-projection.

5. APPLICATIONS

Let us now examine some real molecules and try to understand how our chemical intuition, which has been well developed in R-space, can be extended into P-space.

We generally characterize molecular orbitals as bonding, non-bonding, or antibonding, according to how they contribute to the total force maintaining the atoms at their equilibrium nuclear geometry. The spatial regions of diatomic and triatomic molecules in which charge density contributes to bonding or antibonding effects are shown in figure 5.

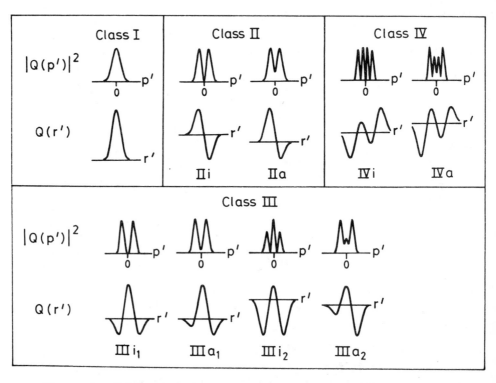

Figure 4. Schematic representation of the different classes of Q_i.

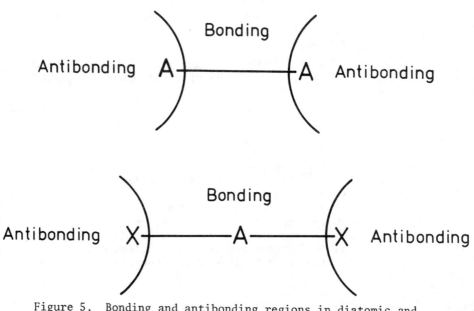

Figure 5. Bonding and antibonding regions in diatomic and linear triatomic molecules.

It is seen that a bonding orbital must increase density in the internuclear region at the expense of the regions outside the ends of the molecule, and antibonding orbitals must do the opposite. Nonbonding orbitals may arise when the charge density is situated mostly on atoms which are far apart relative to the size of the constituent atomic orbitals so that there is little interaction between atoms (type I), or may arise when there is a cancellation of roughly proportionate amounts of bonding and antibonding density (type II).

In the following discussion it will be shown how to recognize these bonding situations in momentum density functions and momentum distributions, although in the latter they are obscured due to spherical averaging and, in experimental measurements, statistical uncertainty.

5.1 The Simplest Molecule: H_2

The momentum distribution[9], and momentum and position density maps[10] for the $1\sigma_g$ orbital of H_2 are shown in figure 6.

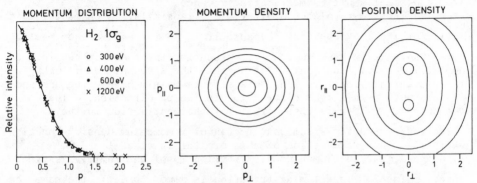

Figure 6. Momentum distributions and momentum and position density maps for the H_2 $1\sigma_g$ orbital. Experimental points and the theoretical momentum distribution are from reference 9. Density map wavefunctions are from reference 10. Contours are at 1%, 3%, 10%, 30% and 80% of the maximum density.

The simplest way to rationalize the shape of these functions is to consider that the position density function is, crudely speaking, an s-type distribution dilated in the $r_\parallel$ direction to span both nuclei.

From our knowledge of atomic density functions (section 2) we understand that s-type functions look similar in both spaces, and it only remains to invoke the Inverse Scale relation to account for the contraction of the momentum density in the $p_\parallel$ direction. This orbital falls into Q-projection class I.

This s-type distribution contracted in the $p_\parallel$ direction is characteristic of all pure σ-bonding valence orbitals, as will be seen in the N_2 $2\sigma_g$ and CO_2 $3\sigma_g$ orbitals.

A more rigorous rationalization of the $1\sigma_g$ density function is of course the same as that given in section 3.3 for the hypothetical σ-bonding orbital, except with 1s functions instead of 2s.

The $1\sigma_g$ momentum distribution is unremarkable: the familiar s-type distribution is seen.

5.2 A Diatomic Molecule: N_2

The electronic configuration of N_2 is:

$$(1\sigma_g)^2(1\sigma_u)^2(2\sigma_g)^2(2\sigma_u)^2(1\pi_u)^4(3\sigma_g)^2$$

and the internuclear spacing is 2.07 a_o. Experimental[11] and theoretical[10] results are given in figure 7.

The N_2 inner shell orbitals serve as an interesting illustration of a very strong interference effect produced when the size of the atomic N1s orbitals is much less than the internuclear spacing R. This effect persists even into the spherically-averaged momentum distributions for two or three oscillations. Unfortunately the $1\sigma_g$ and $1\sigma_u$ orbitals are very nearly degenerate, and the experimentally-measurable combined momentum distribution would be that of the N1s atomic orbital: all geometry effects cancel.

The innermost valence orbital, $2\sigma_g$, is similar in character to the H_2 $1\sigma_g$ orbital, except that, because it is in the n=2 valence shell $\psi(\underline{r})$ changes sign near the nuclei, producing the sharp spikes in the momentum density function. But because the wavefunction is changing rapidly in this low r region, this affects mainly high p regions and is seen only as a slight contraction in the $p_\perp$ direction. The σ-bonding character is responsible for the contraction in the $p_\parallel$ direction, producing a somewhat rectangular momentum density function overall. The Q-projection of this orbital is class I.

Again, the $2\sigma_g$ momentum distribution is of the familiar s-type shape. The $2\sigma_g$ momentum distribution is almost exactly the same shape as the N2s distribution (figure 1), in agreement with the idea that inner valence orbitals have a large degree of atomic character.

The $2\sigma_u$ orbital is strongly antibonding, as seen by the absence of density in or near the $r_\parallel$ =0 plane. By the Symmetry property, this nodal plane must also appear in the momentum density map, and this is the first clue in P-space to the antibonding character of this orbital. The position of $p_{\parallel max}$ at 0.8 a_o^{-1} indicates, according to table 1 a significant amount of position density separated by roughly 4 a_o, which is indeed outside the ends of the molecule in the antibonding region. This orbital is Q-projection class IIi. The momentum distribution is the unremarkable p-type shape.

The $1\pi_u$ orbital is a pure π-bonding case. As with the $2\sigma_g$ case, this orbital retains its overall p-type shape, but suffers a contraction in the $p_\parallel$ direction via the ISR, again characteristic of orbitals which are pure bonding in character. This contraction has a small but noticeable effect on the momentum distribution: although the N2p and N_2 $1\pi_u$ MDs both peak at pmax 0.7 a_o^{-1}, the N_2 $1\pi_u$ MD falls off faster at larger p than does the N2p. Careful examination of figures 1 and 7 will show this. The $1\pi_u$ orbital is class IIi.

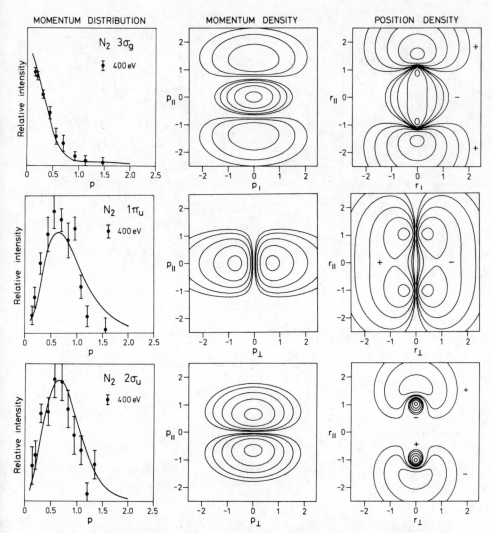

Figure 7. Momentum distributions and momentum and position density maps for all the orbitals of N_2. Experimental points are from reference 11. Theoretical curves are from reference 10. Contours are at 1%, 3%, 10%, 30% and 80% of the maximum density. The sign of the R-space wavefunction is indicated.

The outermost $3\sigma_g$ orbital momentum distribution is unusual in that it falls off with increasing p much faster than either the N2s or N_2 $2\sigma_g$ MDs: the $p_{\frac{1}{2}max}$ value for the $3\sigma_g$ is about 0.4 a_o^{-1}, compared to 0.7 a_o^{-1} for the latter two. The reason for this becomes apparent in the momentum density map which shows two nodal planes at $p_{\parallel} = 0.75 a_o^{-1}$, and only a small amount of density in the lobes at larger $p_{\parallel}$, which, when spherically averaged, appears even smaller. The position density map shows an orbital which is nonbonding (type II), and can overlap in

292

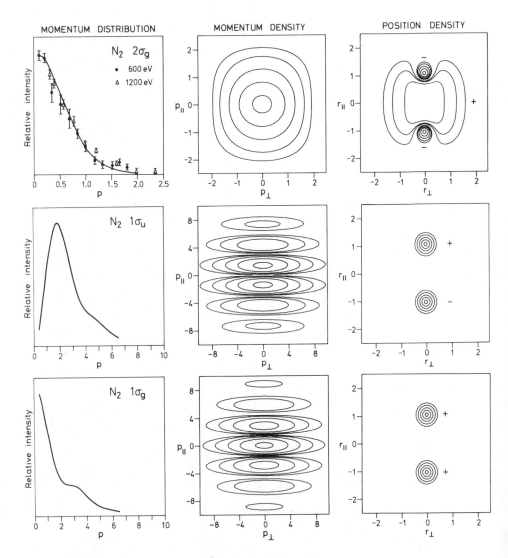

Figure 7 continued

a destructive manner with an FT wave of period around 8 a_o, producing
the nodal planes.   The existence of three lobes of (alternating sign)
position density all separated by about 2 a_o further means that an
FT wave of period 4 a_o will overlap constructively, thus reinforcing
the lobe at $p_{||} \sim 1.5$ a_o^{-1}.   The $3\sigma_g$ orbital is class $IIIi_2$ Q-projection.
     In general, the theoretical wavefunction of Snyder and Basch[10]
gives a very good account of the major features of the electronic
structure in this molecule.

5.3 A Linear Triatomic Molecule: CO_2

     The valence electronic structure of CO_2 is:

$$(3\sigma_g)^2(2\sigma_u)^2(4\sigma_g)^2(3\sigma_u)^2(1\pi_u)^4(1\pi_g)^4$$

and the C-O bond length is 2.2 a_o. Figure 8 gives theoretical[10] and experimental[6,7] results for the $4\sigma_g$, $3\sigma_u$ and $1\pi_g$ MOs of this molecule. These particular orbitals have been chosen as they have interesting new MD shapes, not seen in H_2 or N_2. Complete results are presented elsewhere[6,7].

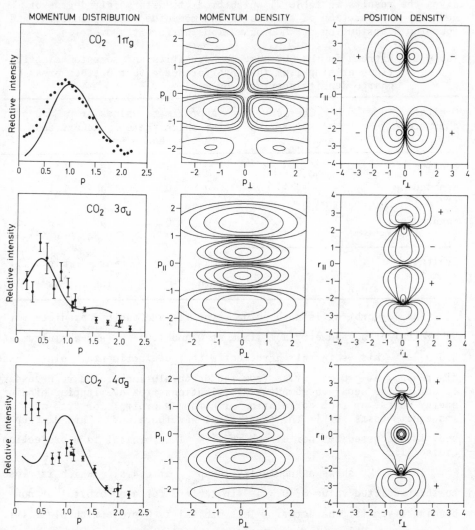

Figure 8. Momentum distributions and momentum and position density maps for the CO_2 $4\sigma_g$, $3\sigma_u$ and $1\pi_g$ orbitals. Experimental points are from references 6 and 7. Theoretical curves are from reference 10. Contours are at 1%, 3%, 10%, 30% and 80% of the maximum density. The sign of the R-space wavefunction is indicated.

Because these CO_2 MOs, even more than in N_2, have lobes of well-localised charge density, with well-defined spatial separation, there is a strong interference effect, and it is possible to predict the shape of the MDs knowing only the molecular geometry. The bonding regions in CO_2 are separated by about 2-3 a_o, and the antibonding regions by about 6 a_o. Nonbonding (type I) density on the oxygen atoms alone is separated by 4.4 a_o. Substituting these values for R in table 1 gives the results in table 2, which predicts with a fair degree of accuracy the location of extrema in the momentum density function and momentum distribution for MOs of various different symmetries.

Table 2. Extrema in the CO_2 momentum densities and momentum distributions arising from the interference effect between areas of charge density separated by R.

Orbital character	Charge separation	Approximate values of $p_\parallel$ and p at which there is an extremum in $\rho(\underline{p})$ and $F_i(p)$			
Bonding	2-3	0	1-1.6	2-3	3-4.7 ...
Nonbonding (type I)	4.4	0	0.7	1.4	2.1 ...
Nonbonding (type II)	2,6	0	0.5	1.0	1.6 ...
Orbital symmetry	σ_g, π_u	max	min	max	min ...
	σ_u, π_g	min	max	min	max ...

The $4\sigma_g$ orbital is nonbonding (type II) and table 2 predicts for an orbital of σ_g symmetry a maximum at 0 and 1.0 a_o^{-1}, and a minimum at 0.5 a_o^{-1}. This is in fair agreement with the experimental points: the discrepancy at $p=0$ may arise from the deconvolution procedure necessary to obtain the separate momentum distributions from overlapping binding energy peaks, or from correlation with the $3\sigma_g$ orbital, or from an inadequate basis set in the theoretical wavefunction. This discrepancy is discussed further elsewhere[6,7]. This orbital is Q-projection class $IIIi_2$.

The $3\sigma_u$ MD shows an unusually low p_{max} at 0.4 a_o^{-1}, which is close to the predicted value of 0.5 a_o^{-1} in table 2 for a σ_u orbital of non-bonding (type II) character. This orbital is Q-projection class IVi.

The large $1\pi_g$ p_{max} at 0.9 a_o^{-1} arises from the π^* symmetry of this orbital: that is, two perpendicular nodal planes. The oxygen 2p atomic orbital constituency of this MO means that $p_{\perp max}$ must be about 0.8 a_o^{-1}. The interference effect dictates that $p_{\parallel max}$ occurs at about 0.6 a_o^{-1} for a nonbonding (type I) π_g orbital (slightly less than $0.7a_o^{-1}$

indicated in table 2 due to the slope of the O2p atomic orbital in this region). Their combined effects put p_{max} at about 0.9 a_o^{-1} radially from the P-space origin, in agreement with the momentum distribution. This large p_{max} occurs for all MOs with this symmetry[6,12,13]. This orbital is class IIIi$_1$.

6. SUMMARY

The three principles regarding the relation between $\psi_i(\underline{r})$ and $\psi_i(\underline{p})$ are here restated:

1. Symmetry Property: The Fourier Transform preserves all aspects of the symmetry of $\psi_i(\underline{r})$ over into $\psi_i(\underline{p})$, and also adds the inversion symmetry element. All nodal symmetry planes present in $\psi_i(\underline{r})$ are also present in $\psi_i(\underline{p})$.

2. Inverse Scale Relation: A dilation of $\psi_i(\underline{r})$ in some dimension results in a proportionate contraction of $\psi_i(\underline{p})$ in that dimension, and vice versa.

3. Interference Effect: If $\psi_i(\underline{r})$ reaches a pronounced extremum at several points separated by $\underline{R}$, then $\psi_i(\underline{p})$ will show a modulation in the direction of $\underline{R}$ with a period $2\pi/R$. If these extrema are of the same sign, the modulation has a maximum at $p_\| = 0$. If the extrema are of alternating sign, the modulation has a minimum at $p_\| = 0$.

With these principles it is possible to explain the major features of the momentum distributions of linear molecules given a reasonably good wavefunction[10] as shown in the cases of H_2, N_2 and CO_2. It is seen that orbital symmetry, molecular geometry and bonding character have pronounced and readily discernible effects on the momentum density function, and hence on the momentum distribution.

The next step will be to study non-linear molecules. Here the situation will be complicated by the fact that there will be interference effects in more than one direction, and density maps in several planes may have to be considered.

REFERENCES

1. I.E. McCarthy and E. Weigold, Physics Reports 27C, 275 (1976).
2. I.E. McCarthy and E. Weigold, Advances in Physics 27, 489 (1976).
3. J.P.D. Cook, C.E. Brion and A. Hamnett, Chem. Phys. 45, 1 (1980).
4. J.A. Tossell, J.H. Moore and M.A. Coplan, J. El. Spect. 22, 61 (1981).
5. I. Suzuki, private communication.
6. J.P.D. Cook, Ph.D. Thesis, unpublished.
7. J.P.D. Cook and C.E. Brion, manuscript in preparation.
8. E. Clementi and R. Roetti, At. Data and Nucl. Data Tables 14, 177 (1974).
9. S. Dey, I.E. McCarthy, P.J.O. Teubner and E. Weigold, Phys. Rev.

Lett. 34, 782 (1975).

10. L.C. Snyder and H. Basch, "Molecular Wavefunctions and Properties", Wiley, New York, 1972.

11. E. Weigold, S. Dey, A.J. Dixon, I.E. McCarthy, K.R. Lassey and P.J.O. Teubner, J. El. Spect. 10, 177 (1977).

12. I.H. Suzuki, E. Weigold, and C.E. Brion, J. El. Spect. 20, 289 (1980).

13. C.E. Brion, J.P.D. Cook, I.G. Fuss and E. Weigold, to be published in J. El. Spect.

RELATIVISTIC (e,2e)

I. Fuss*, R. Helstrom, R. Henderson, J. Mitroy
B.M. Spicer and D. Webb

School of Physics, University of Melbourne,
Parkville, Victoria, Australia 3052.

ABSTRACT

A relativistic plane wave Born approximation is presented and applied to the (e,2e) reaction with the inner and outer shells of heavy elements (Au, Hg and Th). Relativistic effects are shown to be very significant in relation to these states.

The status of a relativistic (e,2e) spectrometer is described and experimental difficulties discussed.

INTRODUCTION

The (e,2e) reaction is an electron impact ionisation reaction in which the two out going electrons are detected in coincidence after angular and energy analysis. The geometry for such a reaction is illustrated in figure 1.

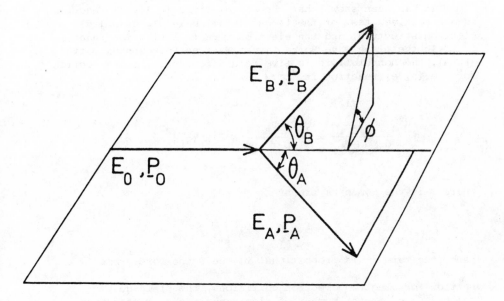

(Figure 1:)   Schematic diagram of the kinematics of an
              (e,2e) reaction

*Present address: Defence Research Centre Salisbury,
                   Salisbury, South Australia 5109

Using the conservation of energy, and ignoring ion recoil energy, the binding energy ε_B of the ejected electron is,

$$\varepsilon_B = E_o - E_A - E_B \qquad\qquad 1.$$

where E_o is the incident electron energy, E_A is the scattered electron energy and E_B is the ejected electron energy. Similarly the ion recoil momentum $\underline{q}$ can be obtained via the conservation of momentum,

$$\underline{q} = \underline{p}_o - \underline{p}_A - \underline{p}_B \qquad\qquad 2.$$

where $\underline{p}_o$ is the incident electron momentum,

$\quad\underline{p}_A$ is the scattered electron momentum, and

$\quad\underline{p}_B$ is the ejected electron momentum.

It has been shown that the nonrelativistic (e,2e) reaction is a sensitive test of the target (atom, molecule or solid) electron structure, and the electron impact ionisation reaction[1-3]. In the plane wave Born, target Hartree Fock approximation (PWBA), the nonrelativistic five fold differential cross section for the (e,2e) reaction is [4,5].

$$\frac{d^5\sigma}{dE_A \, d\Omega_A \, d\Omega_B} = \frac{1}{\pi} \frac{P_A \, P_B}{P_o} \frac{1}{K^4} n_t \left| R_t(q) \right|^2 \qquad\qquad 3.$$

where $\underline{K}$ is the momentum transfer

$$\underline{K} = \underline{p}_o - \underline{p}_A \qquad\qquad 4.$$

hence the term $\dfrac{1}{K^4}$ is proportional to the Rutherford cross section for electron-electron collisions, the state quantum number $t \equiv n, \ell$, the number of electrons in the state t is n_t and $R_t(q)$ is the momentum space radial wave function for that state. Cross sections for electron impact energies from 0.5 to 10 MeV on the K shells Au^{79} and Th^{90} calculated in the PWBA are presented in figures 3 and 4 respectively.

RELATIVITY

As an example of the effect of relativity on the (e,2e) reaction, consider an electron of kinetic energy T_o incident on a stationary free electron, resulting in coplanar symmetric scattering. $T = T_A = T_B$, $\theta = \theta_A = \theta_B$, $\phi = 0^o$ (see figure 1). The angle θ is determined by the conservation of energy $T = T_o/2$ and the conservation of momentum in the direction of the incident electron:

$$p_o = p \cos\theta \qquad\qquad 5.$$

The relativistic calculation gives

$$\theta = \cos^{-1} \left(\frac{T^2 + 2\ T\ mc^2}{T^2 + 4\ T\ mc^2} \right)^{\frac{1}{2}} \qquad\qquad 6.$$

The result of this equation is plotted in figure 2 together with the nonrelativistic calculation which yields $\theta = 45^o$.

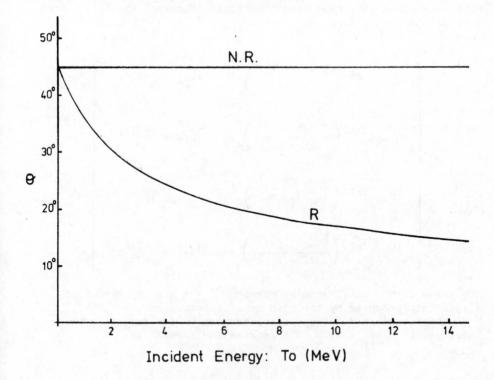

(Figure 2:)   Scattering angle as a function of incident
               electron energy for free electron-electron
               scattering in the symmetric (e,2e) reaction

Thus the angle at which the ion recoil momentum is equal to zero in the (e,2e) reaction, depends on E_o. In particular as E_o increases $\theta \to 0^o$ placing restrictions on the maximum incident energy because of the physical size of the electron detectors.

RELATIVISTIC PLANE WAVE BORN APPROXIMATION

The atomic wave functions are determined from the Hamiltonian[6]

$$H = \sum_i H_D(i) + \sum_{i<j} \frac{1}{r_{ij}} \qquad\qquad 7.$$

where $H_D(i)$ is the Dirac Hamiltonian for the ith electron.

$$H_D(i) = \underline{\alpha}(i) \cdot cp(i) + c^2 \beta(i) - \frac{Z}{r_i} \qquad\qquad 8.$$

The central field solutions to this Hamiltonian are called Dirac Fock wave functions. The single particle states of the Dirac Fock functions have the form

$j = \ell \pm s$

$$\psi_{nj\ell}(\underline{r}) =
\begin{bmatrix}
R_{nj\ell}(r) \left(\dfrac{\ell \pm m + 1/2}{2\ell + 1} \right)^{1/2} Y_{\ell, m - \frac{1}{2}}(\theta,\phi) \\[2ex]
R_{nj\ell}(r) \left(\dfrac{\ell \mp m + 1/2}{2\ell + 1} \right)^{1/2} Y_{\ell, m + \frac{1}{2}}(\theta,\phi) \\[2ex]
- i\varrho_{nj\ell}(r) \left(\dfrac{\ell \mp m + 3/2}{2\ell + 3} \right)^{1/2} Y_{\ell+1, m + \frac{1}{2}}(\theta,\phi) \\[2ex]
\mp i\varrho_{nj\ell}(r) \left(\dfrac{\ell \pm m + 3/2}{2\ell + 3} \right)^{1/2} Y_{\ell+1, m - \frac{1}{2}}(\theta,\phi)
\end{bmatrix} \qquad 9.$$

The terms with orbital angular momentum $\ell \pm 1$ are the result of a spin flip produced by the spin orbit interaction.

The five fold differential cross section for (e,2e) in the relativistic plane wave Born approximation (RPWBA) is

$$\frac{d^5\sigma}{dE_A d\Omega_A d\Omega_B} = \frac{1}{\pi} \frac{P_A P_B}{P_o} \mu \, n_t \frac{(|R(q)|^2 + |\varrho(q)|^2)}{\kappa^4} \qquad 10.$$

where

$$R(q) = \frac{2}{\pi}^{\frac{1}{2}} \int_o^\infty r^2 \, dr \, j_\ell \, (qr) \, R_{nj\ell}(r) \qquad 11.$$

and

$$Q(q) = \frac{2}{\pi}^{\frac{1}{2}} \int_o^\infty r^2 \, dr \, j_{\ell+1}(qr) \, Q_{nj\ell}(r) \qquad 12.$$

are the radial Dirac Fock wave functions in momentum space.

$$\mu = \mu_M \cdot \mu_K \qquad 13.$$

$$\mu_M = \left(1 + \frac{E_A - mc^2}{2mc^2}\right)\left(1 + \frac{E_B - mc^2}{2mc^2}\right) \qquad 14.$$

$$\mu_K = \left(1 + \frac{p_o \cdot p_A}{m^2 c^2}\right)\left(1 + \frac{p_B c}{E_B}\right) \qquad 15.$$

The relativistic term μ_M is magnetic in origin and μ_M is kinematic. The expression

$$\left(\frac{E_O - E_A}{C}\right)^2 \qquad 16.$$

in the "Rutherford" term

$$\kappa^4 = \frac{1}{\left(\kappa^2 - \left(\frac{E_O - E_A}{C}\right)^2\right)^2} \qquad 17.$$

is due to retardation.

The nonrelativistic limit of equation 10 is the PWBA, equation 3.

The effect of including relativity in the PWBA for the (e,2e) reaction is illustrated in figures 3 and 4. These show that for electron impact energies higher than 0.5 MeV relativistic effects significantly change the magnitude of the cross section and its form as a function of energy.

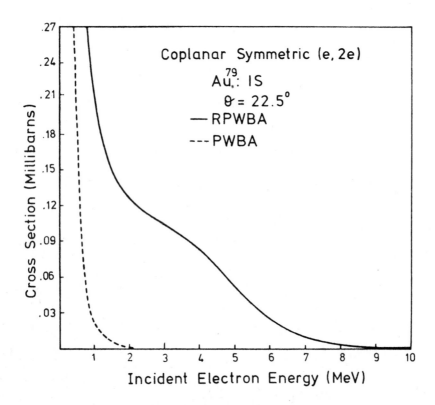

(Figure 3:) The electron impact ionisation cross section for the 1s state of gold calculated using the relativistic plane wave Born approximation RPWBA and the plane wave Born approximation.

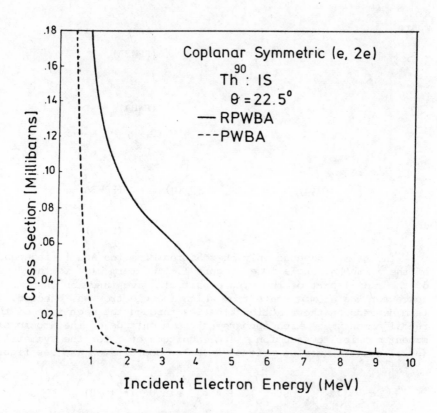

(Figure 4:)   The electron impact ionisation cross section for
the 1s state of thorium calculated using the
relativistic plane wave Born approximation RPWBA
and the plane wave Born approximation.

The (e,2e) cross section in the RPWBA and the PWBA approximations can be broken into the product of two parts

$$\frac{d^5\sigma}{dE_A \, d\Omega_A \, d\Omega_B} = A \cdot \rho(q) \qquad\qquad 18.$$

where

$$A = \frac{1}{\pi} \frac{p_A p_B}{p_o} \cdot \frac{\mu}{K^4} \qquad (RPWBA) \qquad\qquad 19.$$

$$A = \frac{1}{\pi} \frac{p_A p_B}{p_o} \frac{1}{K^4} \qquad (PWBA) \qquad\qquad 20.$$

and

$$\rho(q) = \left| R_t(q) \right|^2 + \left| Q_t(q) \right|^2 \qquad (RPWBA) \qquad\qquad 21.$$

$$(q) \quad \left| R_t(q) \right|^2 \qquad (PWBA) \qquad\qquad 22.$$

The term A depends only on reaction kinetics and is independent of the azimuthal angle ϕ see figure 1. The term $\rho(q)$ is the "square" of the radial part of the single particle wave function in momentum space for the atomic state from which the electron was ejected. This term depends on the reaction kinetics through the magnitude of the ion recoil momentum q which appears as the magnitude of the atomic electron momentum prior to ejection. The magnitude of q for the symmetric (e,2e) reaction may be written in polar coordinates as (see figure 1).

$$q = \left((2 \, p_A \cos\theta - p_o)^2 + 4 \, p_A^2 \sin^2\theta \, \sin^2(\phi/2) \right)^{\frac{1}{2}} \qquad\qquad 23.$$

The magnitude of the ion recoil momentum q depends on the azimuthal angle ϕ and so the square of the radial wavefunction can be measured as a function of momentum q by measuring the (e,2e) cross section as a function of ϕ.

This type of noncoplanar (e,2e) reaction has been studied extensively at nonrelativistic energies and has been shown to accurately measure the square of the wave functions for the outer orbitals of atoms and molecules[2]. It should be possible to make such measurements at relativistic energies of a few MeV and hence measure the square of the wave functions of inner orbitals of heavy elements. These inner orbitals are significantly affected by relativistic effects[6]. Figures 5 and 6 present calculations of cross sections for the K shells of gold and thorium respectively. The two curves presented in each figure are both calculated using relativistic reaction theory,[7] however one curve uses a nonrelativistic Hartree Fock calculation[7] for the atomic wave function and the other a relativistic Hartree Fock calculation[6]. The depression of the relativistic momentum profile towards q = o occurs because of the inward contraction of the relativistic wave function in real space. This effect is more marked in (e,2e) which measures the square of the wave function ρ (q), than in a technique which measures the probability density p.ρ(q) which changes less rapidly as p → 0.

The magnitudes of the cross sections are the same as many nuclear physics cross sections and so should be measurable. Since the K shell measurements can be made using solid targets absolute cross sections should be obtainable[2]. However the shapes of the relativistic and nonrelativistic Hartree Fock wave functions are different and so a relative cross section measurement should be able to distinguish between the two theories.

An intermediate expression can be obtained from equation 10 in which the reaction kinematics are nonrelativistic but relativity is retained in the atomic wave function

$$\frac{d^5\sigma}{dE_A d\Omega_A d\Omega_B} = \frac{1}{\pi} \frac{P_A P_B}{P_o} \frac{1}{K^4} n_t \left(|R_t(q)|^2 + |Q_t(q)|^2 \right)$$

This expression finds application in the study of intermediate energy electron impact ionisation of the outer states of heavy elements e.g. Hg [80]. The outer states of the heavy elements can be significantly affected by the inclusion of relativity in the atomic Hamiltonian. This is the result of the changed atomic potential due to the altered inner electron wave functions and the constraint of orthogonality placed on the independent particle states.

Figures 7 and 8 present momentum space profiles for the principal outer states of Hg[80], respectively $6s_{\frac{1}{2}}$ and the 5d states.

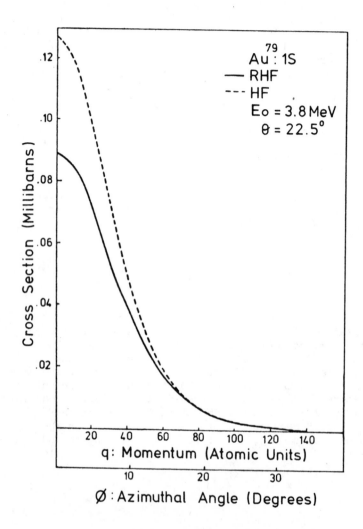

(Figure 5:) The cross section for the (e,2e) reaction with the ls state of gold in the noncoplanar geometry calculated using the relativistic plane wave Born approximation with Dirac Fock ls wave functions RHF and Hartree Fock ls wave functions.

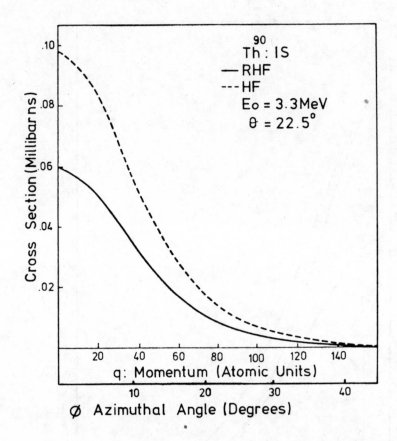

(Figure 6:)   The cross section for the (e,2e) reaction with the
1s state of thorium in the noncoplanar geometry
calculated using the relativistic plane wave
Born approximation with Dirac Fock 1s wave functions
RHF and Hartree Fock 1s wave functions.

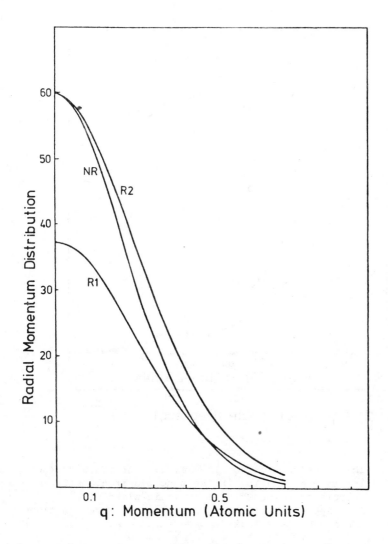

(Figure 7:)   Momentum profiles for the 6s state of mercury
              calculated using Dirac Fock R1 and Hartree
              Fock Hamiltonians NR.   The R2 curve is the R1
              calculation normalised to the NR calculation
              at q = o.

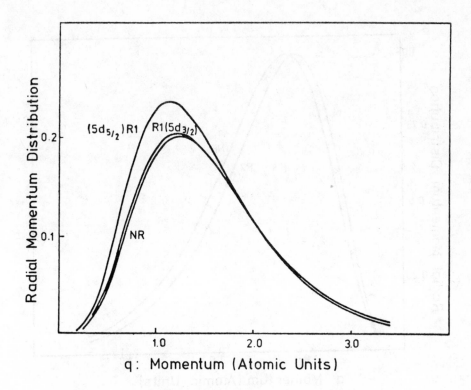

(Figure 8:) Momentum profiles for the 5d state of
mercury calculated using Dirac Fock R1
and Hartree Fock Hamiltonians NR.

The Hartree Fock states are labelled NR and the Dirac Fock with R1.
These figures show that the inclusion of relativistic effects
dramatically changes the momentum profiles of the $6s_{\frac{1}{2}}$ and the $5d_{5/2}$
states. The $5d_{3/2}$ wave function however, is only slightly
modified. The relativistic theory has been normalised (R2) to the
maximum of the nonrelativistic curve in figure 7 ($6s_{\frac{1}{2}}$) and figure 9
($5d_{5/2}$) to display the difference in shape of the momentum profiles,
hence non-absolute measurements should distinguish between the two
theories.

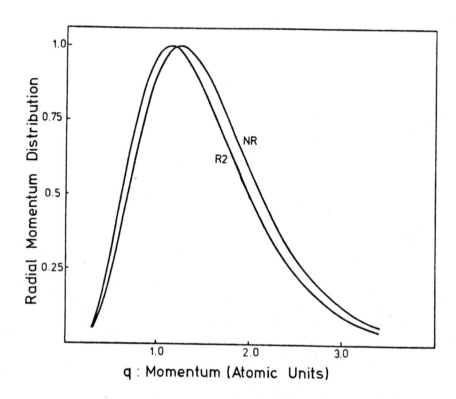

(Figure 9:) Momentum profiles for the $5d_{5/2}$ state of mercury with the Dirac Fock calculation R2 normalised to the Hartree Fock Calculation NR at the profile maxima.

1 - 20 MeV (e,2e) EXPERIMENTS

The existing experimental apparatus is illustrated in figure 10. The source of electrons is a Siemens 30 MeV betatron with a donut equipped with a magnetic shunt to extract the beam. The divergent beam from the betatron is then focused by means of a magnetic lens (quadrupole triplet) through a collimator and bending magnet to a target in a scattering chamber 4 metres away. After passing through the target, the unscattered part of the beam is collected in a Faraday chamber. The current through the chamber to ground is measured using a vibrating reed electrometer. The scattered beams are defined using adjustable 2 centimeter thick copper slits. The resultant beams then pass through double pole magnetic spectometers on to position sensitive detectors in the spectrometer focal plane. The detectors use a carbon monoxide, argon mixture to produce amplification by gaseous ionisation. The position information is obtained using coincidence detection of pulses from a helical delay line, figure 10.

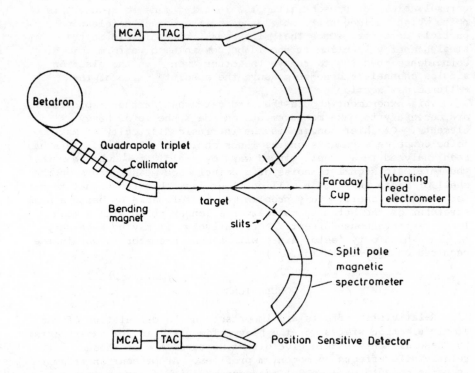

(Figure 10:)

Measurements of nuclear elastic (Coulomb) scattering have verified the betatron energy resolution as $\Delta E_\beta = 20$ keV and that of the electron spectrometers $\Delta E_s \sim 10^{-3} Ep$ where Ep is the pass energy of the spectrometer. The total energy resolution of the (e,2e) spectrometer should thus be $\Delta E \sim 202$ keV. The binding energy of the K shell is ~ 70 keV greater than the other shells in heavy element and so should be resolvable. The elastic scattering also showed that the position sensitive detector enabled the measurement of electron energies in the range from $-2\frac{1}{2}\%$ of the central pass energy (E_c) of the spectrometer to $+2\frac{1}{2}\%$ of E_c.

Further the nuclear scattering showed that for slit sizes ~ 1.0 c.m. the noise was dominated by secondary scattering in the spectrometer with a signal to noise ratio of 4:1 which became progressively worse as the slit size was reduced and the noise became dominated by room radiation effects. The limit of the room radiation on the useable slit size meant the Møller peak was unresolvable. At present attempts are being made to reduce this noise which include extra shielding and the use of a telescope particle detector. Once the Møller peak is resolvable in the singles energy spectrum it should be possible to perform a coincidence spectrum to remove the other events in the electron singles channels and hence measure the absolute value of the Møller cross section.

This experiment should then indicate what further improvements are necessary to make measurements on the K shells of heavy elements. Of chief concern is the intrinsic difficulty of making coincidence measurements on the inner shells of elements. This has been analysed in a simple minded way by Amaldi et al.[9] who show that the intrinsic signal to noise of a coincidence measurements on the K shell of radon Rn^{86} is a million times less than the signal to noise for the outer shell. This problem may be solveable by using a beam stretcher on the betatron to deliver a longer pulse with a much lower instantaneous current. Alternatively, it may be necessary to use a high duty factor linac with a monochromator to obtain the required energy resolution.

CONCLUSIONS

Relativistic effects are important in the calculation of the atomic electron states for the heavy elements. (e,2e) experiments on these elements should give direct measurements of these relativistic effects on momentum profiles. At present an attempt is being made to make (e,2e) measurements with electron impact energies of a few MeV.

REFERENCES

1. R. Camilloni, A. Giardini-Guidoni, R. Tiribelli and G. Stefani, Measurement of quasi free scattering of 9- keV electrons on K and L shells of carbon, Phys. Rev. Lett. 29, 618 (1972).

2. E. Weigold and I.E. McCarthy, (e,2e) Collisions, Adv. in Atom. and Mol. Phys. 14, 127 (1978).

3. A. Giardini-Guidoni, R. Fantoni, R. Camilloni and G. Stefani, (e,2e) Experiments, Comments Atom. Mol. Phys., 10 107 (1981).

4. M. Schulz, Influence of the effective charges on the five fold differential cross section for ionisation of hydrogen and helium, Jrnl. Phys. B: Atom. Molec. Phys. 6 2,580 (1973).

5. E. Weigold, C.J. Noble, S.T. Hood and I. Fuss, Electron impact ionisation of atomic hydrogen: experimental and theoretical (e,2e) differential cross sections, Jrnl. Phys. B: Atom. Molec. Phys., 12, 291 (1979).

6. I.P. Grant, Relativistic calculation of atomic structures, Advance in Phys., 19, 747 (1970).

7. C. Froese Fischer, Comp. Phys. Common 14, 145 (1978).

8. I. Fuss, J. Mitroy and B.M. Spicer, A theory for relativistic (e,2e) reactions, (submitted to Jrnl. of Phys. B).

9. V. Amaldi and F. Ciofi Degli Atti, Value and limitations of quasi free electron scattering experiments on atoms. Nuovo Cimento 64 129, (1970).

PROGRESS REPORT ON (e, 2e) COLLISIONS IN THIN FILMS

J F Williams, S Dey, D Sampson and D McBrinn*

Physics Department, University of Western Australia, Perth
* previous student at Physics Department, Queen's University,
Belfast, Northern Ireland, UK

Introduction

The information extracted from a given scattering experiment depends on the characteristics of the probes and on the nature of its coupling to the target. Neutrons (Egelstaff, 1965), electrons (Raether, 1965), electromagnetic radiation (Wright, 1969) and ion beams (Tolk, et al 1977) have all been used to study condensed matter. This paper concerns electrons probing thin films.

The (e, 2e) collision has been studied for free atoms and molecules principally by Ehrhardt and colleagues at Kaiserslautern for asymmetric kinematics, and by McCarthy, Weigold and colleagues at Flinders and by Guidini et al at Frascati for symmetric kinematics with emphasis on atomic structure.

Amaldi et al (1969) scattered 14.7 KeV electrons through thin (200 Å) films of formvar $(C_5H_7O_2)_n$ to reveal in the (e, 2e) binding energy spectrum well-resolved L and K shells of carbon and K shell of oxygen. No angular correlations were reported. Camilloni et al (1972) used the same apparatus, 9 KeV incident electrons and 200 Å carbon films to measure the ℓ and K shell momentum distributions of carbon atoms. Their work was subsequently verified by Krasilnikova et al (1975) using both carbon and collodion $(C_6H_2O_2 (ONO_2)_3)_n$ films. However their energy resolution of 140 eV was insufficient to obtain any information on the valence states.

Recent work by Persiantseva, Krasilnikova and Neudachin (1979) using 10 KeV electrons transmitted through 200 Å polycrystalline aluminium plus oxide deduced the momentum distribution of the target electron as well as the collective response of the conduction electrons to hole formation. Their energy resolution of 16 eV was sufficient to reveal an anomalously narrow momentum distribution which they attributed to a plasmaron state.

With a gaseous target, the gas atoms are randomly oriented and anisotropic angular correlations are not expected. With a single crystal thin film target, the attractive expectation is that in well defined crystallographic directions the anisotropy of the momentum distributions may be studied. This paper reports a number of experiments which were designed to ultimately lead to such information. Avery (1976) attempted a more ambitious experiment using a modified LEED apparatus to scatter up to 1000 eV from the W surface (001),plane. He observed no true coincidences.

2. Some Theoretical Considerations

Raether (1965) has discussed the theory of characteristic energy losses in solids, which will not be further discussed here although the interpretation of some of the early data is assisted with his work.

0094-243X/82/860314-12$3.00 Copyright 1982 American Institute of Physics

The (e, 2e) coplanar symmetric scattering geometry (McCarthy and Weigold, 1976), where p_o, p_s, p_e and q are the momenta of the incident, scattered, ejected and target electron momenta, leads to the allowed values of q, when parallel to p_o, given by

$$q = [1-2(|p_s|/|p_o|) \cos\theta_s] \, p_o$$

The general (e, 2e) five-fold differential cross-section for a single crystal with elementary cell volume V and containing N electrons is given by

$$\frac{d^5\sigma}{d\Omega_n} = \frac{2mk_s}{(2\pi h)^3} \left(\frac{d\sigma}{d\Omega_s}\right)^{free} N V \ \delta(E_o-E_s-E_e-E_B) \ A(k, E_B)$$

where E_B is the binding energy, $d\sigma/d\Omega$ is the cross section for free scattering of an electron with energy E_o by an electron at rest through an angle θ_s in the laboratory system. $A(k, E_B)$ is the single particle spectral weight function which for a thin aluminium target becomes the square of the solid form factor for the ejected electron

$$A(k, E_B) = |F_{k,f}(q)|^2 = S^2_{k,f} |\phi_k(q)|^2$$

where $S^2_{k,f}$ is the spectroscopic factor proportional to the square of the corresponding fractional parentage coefficient and $\phi_k(q)$ is the wave function of the single particle state from which the electron is ejected. In general $\alpha_k(r)$ satisfies the Bloch relation

$$\phi_k(r + R) = e^{ik \cdot R} \phi_k(r)$$

so that

$$|F_{k,\ell}(q)|^2 = \sum_{B,\ell'} \delta_{q,k+B} | \phi_{k,\ell'} (k + B)|^2$$

where $\phi_{k,\ell}(k + B)$ is the fourier transform of $\phi_\alpha(r)$ and B is a reciprocal lattice vector. For a free electron gas $F_k(q) = ((2\pi)^2/NV)^{\frac{1}{2}} \delta(k-q)$. For a single crystal this function is constant except at the Brillouin zone boundaries where a sharp dip due to the degeneracy results. However for the polycrystalline foils used in our experiment these dips are smoothed by the directional averaging. The result is slowly varying function over the range $o < q < k_f$ which goes to zero for $q > k_f$. The expression also shows that for a fixed binding energy, the cross section will vanish when $k = g$. In a more general analysis that result will be true to within an arbitrary reciprocal lattice vector. By setting q and scanning the binding energy by varying E_o or by setting the binding energy and scanning $\theta_s = \theta_e$, discrete points on the dispersion curve for bands may be mapped out for single crystals. The energy and angular resolutions need to be better than 1 eV and one degree respectively. In the polycrystalline case, only band energies and gaps may be determined requiring a resolution of about 2 eV since the width of the filled valence band for metals is about 3 to 10 eV.

3.   Experimental Method

The experimental apparatus was designed to take measurements in the coplanar symmetric geometry. The spectrometer essentially consists of an electron g⁏ two electron energy analysers with associated retarding electrostatic lens system, a Faraday cup to collect the primary beam and the target.

Electrons are produced by an electron gun that can be operated at 30 kV. A well focussed beam of several microamps was easily obtained.

A 127° electrostatic energy analyser was used in one of the channels to select electrons of energy E_s at θ_s. For the second channel a 180° hemispherical analyser was used to select electrons of energy E_e at θ_e.

The energy resolution is a constant fraction of the electron energy in the analyser which then uses a deceleration lens system with an overall deceleration ratio of 60:1. Triple-element optics data from Harting and Read (1976) were used for the construction of the two-stage circular aperture lens.

After energy analysis the electrons were accelerated from the exit aperture of the analyser and detected by a channel electron multiplier (CEM). The corresponding charge pulse produced was converted to a current pulse which was then amplified, discriminated and counted. Each analyser had a similar pulse handling system. For coincidence experiments the output from one timing discriminator provided the start pulse for the time-to-amplitude converter while the signal from the other discriminator, after being suitably delayed, provided the stop pulse. The time spectrum from the TAC was recorded by a Le Croy 3500 multichannel analyser. A typical spectrum is shown in figure (1).

The coincidence peak corresponds to two electrons arising from the same event and is superimposed on a flat random event background. The time resolution of the circuitry was ~ 20 nsec. The number of true events was evaluated by integrating under the peak and subtracting, after appropriately weighted, an integrated area under a background region.

The targets were self-supporting aluminium and formvar films produced by evaporation and a wet and dry process, varying in thickness from 80 Å - 1000 Å with a 3mm diameter. Aluminium oxidizes easily, so the foils used were a two component system of $A\ell$ and $A\ell_2O_3$. We estimate the oxide coating to be ~ 30-40 Å. The targets were mounted on a slotted rotatable drum holding up to 10 targets. An incident beam of up to 2×10^{-7} amp was used for an $A\ell$ target.

The analyser, Faraday cup and the electron gun were mounted on three independently rotatable concentric tables. The whole spectrometer was enclosed in a mu-metal case to reduce magnetic fields. The chamber was pumped by an oil diffusion pump to ~ 2×10^{-6} torr and finally to a base pressure of ~ 2×10^{-8} torr by a triode ion pump. The system was baked at 150 - 200°C for several hours.

4.   Results and Discussion

(i)   Energy loss measurements: Data on energy loss processes in thin
aluminium foil are presented to highlight the role of multiple
scattering events. $A\ell$ exhibits very sharp volume and surface
plasmon losses. We have made similar energy loss measurements
for formvar where the loss peaks did not exhibit the strength of
those in $A\ell$. The measurements were made to establish the energy

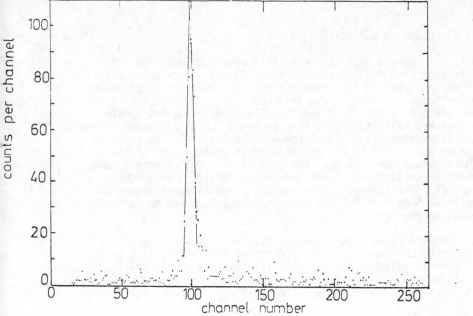

Figure 1. A typical coincidence time spectrum for 6 KeV electrons incident on formvar films (100° thick) in symmetric coplanar geometry at 45°.

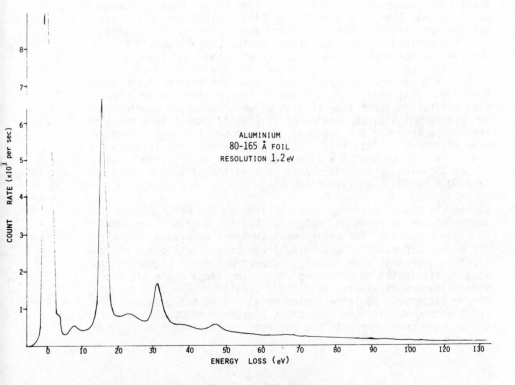

ALUMINIUM
80-165 Å FOIL
RESOLUTION 1.2 eV

Figure 2. Energy loss spectrum due to passage of 3000 eV electrons through a thin (80-165 Å) aluminium foil at θ = 0°.

loss processes in our foils.

Figure (2) shows a typical energy loss spectrum resulting from the passage of 3 KeV electrons through Aℓ foil 100 - 180 Å thick with the analyser in the 'straight through' position, i.e. θ=0°. The sharp peak on the extreme left represents electrons which have been elastically scattered and is called the "zero loss" peak. The measured half-width of this peak was ~ 1.2 eV which was ~ 0.2 eV broader than the instrumental resolution. The peaks at ~ 7.5 eV and ~ 22.5 eV are due to surface plasmon loss at the oxidised aluminium surface. The peak at 15 ± 0.5 eV arises from collective excitations in the bulk of the solid, i.e. volume plasmons. These values agree with the original observations of Powell and Swan (1960). The remaining peaks are due to multiple plasmon energy losses.

The existence of only two plasmon loss peaks of significant strength in the figure and the clarity of the surface loss peaks are taken as evidence for the relative thinness of the foils. For aluminium, the mean free path for plasmon production with E_0 = 3 KeV is ~ 70 Å (Bohm and Pines, 1953).

Figure (3) shows an increase in the number of volume plasmon losses due to the increase in foil thickness. The surface peaks have become less obvious, and at large energy losses the volume peaks are less apparent.

Apart from plasmon losses the energy loss spectrum of aluminium should reveal loss processes due to single electron excitations, such as ionization losses, interband and intraband transitions. Losses corresponding to ionizations of the L-shell should be observed in the range 60 eV to 110 eV but their magnitude is small compared to the plasmon losses and so were not identified. Peaks due to interband transitions lie in the region close to the 'zero-loss' peak. The dielectric theory predicts an interband excitation transition at 1.48 eV which has been studied by optical techniques, Ehrenreich et al (1963). Apart from the high resolution required the intensity of such small energy losses are very dependent on foil thickness. Shklyarevskii and Yarovaya (1964) have shown that the intensity of this loss to the 15 eV plasma oscillation loss decreases by a factor of 100 in going from 60 Å to 500 Å in film thickness.

Figure (4) shows an energy loss spectrum at θ = 45°. A Born approximation calculation indicates that for angles given above $\theta ~ h|\underline{k}_c|/|\underline{p}|$ i.e. for small momentum transfer $h(k) ~ |\underline{p}|\theta$ where for Aℓ $\theta_c ~ 16$ milliradians (0.92°) the cross section for plasmon creation decreases very rapidly. The structure at higher energy loss is attributed to electrons undergoing large angle elastic scattering events as well as plasmon creation. The more rapidly rising background at large angles results from the greater relative probability that electrons scattered through large angles have undergone multiple inelastic collisions.

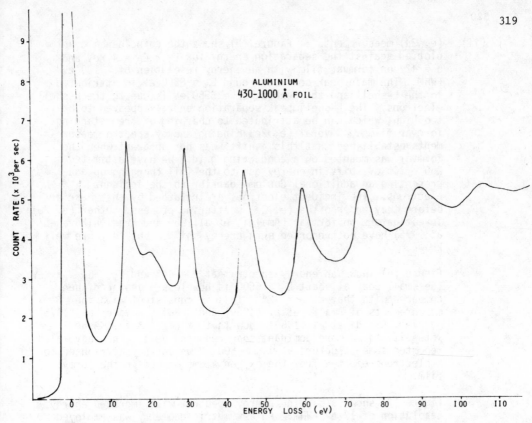

Figure 3. Energy loss spectrum at $\theta = 0°$ for a thicker Al foil (430 - 1000 Å). Incident electron energy = 3000 eV.

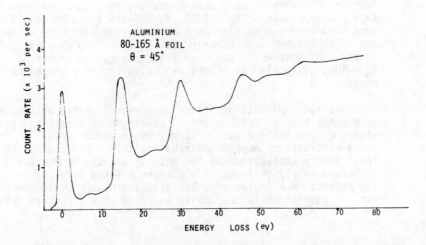

Figure 4. Energy loss spectrum at $\theta = 45°$ for 3000 eV electrons passing through Al foil of 80 - 165 Å thickness.

(ii) <u>(e, 2e) measurements</u> : Figure (5) shows the coincidence count plotted against the separation energy for $E_s = E_e = 4$ KeV and $\theta = 42°$ on formvar film with an energy resolution of ~ 100 eV FWHM. The main peak centred around $E_0 = 8300$ eV is attributed to the L-shell and the peak at $E_0 = 8550$ eV is due to the K-shell electrons. The L-shell peak separation energy appeared to be too high, which can be attributed to charging of free standing formvar films. Several (e, 2e) binding energy spectra measurements established that this 'shift' was not apparent when the formvar was mounted on a conducting grid. We have estimated a 250 - 300 eV shift in energy due to the foil charging up and presenting an additional Coulomb barrier to the incident electrons. The incident electrons are retarded by that barrier before they undergo the (e, 2e) scattering process. Other (e, 2e) investigators on formvar (Amaldi, et al 1969 and Krasilnikova et at 1975) have not reported such energy shifts, but they are well known.

Figure (6) shows an energy spectrum at $\theta = 44°$ and $E_s = E_e = 3$ KeV. The small peak at about $E_0 \approx 6600$ is now less pronounced when compared with that at $\theta = 42°$. This is consistent with the measurements of Amaldi et al (1969) and Krasilnikova et al (1975). However, Amaldi et al (1969) show that at $\theta = 41°$ the K-shell intensity to be more dominant than reported here. They have also reported some structure at about 500 eV which they attributed to K electrons ejected from the oxygen atoms bonded in the formvar film.

Figure (7) shows a binding energy spectrum with improved energy resolution of 17 eV FWHM. An asymmetric geometry was employed for this data collection with $\theta_s = 25°$, $\theta_e = 65°$ and incident energy E_0 varying from 3700 eV to 3900 eV. After taking into account the 250 - 300 eV discrepancy due to surface charging of the film the coincidence binding energy curve is seen to represent loosely bound electrons in the solid rather than electrons ejected from inner shells. The coincidence count rate was zero for incident energies below 3765 eV.

The main features to be observed are the occurrence of three fairly sharp peaks at $E_0 \approx 3810$ eV, 3855 eV and 3885 eV. Also some structure can be seen on the low energy side of $E \approx 3780$ and on the higher side of main peak at $E \approx 3840$ eV. Figure (8) is the continuation of the separation energy spectrum from $E_0 = 4023$ eV to 4123 eV. These data also reveal a series of peaks.

The possible explanations of the observed structure will depend on whether the (e, 2e) process is probing core atomic or molecular states or the solid states. It may be that some or all of the above alternatives may be contributing to the structure. The least likely explanation of the data is to account for the structure solely in terms of electrons ejected from atomic states. The separation energies of these states are very small and clearly our resolution is inadequate to reveal the structure details.

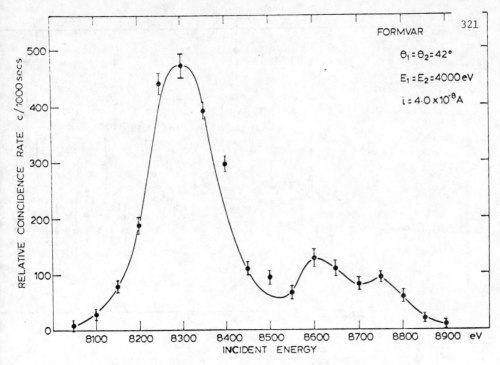

Figure 5. A binding energy curve for 100 Å thick formvar film at $\theta_e = \theta_s = 42°$ and $E_0 = 8000$ eV.

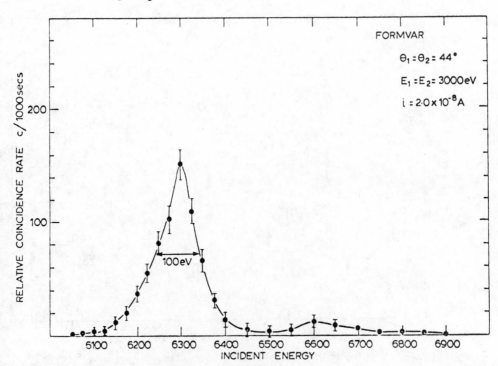

Figure 6. A binding energy spectrum for formvar film 100 Å thick taken at $\theta_e = \theta_s = 44°$ and $E_0 = 6000$ eV.

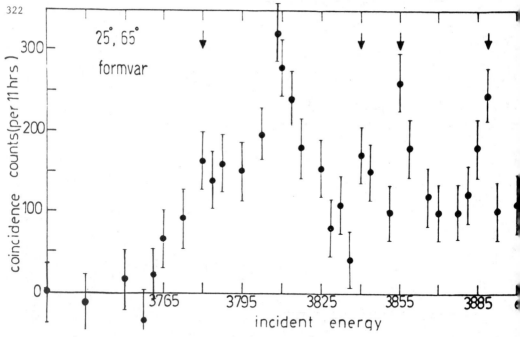

Figure 7. A binding energy curve for 100 Å formvar film at θ_e = 25°, θ_s = 65° from E_o = 3700 eV to 3900 eV.

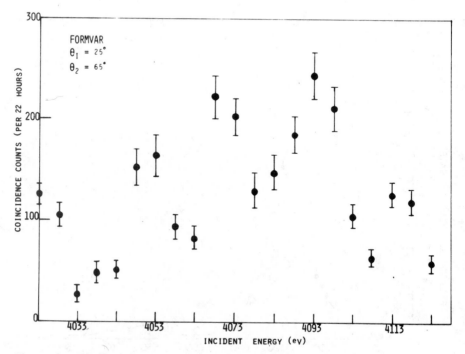

Figure 8. Binding energy scan for a formvar film at θ_e = 25°, θ_s = 65° from E_o = 4023 eV to 4123 eV.

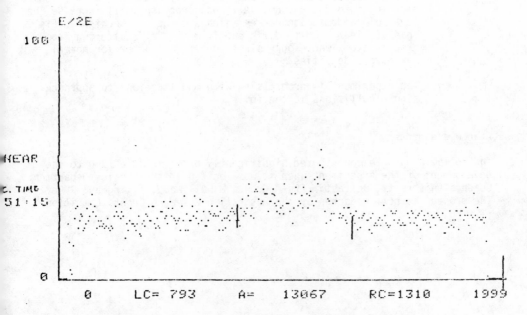

Figure 9. Coincidence time spectra for a 80 - 165 Å thin Al foil
incident electron energy of 6000 eV for symmetric coplanar
geometry.

Secondly, because of the complicated structure of formvar and insufficient energy resolution it seems that the measurements cannot be seriously considered as representing band structure in the solid. The most realistic explanation is that the observed structure is demonstrating the dominant role that plasmon oscillations play in a solid. Another possibility is that each of the observed peaks represents the knock-out of valence shell electrons but are shifted along the energy scale due to the incident energy electrons undergoing multiple plasmon losses discussed earlier. While coincidence spectra were readily obtained for formvar targets it is apparent that simpler targets must be used to permit interpretation of the spectra.

Figure (9) shows two coincidence time spectra for E_0 = 6000 eV electrons on a thin Aℓ foil with an energy resolution of 7 eV. These spectra are from a series of experiments carried out over a range of incident energies corresponding to separation energies between 0 and 400 eV.

The (e, 2e) spectra will contain contributions from both surface and volume scattering. The single electron energy loss spectra from even our thinnest foils (~ 100 Å) show the dominance of volume losses, which is consistent with beam attenuation measurements indicating significant attenuation and multiple scattering. After initial (e, 2e) scattering the two outgoing electrons must undergo subsequent collisions especially with the surface before emerging into the vacuum. All such processes will increase the level of random coincidences which are seen to be high in Fig 9. This is clearly quite different from (e, 2e) scattering from gaseous atoms where only single collisions occur for normal target gas densities.

The experiment is progressing with modifications to approach single collisions in the foils.

Acknowledgements

The following people contributed significantly prior to June 1980 to the development of the experiment when it was located in the Physics Department, Queen's University, Belfast; Drs A Crowe, M Hollywood, T King and R Devlin. The project is now supported by the Australian Research Grants Committee.

References

Ehrhardt, H., Hesselbackher, K.H., Jung, K. and Willman, K., J.Phys.B. <u>5</u>, 1559; <u>B8</u> 1330 (1972).

McCarthy, I. and Weigold, E., Phys. Rep C <u>27</u>, 275 (1976).

Camilloni, F.R., Guidoni, A.G., Tiribelli, R. and Stephani, G., Phys. Rev. Letters <u>29</u> 618 (1972).

Amaldi, U., Egidi, A., Marconero, R. and Pizzella, G., Rev. Sci. Inst. <u>40</u> 1001 (1969).

Raether, H., Springer Tracts in Modern Physics <u>18</u> 84 (1965).

Wright, G.B., Light Scattering Spectra of Solids (Springer-Verlag, N.Y.)(1969)

Egelstaff, P.A. Thermal Neutron Scattering (Academic Press, N.Y.) (1965).

Krasilnikova, N.A., Levin, V.G. and Persiantseva, N.M., Sov. Phys. JETP <u>42</u> 76 (1975).

Persiantseva, N.M., Krasilnikova, N.A. and Neudachin, V.G., Sov. Phys. JETP <u>49</u> 530 (1979).

Bohm, D. and Pines, D., Phys. Rev., <u>92</u> 603 (1953).

Ehrenreich, H., Phillip, H.R. and Segall, B., Phys. Rev. <u>132</u> 1918 (1963).

Shklyarevskii, J.N. and Yarovaya, R., Opt. Spectr. USSR, <u>16</u> 45 (1964).

Powell, C.J. and Swan, J.B., Phys. Rev. <u>118</u> 640 (1960).

Harting, E. and Read, F.H., Electrostatic Lenses (El Sevier Sci. Pub. Co.) (1976).

Tolk, N.H., Tully, J.C., Heiland, W. and White, C.W., eds. Inelastic Ion-Surface Collisions, (Academic Press, N.Y.) (1977).

(e,2e) ON SOLIDS - A PROGRESS REPORT

L. Frost and E. Weigold

Institute for Atomic Studies, The Flinders University of South Australia,
Adelaide, South Australia 5042.

Since the first (e,2e) experiment was performed on the core states in a thin film of carbon[1,2] in 1969, an ongoing goal has been the measurement of separation-energy spectra (showing band gaps) and momentum distributions (showing lattice structure) of electrons in the valence and conduction bands of solids. Theoretical studies by Neudatchin et al.[3], Levin et al.[4] and D'Andrea and Delsole[5] clearly describe the expected dependence of the (e,2e) cross section on the wavefunctions of the electrons in a crystal. Recent transmission measurements on thin amorphous films by Krasil'nikova et al.[6], Persiantseva et al.[7], and Williams et al.[8] have not yet tested these predictions due to a combination of poor energy resolution, poor signal-to-noise ratios and difficulties in sample preparation. In fact the above difficulties essentially obscured all useful details, although these experiments did establish the feasibility of (e,2e) spectroscopy on solids using transmission geometry.

These problems are largely innate to the use of electron beams to probe solid targets. The target density for solids is of course many orders of magnitude greater than that of gaseous targets, which results in severe multiple-scattering and attenuation of the beam for thick targets. For negligible attenuation and multiple scattering, beam energies of $\gtrsim 10$ keV and target thicknesses of $\lesssim 100$Å are required. Uniform self-supporting films of such thickness are difficult to prepare. Sample heating limits beam currents to tens or hundreds of nanoamps per square millimetre, which in turn limits obtainable counting rates and statistical accuracy. Often the energy resolution of the experiment must be deliberately broadened so as to increase the count rate and the signal-to-noise ratio. Finally the random background for a given detected signal is generally greater than that observed using an atomic beam due to the multiplicity of energy-loss processes in solids.

The goal of the development work at Flinders University has been to overcome these problems, using recent technologies and novel means of measuring the kinematic variables of energy and angle. The remainder of this paper outlines some of the phenomena that we expect to measure using the (e,2e) technique and the equipment that we have designed for the task.

The (e,2e) cross section can be written for any target using the plane wave impulse approximation (PWIA) as[9]

$$\sigma(\underline{k}_0,\underline{k}_1,\underline{k}_2) = (2\pi)^4 \frac{k_1 k_2}{k_0} f_{ee} \sum_{av} |F_{if}(\underline{q})|^2 \delta(E_1+E_2+\varepsilon_{if}-E_0)\sigma(\underline{q}+\underline{k}_1+\underline{k}_2-\underline{k}_0) \quad (1)$$

where atomic units have been used and the subscripts 0, 1 and 2 refer to the incident and outgoing electrons. The δ functions ensure energy and momentum conservation, ε_{if} being the separation energy of the electron. It is governed by the dispersion law for each band α,

$$\varepsilon^{(\alpha)} = \varepsilon^{(\alpha)}(\underline{k}) \tag{2}$$

Only one value of the quasimomentum $\underline{k}$ in the first Brillouin zone is allowed for each separation energy of a particular band. The electron-electron Mott scattering cross section f_{ee} is calculated at the energy $E = E_o - \varepsilon_{if}$. The sum and average denotes the usual average over initial degenerate states and sum over final degeneracies. The structure factor $|F_{if}(\underline{q})|^2$ is the absolute square of the momentum space overlap between the initial and final many body states.

For a crystal $|F_{if}(\underline{q})|^2 \sim |\phi_{\underline{k}}^\alpha(\underline{q})|^2$, where the Fourier transform $\phi_{\underline{k}}^\alpha(\underline{q})$ of the band wave function with respect to the recoil momentum $\underline{q}$ is

$$\phi_{\underline{k}}^{(\alpha)}(\underline{q}) = (2\pi)^{-3/2} \int d^3r \, \exp(i\underline{q}.\underline{r}) \psi_{\underline{k}}^{(\alpha)}(\underline{r}) \tag{3}$$

The band wave function satisfies the Bloch condition

$$\psi_{\underline{k}}^{(\alpha)}(\underline{r} + \underline{R}) = \exp(i\underline{k}.\underline{R}) \psi_{\underline{k}}^{(\alpha)}(\underline{r}) \tag{4}$$

where $\underline{R}$ is one of the lattice vectors. In the PWIA the cross section is therefore proportional to

$$|\phi_{\underline{k}}^{(\alpha)}(\underline{q})|^2 = \sum_{\alpha,G} \delta(\underline{q},\underline{k}+\underline{G}) \delta(\varepsilon - \varepsilon^{(\alpha)}(\underline{q}) |\phi_{\underline{k}}^{(\alpha)}(\underline{k}+\underline{G})|^2 \,, \tag{5}$$

where $\underline{G}$ is a reciprocal lattice vector.

The predicted (e,2e) cross section for an oriented crystal of a metal such as aluminium is shown in Figure 1. The Fermi energy and

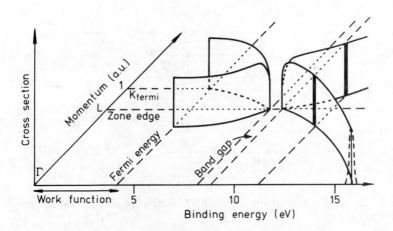

Fig. 1 Schematic drawing of the predicted (e,2e) cross section in the valence region for an oriented crystal of a metal such as aluminium.

Fermi momentum in a given lattice direction are directly measured, as is the band gap at $\varepsilon \sim 7\text{eV}$ where there are no allowed states. The energy dispersion law $\varepsilon(\underline{k})$ for each band is mapped out and the position of the edge of the Brillouin zone is measured. The dotted profile at

15eV shows the expected broadening due to the finite lifetime of a conduction-band hole. The magnitude of the cross-section depends on the details of the electron wavefunction and on the coupling to the lattice. An experimental energy resolution of approximately 0.1eV and momentum resolution of 0.05 au is required to resolve all details.

The separation-energy spectrum at a fixed momentum q should show the upper bands near the vacuum level mentioned above, but in addition there should be peaks due to ionization of localized core states and associated plasmon satellites which may be excited simultaneously[10]. This simple picture may be further complicated in practice by the presence of surface states, and the many inelastic scattering processes which contribute to the background. For example, each core state may have a "tail" extending to larger separation energies due to inelastic scattering in the out-going channels. A successful (e,2e) experiment on crystals must be able to select each band or state with sufficient energy and momentum resolution and at the same time still maintain an adequate signal to provide good statistical accuracy.

The exterior view of the Flinders ultra-high vacuum (e,2e) spectrometer is shown in Figure 2. Basically two electron energy analysers

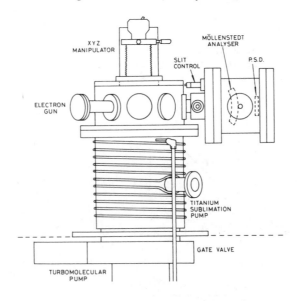

Fig. 2a Side elevation of the Flinders (e,2e) coincidence spectrometer.

are mounted horizontally from the collision chamber at 45° to the electron beam. The chamber is pumped by a titanium sublimation pump and turbomolecular pump. The thin film target at the center of the chamber is oriented using a UHV XYZ manipulator.

Three novel features of the system are illustrated in Figure 3. Firstly the energy analysis is performed by Möllenstedt electrostatic energy analysers[11,12]. These analysers have been designed to be radially and laterally symmetric about the collision centre in order

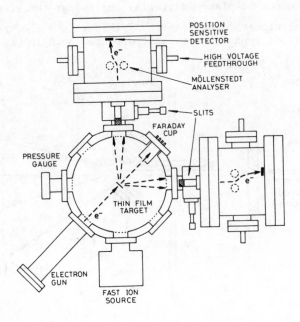

Fig. 2b Top elevation of the Flinders (e,2e) coincidence spectrometer.

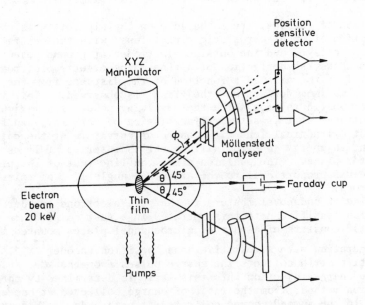

Fig. 3 Schematic drawing of the coincidence spectrometer showing the Möllenstedt energy analysers and position sensitive detectors.

330

to preserve the out-of-plane scattering angle ϕ at the electron detectors. Typical electron trajectories[12] through the electrostatic lens of the Möllenstedt type velocity analyser are shown in Figure 4. The

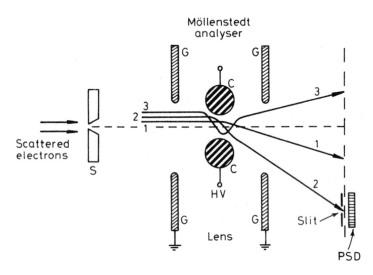

Fig. 4 Schematic electron trajectories through the electrostatic lens of a Mollenstedt velocity analyser. Ray 2 is the normal caustic ray for which the analyser is timed.

operation of the analyser is based upon the high off-axis chromatic aberration of an electrostatic einzel lens, with the central electrode at high voltage, and the outside electrodes at ground potential. The figure shows the deflection of a fine monoenergetic electron beam parallel to the axis as a function of the distance from the axis. For ray 2 a maximum deflection (the first) is observed. This is the so-called caustic ray for which the resolving power is a maximum. Hence the entrance slit is set for transmission of this ray, and for the present cylindrical lens with centre of curvature at the collision region, an energy spectrum will appear as a series of lines on the detector plane. The position along the line (out of the plane of the figure) is proportional to the azimuthal angle $\phi_{1,2}$ of emission of the electrons.

The second novel feature is that the electrons are detected by position sensitive detectors operated in a fast coincidence mode. Each detector consists of a pair of microchannel plates mounted in a chevron configuration with a resistive anode position encoder[13-15]. The resistive strip collects the charge pulses from the MCP. The position of the charge injection into this strip is determined by the charge division method, from the ratio of charge collected at one end of the strip to the summed charge collected at both ends. This arrangement should permit an angular resolution of 0.05° over a total range of out-of-plane azimuthal angle of ±6°. At an incident energy of 20 keV this corresponds to a range of momentum q of ±6 a.u. The MCP detectors will also permit high count rates and fast coincidence timing

(FWHM ~ 2nS).

The third feature of the apparatus is that the analysers can readily be rotated by 90° about the mounting face on the target chamber. Therefore both symmetric coplanar and symmetric noncoplanar geometries are easily achieved. This is equivalent to measuring momenta $\underline{q}$ respectively parallel or perpendicular to the incident beam.

Figure 5 shows the arrangement of the position sensitive detectors

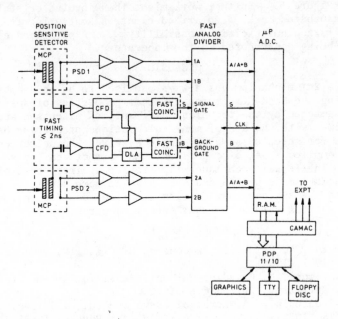

Fig. 5 Schematic drawing of position sensitive detectors and fast timing and position encoding electrons.

and the fast timing and position encoding logic. The experiment is run on-line by a PDP11/10 computer via a CAMAC system. The collection of data requires the detection of both true and accidental coincidences. This is achieved using normal fast timing techniques, with simultaneous, fast analogue division of the resistive anode outputs to determine the position, i.e. out-of-plane azimuthal angles ϕ_1, ϕ_2 of both electrons.

Coincident count rates as a function of ϕ_1, ϕ_2 and ϕ_1-ϕ_2 are stored for both true and background events by a microprocessor and transferred periodically to disc storage on the PDP11 computer.

At any given separation energy the cross section over the whole angular range $\phi = \phi_1$-ϕ_2-π (i.e. whole momentum range $\underline{q}$) can thus be recorded simultaneously. The separation energy is scanned by stepping the incident energy over a programmed range, the energies of the detected emitted electrons being fixed by the Möllenstedt analysers. The equipment has been designed to provide both the required energy and angular (i.e. momentum) resolution with a data acquisition rate at least 10 times that previously attained (with 10 times better resolution).

332

Several experiments suggest themselves. Using simple carbon foils
($\lesssim 200\AA$ thick) the momentum profiles and plasmon satellites of core
states should be measured to check the previous low-resolution work of
Neudachin et al.[3] Band gaps and energy dispersion laws (averaged over
all directions in the polycrystalline sample) could then also be
measured. The next major goal is to progress to oriented single-
crystals of say Al, Cu or W and to obtain results comparable to the
sketch in Figure 1. Further work might then involve crystal surface
states, adsorbates, etc. Polarized electron scattering from magnetic
materials is another exciting possibility[16]. A polarized electron
source is being constructed in the laboratory.

CONCLUSIONS

The (e,2e) mechanism has always promised to be an unambiguous means
of measuring the band-structure and other properties of solids, but it
has been hampered by the technical difficulties discussed above. In
the meantime, with the aid of many calculations predicting band-
structures for selected samples, Angle Resolved Photoelectron Spectroscopy has compiled a set of consistent band-structures for some substances. The (e,2e) technique will provide a direct test of many of
the assumptions made by PES and will also be more sensitive to the bulk
rather than the surface characteristics of solids (due to the longer
mean free path of the much higher energy electrons).

REFERENCES

1. U. Amaldi, A. Egidi, R. Marconero, G. Pizzella, Rev. Sc. Inst.
    40(8) 1001 (1969).
2. R. Camilloni, A. Giardini-Guidoni, R. Tiribelli, G. Stefani,
    Phys. Rev. Lett. 29(10) 618 (1972).
3. V.G. Neudachin, G.A. Novoskol'tseva and Yu. F. Smirnov, Eksp. Teor.
    Fiz. 55, 1039-46 (1968) transl. in Sov. Phys. JETP 28(3) 540 (1969).
4. V.G. Levin, V.G. Neudachin and Yu. F. Smirnov, Phys. Stat. Sol.
    (b) 49, 489 (1972).
5. A. D'Andrea and R. Delsole, Surf. Sc. 71, 306 (1978).
6. N.A. Krasil'nikova, V.G. Levin and N.M. Persiantseva, Zh. Eksp.
    Teor. Fiz. 69, 1562-68 (1975) transl. in Sov. Phys. JETP 42(5)
    796 (1976).
7. N.M. Persiantseva, N.A. Krasil'nikova and V.G. Neudachin, Z. Eksp.
    Teor. Fiz. 76, 1047-57 (1979) transl. in Sov. Phys. JETP 49(3)
    530 (1979).
8. J.F. Williams, S. Dey, D. Sampson and D. McBrinn (preceding paper,
    this workshop).
9. I.E. McCarthy and E. Weigold, Phys. Rep. 27C, 275 (1976).
10. V.G. Neudatchin and F.A. Zhivopisteo, Phys. Reo. Lett. 32, 995
    (1974).
11. G. Müllenstedt, Optik 5, 499 (1949).
12. H.F. Wellenstein, J. Appl. Phys. 44, 3669 (1973).
13. W. Parkes, K.D. Evans and E. Mathieson, Nuc. Instr. Meth. 121,
    151 (1974).
14. E. Mathieson, K.D. Evans, W. Parkes and R.F. Christie. Nucl. Instr.
    Meth. 121, 139 (1974).
15. S.T. Hood and E. Weigold, J. Electron Spectrosc. 15, 237 (1979).
16. L. Vriens, Phys. Rev. B4(9) 3088 (1971).

WORKSHOP SUMMARY

I.E. McCarthy

Institute for Atomic Studies, The Flinders University of South Australia,
Bedford Park, S.A. 5042, Australia.

This workshop has covered nearly every imaginable way of getting at
atomic and nuclear momentum distributions. It has demonstrated many
similarities between the two fields in experimental and theoretical
technology, some similarities and some differences in outlook and expectation. I first experienced the latter some years ago, before I changed
my Physical Review subscription from C to A. I gave a talk at a well-
known atomic physics center for a reasonably-successful optical model
for atomic scattering, which used an imaginary potential strength that
was freely adjusted to fit the total reaction cross section. I was
soundly criticized for using a parameter and it was suggested that it
would have been better to have used the Born approximation. When I
objected that the Born approximation is wrong I was told that this
doesn't matter, at least it is ab initio.

This illustrates what is probably the basic difference between the
two fields, caused by the difference in the degree of understanding of
the two-body force. Atomic and molecular physicists use $1/r$. The
question of the possible effect of other particles on the force such as
Pauli blocking causing renormalization does not seem to occur. I
remember being told once by Bob Nesbet that he had estimated this
effect at about 4%. The nuclear force suffers from technical complications such as the coupling between spin channels caused by the tensor
force and the coupling at higher energies to nucleon-isobar (such as
delta) channels. This is not as important as the fact that it is
basically unknown (although accurately parametrized for relevant
energies). Not only is the two-body force unknown, but there is no
real answer to the question of n-body forces in the n-body problem.
Certainly accurate triton calculations by Afnan and others show that
the best two-body forces make errors of 1 MeV in the 8 MeV binding
energy.

The outlook of atomic physics generated by the relative understanding of the basic force is that calculations should be ab initio
(in atomic language) or microscopic (in nuclear language). We are
really testing our ability to handle (experimentally and theoretically)
Coulomb many-body problems. This workshop has shown how close we
come in some cases and how far we have to go in others. In nuclear
physics we have a much wider diversity of phenomena leading to much
effort in their classification and relationship, often less-precise
experiments, less concentration on one particular aspect such as
momentum distributions, and the added interest of trying to unravel
the mysteries of the two-body force.

The subject matter of the workshop can be summarized by the
following figure in which the entrance channel consists of a particle
of momentum $\vec{k}_o$ interacting with an (n+1)-particle system through a
distorting interaction D_o. It has an interaction T with a target
particle resulting in a final state in which particles of momenta $\vec{k}_A$
and $\vec{k}_B$ interact with the residual n-particle system by distorting inter-

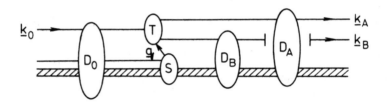

actions D_A and D_B. The structure-dynamics vertex $S(\vec{q})$ contains the subject of the workshop, the momentum distribution $f(\vec{q})$. The reactions may be regarded as ways of getting at this. The diagram represents essentially a three-body or binary-encounter view of the reaction which is not the whole story, but can be shown in many cases to give an excellent understanding, particularly if we extend the meaning of the two-body operators D and T to include coupling of elastic and inelastic (particularly resonant) channels in the relevant two-body subsystems, allowing so-called two-step processes.

Erich Weigold first showed us that for high-energy (e,2e) on the hydrogen atom most of the diagram can be ignored in a chosen kinematic region and, since the structure of the ground state is trivial, we can get directly at the dynamics, obtaining the well-known electron momentum distribution within experimental error.

At the other extreme we had the discussions by Bruce Barrett and Ian Morrison on the amazing relationship of phenomena introduced into nuclear, atomic and molecular structure by recognizing higher symmetries, in particular those of the interacting boson model. To get absolute numbers for a wide variety of reactions we need only a few quantities, which contain the bound-state dynamics, essentially the momentum distributions. However Ken Amos and Ian Morrison showed us that present inelastic scattering experiments can be fitted with essentially no precise knowledge of the dynamics, for example one does equally well using harmonic-oscillator or Woods-Saxon single-particle wave functions.

An experiment that contains a wealth of illustration of most of the points of the workshop is the (e,2e) reaction to the doublet S (3s-hole) states of the argon ion (figures 1 and 3 in my talk at this meeting). The structure aspect is represented by the relative cross sections (fig. 1) of the few states seen in the experiment (spectroscopic factors). The dynamics is represented by the momentum profiles (fig. 3), which are all the same, regardless of incident energy, for $q \lesssim 1$ a.u. This confirms that the energy-dependent operators D_o, D_A, D_B and T are unimportant in the understanding of the reaction in this kinematic range. The fact that the dynamics is well-described by the 3s Hartree-Fock wave function in this range illustrates how hard we have to work to describe it. For $q \gtrsim 1$ a.u. the experimental data become energy dependent and there are more high-momentum components than given by the simple undistorted reaction theory with one-body bound-state dynamics. Clearly this kinematic region contains information about distortion effects and possible effects of bound-state correlations, which must be unravelled by more-sophisticated theory allied to possible different experiments designed to bring out one or more of these effects.

The same ideas apply to the discussions of the nuclear knockout reactions (p,2p), (α,αp), (π,πp), (p,pn) and (α,2α) by Nick Chant and Del Devins. Here the mechanism is well-understood in certain kinematic ranges where the factorization of the matrix elements of T from the reaction amplitude is confirmed to be valid both for polarized and unpolarized incident beams in the case of (p,2p). However the experiments have not reached the stage of identifying dynamical details. The spectroscopic factors depend on distorting parameters. For example large differences occur if a relativistic equation is solved to obtain the optical model wave functions used in the representation of the distortion operators D. This highlights the need for greater understanding of the relationship between the distorting and knockout parts of fig. 1. The bound-state dynamics is adequately described by Woods-Saxon wave functions. The experiment is not done at the level of detail that discriminates for example between these and Hartree-Fock functions calculated by one of the nuclear methods described by Bruce Barrett. Del Devins' suggestion that kinematic ranges sensitive to inelastic distorting (two-steps) processes should be investigated is a step in the direction of understanding distortion. None of these uncertainties can account for the enhancement of the spectroscopic factors by a factor of 100 in the (α,2α) reaction, which is otherwise understood at the same level as the other nuclear knockout reactions. It seems that we have at last the long-sought confirmation of the condensation of nuclear matter to α-particles in the far nuclear periphery.

We now turn to the mechanism of the high momentum-transfer (e,2e) reaction. First a good understanding of absolute cross sections for a wide energy and momentum range has been achieved by the Frascati group for Xenon using consistent distorting parameters in the simplified averaged-eikonal approximation. More-accurate absolute cross sections obtained by the FOM group for helium are well-reproduced by a distorted-wave model. The enhancement of the momentum profile for q up to 3 a.u. has been well-described in calculations by Cliff Noble for neon using distorting potentials calculated from an ab-initio optical model that includes excitation and breakup effects in the closure approximation. It seems that for this range of q the Hartree-Fock dynamics still holds. Good fits to relative cross sections for hydrogen are obtained with a similar calculation for a wide range of symmetric and asymmetric kinematics. These calculations all represent T by the half-shell factorization approximation using the Mott t-matrix. The factorization has been confirmed by the Frascati group. The conclusion that the plane-wave impulse approximation adequately describes $S(q)$ for $q \lesssim 1$ a.u. in noncoplanar symmetric geometry is based on these distorted-wave calculations as well as the independence of relative cross sections for changes of incident energy. The plane-wave approximation does not describe absolute cross sections well, but this is not necessary for extraction of $S(q)$, provided all the reaction strength is observed and assigned to symmetry classes. The sum rule for spectroscopic factors then fixes the structure information. These conditions hold for the 3-s hole representation of the argon ion for example.

Helmut Ehrhardt showed us a very different picture for the reaction mechanism of (e,2e) when one of the final-state electrons is slow. Here it is absolutely necessary to use distortion D_B for the slow

electron wave and the first-order approximation to T is adequate.   This
is the first Born approximation.   Angular profiles for the slow electron
are not centered correctly in this approximation.   Great improvement is
obtained in calculations by Byron, Joachain and Piraux which use distortion D_0 described by the second Born approximation with closure.   Larry
Mendelsohn and Russell Bonham showed that an effective hydrogenic model
with a charge parameter Z^* is an adequate description of asymmetric
(e,2e) and Compton scattering.   In no case is there evidence that
asymptotic screening is relevant in the reactions, so that new insight
being gained into the three-body Coulomb boundary condition (for example
by Altick) is probably not tested by these experiments.

We now return to the problem of the high-q bound-state dynamics.
There is an alternative reaction that sorts this out very well.   We
replace the scattered particle in fig. 1 by a photon.   Because of the
restrictions of two-body kinematics the observable range of q is closely
tied to the energy of the photon.   In the nuclear reactions described
by June Mathews the range observed is q > 300 MeV/c.   This is just the
range where simple one-body wave functions such as harmonic oscillator
functions give very different dynamics from the functions calculated by
the various Hartree-Fock schemes using the nuclear-matter G-matrix as
the two-body force.   The trends of the more-sophisticated calculations,
which were described by Bruce Barrett, are very similar to the enhanced
cross sections observed in the experiment.   The range $q \gtrsim 300$ MeV/c
is quite-well described by the (e,e'p) reaction.   A similar situation
was described by Frank Larkins for the photoelectron reaction on helium
atoms at high energy.   Here there is strong evidence, gained from
(e,2e) for $q \lesssim 2$ a.u. and from (γ,e) for $q \gtrsim 2$ a.u. that the dynamics
and structure of the ground state must be understood far beyond the
Hartree-Fock level to describe the forbidden excitation of the n=2
states of the helium ion.

There were two very interesting talks on the understanding of
individual vertices in fig. 1.   Heinz von Geramb described the microscopic calculation of proton elastic scattering, which uses the vertex
T.   For momentum transfer K up to 600 MeV/c excellent description of
the cross section is obtained using the density-dependent G-matrix for
a well-parametrized two-nucleon potential.   For K up to 1200 MeV/c this
model gives errors up to a factor of 2 and qualitatively-different cross-
section shapes.   The situation is dramatically improved by coupling
the delta channel into T.   This is an excellent illustration of the
improved understanding of nuclear physics to be obtained by going beyond
the many-nucleon description.

Andris Stelbovics described a microscopic (ab initio) calculation
of electron-atom scattering which takes all reaction channels into
account explicitly.   The continuum is described by an approximation
which gives an optical potential that can be checked in a certain limit;
namely it correctly reproduces the total ionization cross section.   The
effect of exciting unobserved channels may be described by contributions
to the optical potential or by explicit coupling.   All observed channels
are treated by explicit solution of the coupled-channels equations in
momentum space.   This calculation seems from preliminary results to
fulfil the atomic physics expectation of calculating the results of
detailed experiments ab initio.   A very-detailed test will be provided
by the amplitudes for coherent excitation of states with different

magnetic quantum numbers that are measured in the elegant quantum-beat experiments described by Peter Teubner. The microscopic atomic scattering T-matrix elements can of course be used to describe the D-vertices in (e,2e), completing the circle of self-consistency between scattering and knockout calculations.

A very elegant and successful calculation of self-consistent distortion and knockout was described by Iraj Afnan. The distortion vertices in the reaction $\pi d \to NN$ are those for $\pi d \to \pi d$ and $NN \to NN$. All three reactions are described by the solution of one set of coupled integral equations, with essentially completely accurate prediction of all relevant experimental data.

All the calculations of reactions described above use bound-state calculations as input and may be regarded as tests of validity of these calculations. Bound-state calculations in atomic and molecular physics were described by Graham Chandler, Ron Glass, Frank Larkins and Joel Epstein. For nuclei they were described by Bruce Barrett and Ian Morrison.

Alan Minchinton showed a convincing weight of evidence that the low-q dynamics of molecules is correctly calculated by Hartree-Fock methods and observed by noncoplanar-symmetric (e,2e). Improvement is obtained by direct calculation of the overlap amplitude $S(\vec{q})$ by perturbation expansion of the one-electron Green's function. John Cook introduced us to the direct understanding of molecular orbitals in momentum space and Jack Moore showed us how to extract numbers such as autocorrelation functions from (e,2e) data that can be related to co-ordinate space. Frank Larkins suggested that the influence of ground-state correlations may be observable in more-precise (e,2e) experiments at very-low q. Ian Fuss showed that relativistic effects can be important in (e,2e) even for reactions on outer valence shells at energies of the order of 1 keV for larger atoms.

The question of the configuration structure of atomic wave functions is very-well probed by (e,2e) as shown in the example of the 3s-hole structure of the argon ion. Here we have one of the big contrasts between nuclear physics, where vast amounts of data are correlated by an excellent understanding of structure, and atomic physics, where it is very difficult even to get qualitatively-correct description of the argon-ion example. This was emphasized by Ron Glass for argon, krypton and xenon. It seems that a very large configuration basis is needed for atomic calculations (including variational pseudostates to mimic states outside the basis), just as it is in the nuclear shell model, and that it would help considerably if a much larger basis were used than is possible at present. This may be made possible by the recognition of higher symmetries, such as those of the interacting boson model in nuclear physics, which enable a valid truncation of the basis. Here atomic physics is on the threshold of a new challenge.

Molecular wave functions are understood in terms of a knowledge of the nuclear positions. Gary Willett described a very elegant co-incidence method of studying different rearrangements of nuclei in the dissociation of molecules.

Finally we were given a glimpse into the future of momentum studies, which could lie in the field of solid-state physics. Robert Leckey showed that electron bands can be mapped by angle-resolved photoelectron reactions. Jim Williams showed that existing (e,2e) technology is already capable of identifying different electronic states in thin films.

Lindsay Frost described the coming generation of (e,2e) experiments on solids which, in the light of the former information, promise detailed observations of bands and their spectral functions.

WORKSHOP SUMMARY ON (e,2e) AND RELATED SPECTROSCOPIES

R.A. Bonham,
Department of Chemistry, University of Indiana.

Numerous new (e,2e) studies from a number of different laboratories have appeared in the literature since the 1976 momentum wave function conference at Indiana University. In fact the sheer number of studies presented at this conference is staggering. The main impression which one gets from a perusal of this work is that when careful Hartree-Fock level wave functions are available good agreement is obtained with experiment in all cases. Some deviations between experiment and theory have been noted but the emphasis has been placed on the agreement. This has occurred, I think, because the conservative scientific community had to be convinced that both theory and experiment had a sound basis in fact. During this period a number of fundamental theoretical calculations (McCarthy and Noble) and crucial experimental tests (Weigold on atomic hydrogen, De Heer et al. on the absolute e,2e cross section in He and Giardini-Guidoni and co-workers on factorization tests) have helped to validate the method.

It is my feeling that the (e,2e) method as presented at this conference has reached a cross roads. As Prof. Pople put it at the recent Flinders conference on atomic and molecular reactions and structure: What new has been learned from (e,2e) studies? Actually a significant body of new information dealing with electronic structure has appeared (Weigold and McCarthy on correlation effects in atoms and molecules, Brion on the structure of lone pair orbitals in NH_3 and PH_3 and on

settling disputed assignments in ESCA spectra) but by and large the emphasis so far seems to have been placed on agreement to a greater extent that disagreement with existing theories. I would suggest that the (e,2e) spectroscopists have surely convinced even the most pessimistic observer that there is indeed substance to their approach and it is now time to follow the lead of the nuclear experimentalists who, as this conference has clearly shown, are focusing their energies on those experiments and experimental regions of energy and momentum where theory and experiment do not agree. After all, in atomic and molecular science there is little room left in this day of ever faster and cheaper computers for any experimental method which can only churn out results which can be reproduced by Hartree-Fock calculations. I would go even further and suggest that the real power and utility of the (e,2e) and other methods for determining momentum distributions will lie in helping many body theorists develop reliable approximations for treating complex atomic and molecular systems including possibly solids and surfaces. I seriously doubt that such studies will have any practical impact on organic or inorganic chemistry except to provide a sound physical basis to already existing working models of chemical behavior. The kind of interaction that I envision will not be an easy one especially for the experimenter. To go beyond the Hartree-Fock level requires increased precision and experiments designed to investigate both the low q and high q regions of the (e,2e) spectrum. It is reasonably certain that the influence of reaction mechanisms which now seem to be understood at the 1% level for the symmetric noncoplanar geometry will rear their ugly heads again if 0.1% precision is attained. This has happened in the case of (e,e') compton measurements (Mendelsohn and Wellenstein)

and one has every reason to believe it will crop up again in (e,2e)
studies.   In fact there are current problems in comparing the (e,2e)
theory (V. Smith) with experiment (Coplan and Moore) on an absolute
basis for molecular hydrogen in spite of good agreement in the case of
He.   This is very reminiscent of the situation for H_2 with Compton
studies in 1974 where good agreement for He had also been obtained.
My own experience is that greater understanding of the underlying
physical mechanism of a process has nearly always been attained from
resolving discrepancies such as these between theory and experiment
than from situations where theory and experiment agreed from the start.

For all the above reasons it appears to me to be essential that
(e,2e), Compton, (e,e') and positron annihilation experimentalists work
closely with reaction mechanism theorists on the one hand and many body
theorists on the other.

This triumvirate cannot, however, exist as a single closed loop
system.   For reaction mechanisms of the complex type being developed
here at Flinders (McCarthy and Stelbovics) to be worked out properly,
they must be tested against experiments that show maximum sensitivity
to perturbing effects.   This means that any valid mechanistic theory must
be capable of explaining low energy e,2e experiments of diverse geometry
(Ehrhardt) and low energy (e,e') studies including the sensitive measurements of cross section ratios provided by the (e,γ) (Teubner) coincidence experiments.   Further the many body theorists, as represented
at this workshop by Larkins, Chandler and Glass, must test their codes
by computing a wide range of experimentally observable quantities such
as photoionization β values (Leckey), charge and spin densities and
potential surfaces (Epstein) capable of explaining the myriad and detail
of chemical reactions as revealed by (e,e' + ion) studies (Willet).
In addition the more classical areas of spectroscopy, molecular structure
and the measurements of electric moments will play an important role.
In Fig. I I have attempted to outline how this interaction might look
in a flow diagram.

The box marked applications raises a question.   I have pressed the
thesis that the main contribution of (e,2e) and related methods will lie
in their serving as a guide to the development of viable approximation
methods in many body theory.   Hence in some sense I am relying on the
development of such theories to result in the major applications of
the study of momentum distributions.   Here of course I could be quite
wrong or at least partly so.   In some areas such as surface science it
may turn out that momentum distribution measurements may yield valuable
applied information directly.   It is also true that with the current
difficulties in obtaining reliable CI calculations that chemical reac-
tivity due to electron correlation effects may be sorted out by the
application of (e,2e) methods before theoretical tools of sufficient
sophistication are developed.

Because of the strong emphasis on e,2e studies in this workshop it
seemed worthwhile to speculate on where these studies will lead in the
coming years.   In Fig. II a flow chart summarizes experiments reported
on or proposed at this conference.   In addition the squares with dotted
boundaries indicate areas of possible application of (e,2e) methods
which were not discussed in this conference.

The results of scattering on thin films (Williams) seemed to
indicate that plasma modes probably do not decay by electron ejection.
It appears that significant progress is being made in thin film (e,2e)

Fig. I. The Interaction Between Momentum Distribution Measurements and other Areas of Science.

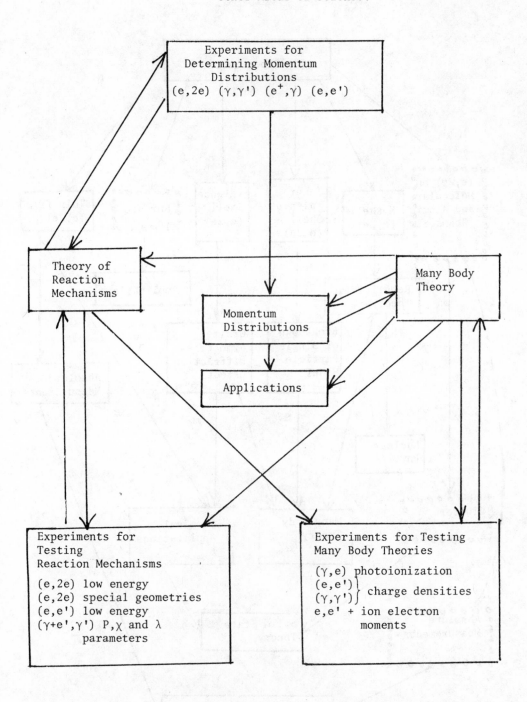

Fig. II. Applications of the (e,2e) Method

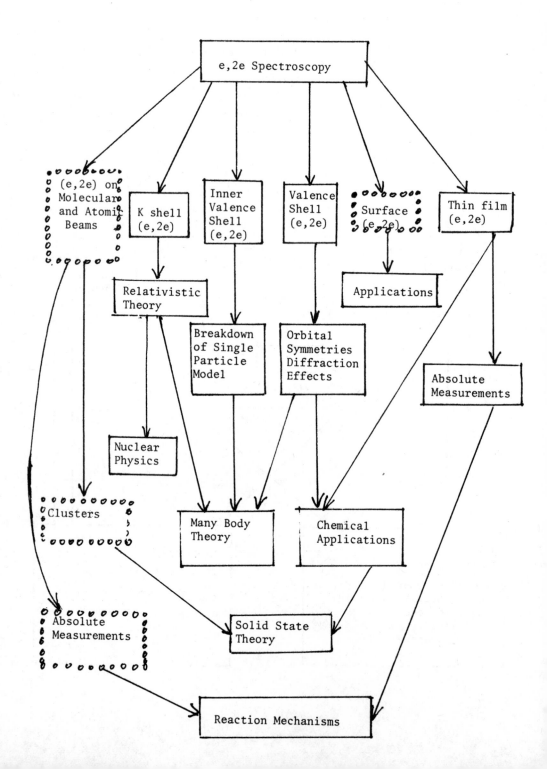

(Williams) and proposed experiments using novel techniques show a great deal of promise (Frost-Weigold). The study of molecules by (e,2e) is progressing at a very rapid rate. Numerous new results were presented (Minchinton and Moore). Finally diffraction features in (e,2e) as an additional tool in orbital symmetry assignments was pointed out (Cook).

The importance of obtaining absolute measurements to make sure that scattering mechanism effects were not present in the data was obvious from many of the nuclear studies presented. Possible ways to do this may be through thin film measurements as a function of film thickness or by use of carefully defined molecular beams.

There seemed to be a certain air of hopelessness concerned with (e,2e) of unresolved inner valence structure. It is now known that in second row and heavier elements serious breakdown of the independent particle model occurs in this region. It should be kept in mind that even if (e,2e) studies cannot be cleanly interpreted in this region they still furnish the most detailed data for comparison with theory that is currently available. Close collaboration between (e,2e) experimentalists and theorists working with Green's function and other approaches should be profitable for unravelling the mysteries of the inner valence region. An exciting attempt to obtain (e,2e) data on tightly bound core electrons using 5 MeV electrons was outlined by the Melbourne nuclear group (Spicer and Fuss). Preliminary calculations indicate that relativistic effects should be observable although some question remains as to whether coincidence rates will be high enough to make the experiment feasible.

WORKSHOP PARTICIPANTS

Afnan, Iraj	School of Physical Sciences, Flinders University, Adelaide, South Australia.
Amos, Ken	Dept. of Physics, Melbourne University, Parkville. Victoria.
Barrett, Bruce	Department of Physics, University of Arizona, Tucson, Arizona, U.S.A.
Bonham, Russ	Department of Chemistry, Indiana University, Bloomington, Indiana, U.S.A.
Chandler, Graham	Department of Chemistry, University of Western Australia, Nedlands, Western Australia.
Chant, N.S.	Department of Physics, University of Maryland, College Park, Maryland, U.S.A.
Collins, S.F.	Physics Department, Melbourne University, Parkville, Victoria.
Cook, John	School of Physical Sciences, Flinders University, Adelaide, South Australia.
Devins, D.	Department of Physics, University of Indiana, Bloomington Indiana, U.S.A.
Dodd, L.	Dept. of Mathematical Physics, University of Adelaide, Adelaide, South Australia.
Ehrhardt, H.	Fachbereich Physik, Universität Kaiserslautern, 6750 Kaiserslautern, Federal Republic of Germany.
Epstein, J.	Research School of Chemistry, A.N.U., Canberra, A.C.T.
Frost, Lindsay	School of Physical Sciences, Flinders University, Adelaide, South Australia.
Furst, J.	School of Physical Sciences, Flinders University, Adelaide, South Australia.
Fuss, Ian	Defence Research Centre, Salisbury, South Australia.
Glass, R.	Department of Chemistry, University of Western Australia, Nedlands, Western Australia.
Golden, D.E.	Department of Physics and Astronomy, University of Oklahoma, Norman, Oklahoma, U.S.A.
Larkins, F.	Dept. of Chemistry, Monash University, Clayton, Victoria.
Leckey, R.C.G.	Physics Department, La Trobe University, Bundoora. Victoria.
Lohmann, B.	School of Physical Sciences, Flinders University, Adelaide, South Australia.
Matthews, June	Physics Department, Massachusetts Institute of Technology, Cambridge, Massachusetts, U.S.A.
McCarthy, Ian	School of Physical Sciences, Flinders University, Adelaide, South Australia.
Mendelson, L.B.	Department of Physics, Brooklyn College, Brooklyn, New York, U.S.A.
Minchinton, Alan	School of Physical Sciences, Flinders University, Adelaide, South Australia.
Mitroy, J.	Physics Department, Melbourne University, Victoria.
Moore, J.	Department of Chemistry, University of Maryland. College Park, Maryland U.S.A.

Morrison, I.	Dept. of Physics, Melbourne University, Victoria.
Peel, J.B.	Department of Chemistry, La Trobe University, Bundoora, Victoria.
Riley, J.	School of Physical Sciences, Flinders University, Adelaide, South Australia.
Shute, G.	Physics Department, Melbourne University, Parkville, Victoria.
Standage, M.	School of Science, Griffith University, Nathan, Queensland.
Stelbovics, A.	School of Physical Sciences, Flinders University, Adelaide, South Australia.
Tassie, Lindsay	Research School of Physical Sciences, A.N.U., Canberra, A.C.T.
Teubner, Peter	School of Physical Sciences, Flinders University, Adelaide, South Australia.
von Geramb, H.V.	Theoretische Kernphysik, Universität Hamburg, Hamburg, Federal Republic of Germany.
Wastell, J.M.	Physics Department, University of Melbourne, Parkville, Victoria.
Weigold, Erich	School of Physical Sciences, Flinders University, Adelaide, South Australia.
Willett, G.	Dept. of Theoretical Chemistry, University of Sydney, Sydney, New South Wales.
Williams, J.F.	Physics Department, University of Western Australia, Nedlands, Western Australia.

AIP Conference Proceedings

		L.C. Number	ISBN
No.1	Feedback and Dynamic Control of Plasmas	70-141596	0-88318-100-2
No.2	Particles and Fields - 1971 (Rochester)	71-184662	0-88318-101-0
No.3	Thermal Expansion - 1971 (Corning)	72-76970	0-88318-102-9
No.4	Superconductivity in d-and f-Band Metals (Rochester, 1971)	74-18879	0-88318-103-7
No.5	Magnetism and Magnetic Materials - 1971 (2 parts) (Chicago)	59-2468	0-88318-104-5
No.6	Particle Physics (Irvine, 1971)	72-81239	0-88318-105-3
No.7	Exploring the History of Nuclear Physics	72-81883	0-88318-106-1
No.8	Experimental Meson Spectroscopy - 1972	72-88226	0-88318-107-X
No.9	Cyclotrons - 1972 (Vancouver)	72-92798	0-88318-108-8
No.10	Magnetism and Magnetic Materials - 1972	72-623469	0-88318-109-6
No.11	Transport Phenomena - 1973 (Brown University Conference)	73-80682	0-88318-110-X
No.12	Experiments on High Energy Particle Collisions - 1973 (Vanderbilt Conference)	73-81705	0-88318-111-8
No.13	π-π Scattering - 1973 (Tallahassee Conference)	73-81704	0-88318-112-6
No.14	Particles and Fields - 1973 (APS/DPF Berkeley)	73-91923	0-88318-113-4
No.15	High Energy Collisions - 1973 (Stony Brook)	73-92324	0-88318-114-2
No.16	Causality and Physical Theories (Wayne State University, 1973)	73-93420	0-88318-115-0
No.17	Thermal Expansion - 1973 (lake of the Ozarks)	73-94415	0-88318-116-9
No.18	Magnetism and Magnetic Materials - 1973 (2 parts) (Boston)	59-2468	0-88318-117-7
No.19	Physics and the Energy Problem - 1974 (APS Chicago)	73-94416	0-88318-118-5
No.20	Tetrahedrally Bonded Amorphous Semiconductors (Yorktown Heights, 1974)	74-80145	0-88318-119-3
No.21	Experimental Meson Spectroscopy - 1974 (Boston)	74-82628	0-88318-120-7
No.22	Neutrinos - 1974 (Philadelphia)	74-82413	0-88318-121-5
No.23	Particles and Fields - 1974 (APS/DPF Williamsburg)	74-27575	0-88318-122-3
No.24	Magnetism and Magnetic Materials - 1974 (20th Annual Conference, San Francisco)	75-2647	0-88318-123-1
No.25	Efficient Use of Energy (The APS Studies on the Technical Aspects of the More Efficient Use of Energy)	75-18227	0-88318-124-X

Date Due